Lecture Notes in Computer Science 6177

Commenced Publication in 1973
Founding and Former Series Editors:
Gerhard Goos, Juris Hartmanis, and Jan van Leeuwen

Editorial Board

Christina J. Hopfe   Yacine Rezgui
Elisabeth Métais   Alun Preece
Haijiang Li (Eds.)

Natural Language Processing and Information Systems

15th International Conference on Applications
of Natural Language to Information Systems, NLDB 2010
Cardiff, UK, June 23-25, 2010
Proceedings

 Springer

Volume Editors

Christina J. Hopfe
Cardiff University, School of Engineering, UK
E-mail: HopfeC@cardiff.ac.uk

Yacine Rezgui
Cardiff University, School of Engineering, UK
E-mail: RezguiY@cardiff.ac.uk

Elisabeth Métais
Centre National des Arts et Métiers, France
E-mail: Elisabeth.Metais@cnam.fr

Alun Preece
Cardiff University, School of Computer Science, UK
E-mail: A.D.Preece@cs.cardiff.ac.uk

Haijiang Li
Cardiff University, School of Engineering, UK
E-mail: LiH@cardiff.ac.uk

Library of Congress Control Number: 2010928431

CR Subject Classification (1998): I.2.7, I.5, H.2, J.3, H.2.8, I.2.6

LNCS Sublibrary: SL 3 – Information Systems and Application, incl. Internet/Web and HCI

ISSN   0302-9743
ISBN-10   3-642-13880-2 Springer Berlin Heidelberg New York
ISBN-13   978-3-642-13880-5 Springer Berlin Heidelberg New York

springer.com

© Springer-Verlag Berlin Heidelberg 2010
Printed in Germany

Typesetting: Camera-ready by author, data conversion by Scientific Publishing Services, Chennai, India
Printed on acid-free paper   06/3180

Preface

The 15th International Conference on Applications of Natural Language to Information Systems (NLDB 2010) took place during June 23–25 in Cardiff (UK). Since the first edition in 1995, the NLDB conference has been aiming at bringing together researchers, people working in industry and potential users interested in various applications of natural language in the database and information system area. However, in order to reflect the growing importance of accessing information from a diverse collection of sources (Web, Databases, Sensors, Cloud) in an equally wide range of contexts (including mobile and tethered), the theme of the 15th International Conference on Applications of Natural Language to Information Systems 2010 was "Communicating with Anything, Anywhere in Natural Language."

Natural languages and databases are core components in the development of information systems. Natural language processing (NLP) techniques may substantially enhance most phases of the information system lifecycle, starting with requirement analysis, specification and validation, and going up to conflict resolution, result processing and presentation. Furthermore, natural language-based query languages and user interfaces facilitate the access to information for all and allow for new paradigms in the usage of computerized services.

Hot topics such as information retrieval (IR), software engineering applications, hidden Markov models, natural language interfaces and semantic networks and graphs imply a complete fusion of databases, IR and NLP techniques.

Among an increasing number of submitted papers, the Program Committee selected 14 papers as full papers, thus coming up with an acceptance rate of less than 30%. These proceedings also include 19 short papers that were presented at the conference and two invited talks from industry, one given by James Luke (IBM) and the other one given by Enrique Alfonseca (Google).

This conference was possible thanks to the support of three organizing institutions: Cardiff University, School of Engineering and School of Computer Science and Informatics, UK and Centre National des Arts et Métiers (CNAM), France. We also wish to thank the entire organizing team including secretaries, researchers and students who put their competence, enthusiasm and kindness into making this conference a real success. Special thanks are due to the Welsh Assembly Government (WAG). WAG was the main sponsor of this year's conference.

June 2010

Christina J. Hopfe
Yacine Rezgui
Elisabeth Métais
Alun Preece
Haijiang Li

Conference Organization

Conference Co-chairs

Yacine Rezgui   Cardiff University, UK
Elisabeth Métais   CNAM, France

Program Co-chairs

Alun Preece   Cardiff University, UK
Christina J. Hopfe   Cardiff University, UK

Conference Publicity Chair

Nadira Lammari   CNAM, France

Program Committee

Akhilesh Bajaj	University of Tulsa, USA
Alexander Gelbukh	Mexican Academy of Sciences, Mexico
Alexander Hinneburg	University of Halle, Germany
Alfredo Cuzzocreaq	University of Calabria, Italy
Andreas Hotho	University of Kassel, Germany
Andrés Montoyo	Universidad de Alicante, Spain
Antje Düsterhöft	Hochschule Wismar, Germany
Bernhard Thalheim	Kiel University, Germany
Cedric du Mouza	CNAM, France
Christian Kop	University of Klagenfurt, Austria
Christian Winkler	University of Klagenfurt, Austria
Christina J. Hopfe	Cardiff University, UK
Deryle Lonsdale	Brigham Young Uinversity, USA
Elisabeth Métais	CNAM, France
Epaminondas Kapetanios	University of Westminster, UK
Fabio Rinaldi	University of Zurich, Switzerland
Farid Meziane	Salford University, UK
Frederic Andres	University of Advanced Studies, Japan
Georgia Koutrika	Stanford University, USA
Grigori Sidorov	National Researcher of Mexico, Mexico
Günter Neumann	DFKI, Germany
Günther Fliedl	University of Klagenfurt, Austria
Hae-Chang Rim	Korea University, Korea
Harmain Harmain	United Arab Emirates University, UAE
Heinrich C. Mayr	University of Klagenfurt, Austria

Helmut Horacek	Saarland University, Germany
Hiram Calvo	National Polytechnic Institute, Mexico
Irena Spasic	Manchester Centre for Integrative Systems Biology, UK
Isabelle Comyn-Wattiau	CNAM, France
Jacky Akoka	CNAM, France
Jana Lewerenz	Capgemini sd&m Düsseldorf, Germany
Jian-Yun Nie	Université de Montréal, Canada
Jing Bai	Yahoo Inc., Canada
Jon Atle Gulla	Norwegian University of Science and Technology, Norway
Juan Carlos Trujillo	Universidad de Alicante, Spain
Jürgen Rilling	Concordia University, Canada
Karin Harbusch	Universität Koblenz-Landau, Germany
Krishnaprasad Thirunarayan	Wright State University, USA
Leila Kosseim	Concordia University, Canada
Luis Alfonso Ureña	Universidad de Jaén, Spain
Luisa Mich	University of Trento, Italy
Magdalena Wolska	Saarland University, Germany
Manolis Koubarakis	University of Athens, Greece
Markus Schaal	Bilkent University, Turkey
Max Silberztein	Université de Franche-Comté , France
Mokrane Bouzeghoub	Université de Versailles, France
Nadira Lammari	CNAM, France
Odile Piton	Université Paris I Panthé on-Sorbonne, France
Panos Vassiliadis	University of Ioannina, Greece
Paul Johannesson	Stockholm University, Sweden
Paul McFetridge	Simon Fraser University, Canada
Philipp Cimiano	CITEC, University of Bielefeld
Pit Pichappan	Annamalai University, India
Rafael Muñoz	Universidad de Alicante, Spain
René Witte	Concordia University, Canada
Roger Chiang	University of Cincinnati, USA
Roland Wagner	University of Linz, Austria
Samia Nefti-Meziani	University of Salford, UK
Samira Si-Said Cherfi	CNAM, France
Sophia Ananiadou	NaCTeM, Manchester Interdisciplinary Biocentre, UK
Stéphane Lopes	Université de Versailles, France
Udo Hahn	Friedrich-Schiller-Universität Jena, Germany
Veda Storey	Georgia State University, USA
Vijay Sugumaran	Oakland University Rochester, USA
Yacine Rezgui	University of Salford, UK
Zoubida Kedad	Université de Versailles, France

Organizing Committee Chair

Christina J. Hopfe1   Cardiff University, UK

Organizing Committee

Haijiang Li   Cardiff University, UK
John Miles   Cardiff University, UK
Elisabeth Métais   CNAM, France
Alan Preece   Cardiff University, UK
Yacine Rezgui   Cardiff University, UK

Table of Contents

Domain Modelling

Information Extraction

Semantic Networks & Graphs

An Approach for Adding Noise-Tolerance to Restricted-Domain Information Retrieval

Katia Vila[1], Josval Díaz[1], Antonio Fernández[1], and Antonio Ferrández[2]

[1] University of Matanzas, Department of Informatics
Varadero Road, 40100 Matanzas, Cuba
{kvila,antonio}@dlsi.ua.es
[2] University of Alicante, Department of Software and Computing Systems
San Vicente del Raspeig Road, 03690 Alicante, Spain
{josval.diaz,antonio.fernandez}@umcc.cu

Abstract. Corpus of Information Retrieval (IR) systems are formed by text documents that often come from rather heterogeneous sources, such as Web sites or OCR (Optical Character Recognition) systems. Faithfully converting these sources into flat text files is not a trivial task, since noise can be easily introduced due to spelling or typeset errors. Importantly, if the size of the corpus is large enough, then redundancy helps in controlling the effects of noise because the same text often appears with and without noise throughout the corpus. Conversely, noise becomes a serious problem in restricted-domain IR where corpus is usually small and it has little or no redundancy. Therefore, noise hinders the retrieval task in restricted domains and erroneous results are likely to be obtained. In order to overcome this situation, this paper presents an approach for using restricted-domain resources, such as Knowledge Organization Systems (KOS), to add noise-tolerance to existing IR systems. To show the suitability of our approach in one real restricted-domain case study, a set of experiments has been carried out for the agricultural domain.

Keywords: information retrieval, noise-tolerance, restricted domain.

1 Introduction

Human beings must continuously face up to noise when they write or read documents. Noise in text is widely defined as [14] "any kind of difference in the surface form of an electronic text from the intended, correct or original text". Hence, noise is likely to cause erroneous results (due to spelling mistakes, typeset errors or special characters) in applications that process electronic texts in an automatic manner, such as Natural Language Processing (NLP) applications [3,21]. Interestingly, dealing with noise is of most importance for Information Retrieval (IR) systems, since they are usually at the core of most NLP applications, such as Question Answering (QA), Information Extraction (IE), etc.

IR is the task of searching for information required by users in a huge repository of documents or corpus. IR systems are based on comparing text strings

C.J. Hopfe et al. (Eds.): NLDB 2010, LNCS 6177, pp. 1–12, 2010.

between the user's query and the corpus where the answer should be found. Specifically, from a user's query, an IR system returns a list of relevant documents which may contain the answer of the query [2]. Therefore, noise can appear in (i) the *query*, since its terms can be directly written in an incorrect manner [1,21]; or (ii) the *corpus*, since it must be automatically processed to obtain a set of text files as input of the IR system, from the Web, from PDF (Portable Document Format) files or even from OCR (Optical Character Recognition) or ASR (Automatic Speech Recognition) tools [23].

Surprisingly, current IR approaches have shown little interest in explicitly dealing with noise in the corpus, thus only focusing on dealing with noise in the query terms (examples could be found in [4,16]). At first sight, this seems a reasonable situation, since corpus consists of huge amounts of redundant documents in which the expected answer of a query is often repeated in quite a lot documents, with and without noise. Redundant corpus thus avoids that IR systems are affected by noise problems. Unfortunately, this is only true for open-domain corpus, since corpus from restricted domains are usually rather small and, therefore, with little or no redundancy [18]. Non-redundant corpus thus makes that the IR system may find the answer in very few documents, which probably contain noise and the answer will be never found. Consequently, this scenario hampers the use of IR systems in real-world situations where (i) a restricted-domain corpus is used, and (ii) noise is unavoidable.

Bearing these considerations in mind, a novel approach is presented in this paper in order to add noise-tolerance to any IR system when restricted-domain corpus is used. The basis for our approach is the use of the often available restricted-domain Knowledge Organization Systems[1] (KOS), such as thesaurus or ontologies. Our motivation is that important terms in restricted-domain corpus belong to a specific and small vocabulary which can be obtained from a KOS. Therefore, noise in the corpus can be avoided by comparing terms from the KOS and terms from the corpus. To this aim, in this paper the Extended Edit Distance (DEx) [7] has been adapted in order to consider multi-words, which commonly appear in most of restricted-domain corpus (e.g. in scientific domain such as "calcium hydroxide" or "adrenal cortex hormones") when terms are being compared.

The main benefit of our approach is that the performance of the IR system is kept although a noisy restricted-domain corpus is used, as is shown in several experiments that have been carried out. These experiments show the suitability of our approach in a real restricted-domain case study: an IR system for one scientific journal of the agricultural domain.

Once our work has been motivated above, related work is briefly described in Sect. 2. Our approach for adding noise-tolerance to IR systems is described in Sect. 3, while in Sect. 4 a set of experiments are discussed. Our conclusions and future work are sketched in Sect. 5.

---

[1] Knowledge Organization Systems include a variety of schemes that organize, manage, and retrieve information. This term is intended to encompass all types of schemes for promoting knowledge management [11].

2   Related Work

So far, few methods have been proposed to deal with noise in the corpus [21]. These methods focused on preprocessing and filtering noise [13,5]. In contrast, our approach agrees with Esser [6] who advocated the addition of a fault-tolerant behavior to the system, in such a way that the noisy corpus can be used directly independently of the kind of noise. In this sense, some events have highlighted the importance of query expansion for dealing with noise in IR systems, such as TREC Confusion Track [12] or Legal Track [20]. They advocate that query must be expanded with new terms which are obtained by adding common corruption errors previously found in the corpus or obtained from lists of pair of correct and incorrect words.

To sum up, all these approaches depend on a deep knowledge about how noise can appear in the corpus in order to detect patterns in spelling or typeset errors to help to detect noise. Therefore, current approaches are not suitable for restricted domain corpus since they are small, thus each kind of errors can occur very few times and error patterns are difficult to find.

Our approach overcomes this drawback since we take advantage of the highly available KOS for restricted domains to compare terms from the noisy corpus and important terms of the domain in such a way that a previous analysis of the errors occurring in restricted-domain terms of the corpus is unneeded.

Finally, it is worth pointing out that although the negative impact of noise is lately attracting attention of researchers [17], surprisingly dealing with noise in restricted-domain corpus has not been widely considered so far.

3   Adding Noise-Tolerance to an IR System

In this section, our approach for adding a noise-tolerant behavior to a restricted-domain IR system is described. As the most important query terms in restricted-domain IR systems are those related to the domain, noise affecting these restricted-domain terms in the corpus makes the precision of the IR system decreased. Consequently, restricted-domain IR systems must be aware of noise in restricted-domain terms appearing in the corpus. To this aim, our approach compares terms in a KOS with terms in the corpus by means of an extended edit distance algorithm.

3.1   Extended Edit Distance Algorithm

In our approach, distance between terms has been calculated on the basis of a new edit distance called the Extended Edit Distance (DEx:"Distancia de Edición eXtendida"). In our previous work [7], DEx shows excellent results when it is compared with other distances as [15,19,24]. Some features make DEx appropriate for our purposes: (i) distances are calculated on the original words, i.e. without applying lemmatization, which is useful in noisy words because they cannot be lemmatized; and (ii) penalties are applied according to the kind, position and character involved in the modification operation (insertion, deletion, substitution or non operation), thus being useful for dealing with noise patterns that frequently appear in the corpus (highly repetitive noise). How to calculate DEx is briefly described as follows (we refer reader to [7] for a further description):

1. Making the ranking out of characters of all the terms contained in a KOS.
2. Both corpus terms and terms from the KOS are tokenized.
3. A matrix is created and filled according to Levenshtein's edit distance [15]. Levenshtein's matrix is used to determine the modification operation chain. Terms are assumed to be independent entities to be compared with DEx. A dynamic threshold is established from the evaluation of the lowest ranking's character (calculated in step 1) from the KOS just in the middle of the operation chain. By doing so, it is decided the similarity of the compared strings: if the DEx is lower than the threshold then it is assumed that tokens are similar, otherwise they are different.
4. Finally, DEx values between compared tokens are stored in other matrix.

As previously-stated, to be considered in a restricted-domain, DEx has to be aware of multi-words, since they often appear in restricted-domain KOS. The following formula (MD: Multi-words Distance in Eq. 1) is based on calculating DEx for each word that appears in multi-words.

$$MD = \sum_{i=1}^{L} (V(O_i) * CP + CD * DE_{x_i} * 100) * 2^{-i}$$

$$where:$$
$$V(O_i) = \begin{cases} 0 \ if \ O_i = o \\ 1 \ if \ O_i \neq o \end{cases} \tag{1}$$

Descriptions of parameters and constants used are as follows:

O_i: Modification operations' chain.
L: Length of the operations' chain.
CP: Degree of affectation due to the position (0.95). This is a constant that generates a cost according to the position in which the operation appears. This value was empirically obtained.
CD: Degree of affectation due to DEx (0.05). This is a cost applied to DEx. Its value corresponds to the remaining percentage of CP. CD is used to establish an order between similar comparisons.
DE_{x_i}: DEx between words at the position i of the multi-words that are been compared.
2^{-i}: This element is used to give a weight to the position where the transformation appears among compared words. Thanks to the exponential behavior of this element, those transformations that appear at the left-hand side of the word are more strictly penalized.

Finally, is necessary to note that the higher the result of MD is, the higher the distance between words is.

3.2   Our Approach

An overview of our approach is depicted in Fig. 1. Our approach has two main stages: one at indexing time (offline) and other at query time (online).

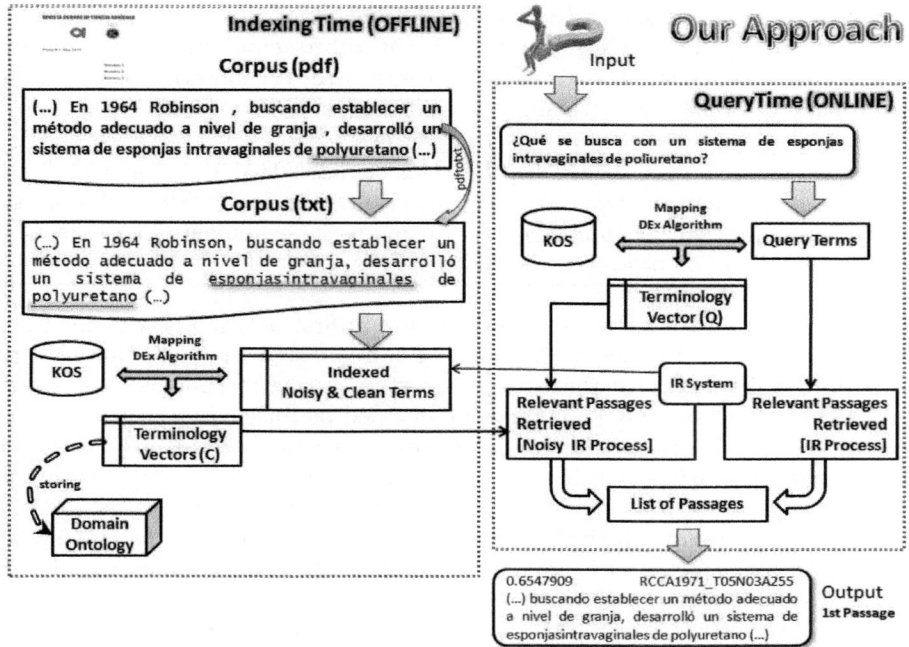

Fig. 1. Overview of our approach for adding noise-tolerance to an IR system

At indexing time (see the offline process at left-hand side of Fig. 1), the IR system converts the corpus into flat files and their terms are indexed independently of the noise, by computing the frequency of these terms by document. Then, every indexed term is mapped to the terms in the domain-specific KOS by using MD algorithm previously detailed. For each indexed term, their mapped terms and corresponding distances are kept in terminology vectors (C). A terminology vector can be defined as follows: let T be the set of n terms from the KOS mapped with the MD algorithm. $t_r \in T$ denotes the term r in the set of terms. Then, the terminology vector that represents the term t_s is defined as the vector $V_{t_s} = [(t_1, w_1), (t_2, w_2), ..., (t_n, w_n)]$ where w_r denotes the distance between t_s and t_r.

These terminology vectors are stored in a domain ontology by following our proposal [22]. An example of spacing error is the noisy term "esponjasintravaginales" (instead of "esponjas_intravaginales"), extracted from the corpus, whose terminology vector is: $C_{esponjasintravaginales} = [(esponjas, 0.13), (esponjilla, 0.18), (esponja\_vegetal, 0.25), (...)]$.

The second stage takes place in the question analysis process at query time (see the online process at right-hand side of Fig. 1). This stage consists of two independent retrieval processes. The first one uses the IR system with the original query terms in order to return the passages with more probability of having the answer, as usually. For example, if the retrieval system tries to answer the following question: *"Qué se busca con un sistema de esponjas intravaginales de poliuretano?"* (*"What is searched by using an intravaginal sponge system made of polyurethane?"*), this first retrieval process returns the following passage (with a probability of 0.34 of containing the right answer).

Input query: 174146 0.34165043 RCCA1989_T23N03A241 "... iguales para los tres sistemas. Sc concluye que en el sistema de dos cuartones las vacas tuvieron que realizar un mayor esfuerzo cn busca del alimento, por la menor disponibilidad dc pastos cn este sistema ..."

The reason for the retrieval of this passage in first position is that one of the weightiest term of the query ("sistemas") is frequently repeated. However, the low probability indicates that other important terms ("esponjas", "intravaginales", and "poliuretano") could not be mapped because corpus contains noise.

In the second process, also noisy weightiest question's terms are mapped to the terms in the KOS by using MD in order to obtain their corresponding terminology vector (Q), as at indexing time. Then, terminology vectors C and Q are used to properly substitute those terms in the question by the corresponding terms that appear with noise in the corpus: the term i represented by the terminology vector $Q_i \in Q$ can be substituted by the term j represented by the terminology vector $C_j \in C$ provided that $Q_i \neq C_j$ and one of the following two empirically-obtained rules are fulfilled[2]:

- There exist three or more terms in Q_i that are contained also in C_j whose distances are between 0 and 0.37.
- There exist less than three elements in Q_i that are contained also in C_j whose distances are all equal to 0.

Obviously, this second process really consists on launching an independent retrieval process per each substituted term without applying any penalty to the probability of the retrieved passages to contain the right answer. Finally, returned passages by both retrieval processes are merge to find the passage that contains the answer. This merging reorders the passages according to their weights, without doing any modification or penalization.

As an example of this second retrieval process, the query term "esponja" would have the following vector: $Q_{esponjas} = [(esponjas, 0), (esponjilla, 0.1), (esponja\_vegetal, 0.25), (...)]$. Therefore, we can conclude that "esponjas" can be substituted by "esponjasintravaginales", since there exist three terms ("esponja",

---

[2] It is worth clarifying that thresholds are obtained from a previous experimentation (more details in Sect. 4.4) whose aim was to evaluate effectiveness of MD. However, these thresholds can be adjusted according to the application domain.

"esponjilla", and "esponja vegetal") in $Q_{esponjas}$ that are contained also in $C_{esponjasintravaginales}$ whose distances are between 0 and 0.37.

In doing so, the IR system does obtain the right answer with a weight of 0.65 in the passage shown at the bottom of Fig. 1. It is worth highlighting that this passage was not even returned by the baseline IR system. Moreover, the greatest value of the weight of the returned passages by the baseline IR system was 0.34, so the passage returned by the IR system with noise tolerance contains the right answer and takes the first position in the new ranking.

4   Experiments and Results

This section describes the experiments performed to show the suitability of our approach for retrieval the right information from an agricultural-domain noisy corpus. The first experiment aims to determine values of several performance measures when a clean corpus (upper bound values) and a noisy corpus (lower bound values) are used in a baseline IR system. This experiment also shows that the lower bound values drastically decreases comparing with the upper bound values. The second experiment aims to show that our approach is useful for increasing the lower bound values of the baseline IR system until similar values to the upper bound values when a noisy corpus is used. For the sake of reproducibility, before presenting the experiments, we briefly describe the required resources and how the experiments have been prepared.

4.1   Resources

Corpus. Our corpus is the Cuban Journal of Agricultural Science (RCCA, *Revista Cubana de Ciencia Agrícola*)[3]. It was created in both English and Spanish in 1966. At this moment the RCCA Journal has published 43 volumes, each one with an average of three or four numbers, that makes a total of 140 numbers and 2000 articles (28.65 MB as PDF files). This journal comprises topics related to agricultural science, such as Pastures and Forages or Animal Science. In this paper, we use the Spanish part of this journal as corpus.

RCCA journal accomplishes the three conditions stated by Minock in [18] for being a restricted domain corpus: (i) it is *circumscribed* because user queries are only related to agricultural science, (ii) it is *complex* since it contains a plethora of agricultural-specific terms, and (iii) it is *practical* because all agricultural researchers should be interested in it.

Importantly, as the RCCA publishes papers from 1966, a lot of papers have been digitalized, which may include even more noise in the corpus when they are converted to flat files: 1479 papers published between 1967-2000 were scanned and stored as PDF files which require OCR tools to extract the text documents, which represents a significant percentage (73.95% of total). Therefore, we can state that our experiments are done with *real* noisy data instead of introducing simulated data corruption which makes our case study highly representative in order to evaluate our approach.

---

[3] http://www.ica.inf.cu/productos/rcca/

IR system. The baseline IR system used to conduct the experiments is called JIRS (JAVA Information Retrieval System) [8,9]. This retrieval system has a natural language question as an input (instead of keywords) and a list of passages containing the answer as an output, which are decreasingly ordered according to the probability by using a distance density n-gram model.

AGROVOC. In our case study of the agricultural domain, we have used the AGROVOC[4] thesaurus as KOS. The AGROVOC thesaurus has a total number of 16700 descriptors, and 10758 non-descriptors, which are specific descriptors and terminological terms used in agricultural science. Shortly, AGROVOC is a multilingual structured controlled vocabulary used for indexing and retrieving data in agricultural information systems.

4.2   Experiments Preparation

Required steps performed before carrying out the experiments are briefly described in this section.

Converting corpus from PDF to text files. As some authors have pointed out [10], automatic processing and data extraction of PDF files is not a trivial task since most of them have little structural information. It is even worst in the scientific domain where character codification problems or lay-out problems (e.g. double-column format, page head and foot notes, table or formulæ conversion, etc.) often appear, thus introducing in the converted document many syntactic and semantic errors. For our experiments, the well-known tool `pdftotext` has been used to deal with this kind of problems by means of applying the following parameters: [-enc encoding] to correctly obtain the character codification, [-nopgbrk] to avoid mixing head and foot notes with the main text and [-layout] to keep the original text structure.

Creating test queries. A total of 150 questions were used in the experiments. They were formulated in natural language instead of a list of keywords because they will be used in QA systems in our future work (see Sect. 5). These questions are in Spanish to be able to interact with the Spanish corpus of the RCCA journal. Some sample questions are: *Qué es la necrosis cerebrocortical? (What is the cerebrocortical necrosis?)* or *Qué produce la cytophaga? (What is produced by the cytophaga?)*. These questions were elicited by interviewing agricultural domain experts of the RCCA journal. The answer of 56 of these questions appears free of noise in the corpus, while the answer of 94 of them is affected by noise.

Removing noise from a small piece of the corpus. In order to determine the upper bound of performance of the baseline IR system, noise has been manually removing from the 150 files of the corpus where answers of the previously described test questions appear.

---

[4] http://www.fao.org/agrovoc/

4.3   Experiments

In this section, experiments carried out are described in detail. A summary statistics and results of the experiments are shown in Tab. 1.

First experiment: determining bound values. This experiment aims to obtain the maximum and the minimum values of several performance measures when JIRS (i.e. our baseline IR system) is used. These values will be used to show the suitability of our approach later on, in the second experiment.

In Tab. 1(a), some of the results of this experiment are shown. As a result of applying the indexing documents process to all the corpus, 432,997 passages and 180,460 domain terms were obtained. When applying this process to the 150 preprocessed documents, 6,795 passages and 10,437 domain terms (1894 out of them containing noise). Therefore, around 18% of terms had noise.

The experiment has been conducted by using the 56 questions which are not affected by noise and the 94 questions whose answer is affected by noise. The amount of retrieved passages and relevant retrieved passages for all questions are also shown in Tab. 1(a). It is worth noting that we decided to return 25 passages per question in order to properly analyze the results and the ranking of the right answer.

This experiment consists of two parts: first the best performance for JIRS is obtained by using the corpus that has been preprocessed to partially remove noise (PCB). Secondly, the worst performance is obtained by using the noisy corpus (NCB). We can conclude that, in order to consider our proposal suitable, upper and lower bound of relevant retrieved documents for all questions should be between 79 and 130.

Others results obtained in this experiment for each corpus are shown in Tab. 1(b). We calculated the following measures: precision, recall, F1[5], Mean Average Precision (MAP) and Mean Reciprocal Rank[6] (MRR). The values in this table show that the noise widely affects results returned by the IR system (e.g. $MRR(PCB) = 0.77$ vs. $MRR(NCB) = 0.22$).

Second experiment: evaluating our approach. The aim of this experiment is to evaluate the effectiveness of our approach by comparing the results obtained by our approach with the previously-obtained results in an explicit and exhaustive manner. Expected results must be found between the values obtained from the previous experiment and, therefore, the nearer the results will be from the ones obtained in the first part of our previous experiment, the better our approach will work.

Results of this experiment ($MRR(NCA) = 0.71$) are shown in the NCA column of Tab. 1(b). These results considerably improve the baseline values obtained in the previous experiment by using the noisy corpus ($MRR(NCB) = 0.22$), while they are near the optimal results returned by the IR system when a clean corpus is used ($MRR(PCB) = 0.77$).

---

[5] $F1 = 2 * \frac{precision*recall}{precision+recall}$

[6] MRR is defined with $MRR = \frac{1}{Q}\sum_{i=1}^{Q}\frac{1}{rank_i}$, where Q is the number of questions and $rank_i$ is the rank of the first relevant passage retrieved for query i.

Table 1. Summary statistics and results of experiments

(a) Summary Statistics.

Characteristics of the collection	
Number of Documents in the Collection:	2000
Number of Passages in the Collection:	432997
Number of Questions:	150
Total number of passages over all questions	
Retrieved (25 per question):	3750
Relevant:	150
Relevant Retrieved (RR)	
Preprocessed Corpus + Baseline (PCB)	130
Noisy Corpus + Baseline (NCB)	79
Noisy Corpus + Approach (NCA)	110

(b) Results.

Passage Level Averages						
RR (# pass)				Precision		
PCB	NCB	NCA		PCB	NCB	NCA
110	20	105	At 1 pass	0.73	0.13	0.70
118	25	108	At 2 pass	0.39	0.08	0.36
122	28	110	At 3 pass	0.27	0.06	0.24
125	58	110	At 5 pass	0.16	0.07	0.14
130	71	110	At 10 pass	0.08	0.04	0.07
130	75	110	At 15 pass	0.05	0.03	0.04
130	79	110	At 20 pass	0.04	0.02	0.03
non-interpolated MAP				0.77	0.22	0.71
Other Measures						
Overall Recall				0.86	0.52	0.73
F1				0.06	0.04	0.06
MRR				0.77	0.22	0.71

4.4 Discussion

Our first experiment shows that noise affects the baseline IR system, as is shown by the upper and lower bound values. Our second experiments shows that our approach performs well when the corpus is noisy, since obtained results are close to the upper bound values calculated in the first experiment.

It is worth highlighting that values of MAP and F1 in Tab. 1(b) are similar for our approach and for the baseline IR with cleaned corpus, while the difference in the overall recall is only 0.13 and the difference in MRR is 0.06. A deep analysis of all the evaluated measures shows that, in our approach ("NCA"), recall is affected since 20 relevant passages are retrieved less. The main reason is that some noisy terms have not their counterparts in AGROVOC due to the fact that they were too deformed or they are not in the thesaurus. However, the weighted harmonic mean (F1) of precision and recall reach the same result in both experiments.

Effectiveness of our algorithm for carrying out the thesaurus mapping is also measured. A previous study showed us that around 70% of terms are noisy and 45% are multi-words from 500 terms from the corpus and the collection of questions. Only 200 out of 500 terms did not find the corresponding mapped term in AGROVOC: 80 due to an incorrect mapping and 120 due to be missed in AGROVOC. It is then required to carry out a study about how a wrong mapping of terms affects to IR process.

Finally, it is worth noting that results are similars in simple words and multi-words, which show the feasibility of our novel *MD* algorithm to deal with restricted-domain corpus.

5 Conclusions and Future Work

Real-world data are inherently noisy, so techniques to process noise are crucial in IR systems to obtain useful and actionable results [14].

Due to the huge amount of redundancy inherent to open-domain corpus, they are insensitive to noise. Nevertheless, corpus in restricted domains are usually rather smaller and, therefore, with little or no redundancy. Consequently, IR systems using restricted-domain corpus are likely to fail.

In order to overcome this problem, in this paper, we show how noise tolerance can be added to the retrieval process for restricted domains. Our approach is based on comparing terms from the noisy corpus and important terms of the restricted-domain obtained from a KOS. This comparison is carried out by a novel algorithm that adapts a edit distance (MD) to consider multi-words.

The main benefit of our approach is that the performance of a IR system is kept although noisy restricted-domain corpus are used, as is shown in several experiments that have been carried out within the agricultural domain.

Our future work is focused on extending the proposed experiments to improve the MD algorithm to be used in other NLP tasks such as QA. Finally, as we believe that our approach is suitable for dealing with noise in the question, we plan to do several experiments in this sense.

Acknowledgments. This research has been partially funded by the Valencia Government under project PROMETEO/2009/119, and by the Spanish government under project Textmess 2.0 (TIN2009-13391-C04-01).

References

1. Aunimo, L., Heinonen, O., Kuuskoski, R., Makkonen, J., Petit, R., Virtanen, O.: Question answering system for incomplete and noisy data. In: Sebastiani, F. (ed.) ECIR 2003. LNCS, vol. 2633, pp. 193–206. Springer, Heidelberg (2003)
2. Baeza-Yates, R.A., Ribeiro-Neto, B.A.: Modern Information Retrieval. ACM Press / Addison-Wesley (1999)
3. Bourdaillet, J., Ganascia, J.-G.: Alignment of noisy unstructured data. In: IJCAI 2007 Workshop on Analytics for Noisy Unstructured Text Data (January 2007)
4. Chen, Q., Li, M., Zhou, M.: Improving query spelling correction using web search results. In: Proceedings of the 2007 Joint Conference on Empirical Methods in Natural Language Processing and Computational Natural Language Learning (EMNLP-CoNLL), Prague, Czech Republic, June 2007, pp. 181–189. Association for Computational Linguistics (2007)
5. Clark, A.: Pre-processing very noisy text. In: Proceedings of Workshop on Shallow Processing of Large Corpora, Corpus Linguistics (2003)
6. Esser, W.M.: Fault-tolerant fulltext information retrieval in digital multilingual encyclopedias with weighted pattern morphing. In: McDonald, S., Tait, J.I. (eds.) ECIR 2004. LNCS, vol. 2997, pp. 338–352. Springer, Heidelberg (2004)
7. Fernández, A.C., Díaz, J., Fundora, A., Muñoz, R.: Un algoritmo para la extracción de características lexicográficas en la comparación de palabras. In: IV Convención Científica Internacional de La Universidad De Matanzas CIUM 2009 (2009)
8. Gómez, J.M.: Recuperación de Pasajes Multilingüe para la Búsqueda de Respuestas. Phd. thesis, Departamento de Sistemas Informáticos y Computación, Universidad Politécnica de Valencia, Valencia, Spain (2007)
9. Gómez, J.M., Rosso, P., Sanchis, E.: Jirs language-independent passage retrieval system: A comparative study. In: 5th Internacional Conference on Natural Language Proceeding (ICON 2007), MaCMillan Publisher, Basingstoke (2007)
10. Hassan, T., Baumgartner, R.: Intelligent text extraction from pdf documents. In: CIMCA/IAWTIC, pp. 2–6 (2005)

11. Hodge, G.: Systems of Knowledge Organization for Digital Libraries: Beyond Traditional Authority Files. In: The Digital Library Federation Council on Library and Information Resources (2000)
12. Kantor, P.B., Voorhees, E.M.: The trec-5 confusion track: Comparing retrieval methods for scanned text. Inf. Retr. 2(2/3), 165–176 (2000)
13. Khadivi, S., Ney, H.: Automatic filtering of bilingual corpora for statistical machine translation. In: Montoyo, A., Muñoz, R., Métais, E. (eds.) NLDB 2005. LNCS, vol. 3513, pp. 263–274. Springer, Heidelberg (2005)
14. Knoblock, C.A., Lopresti, D.P., Roy, S., Venkata Subramaniam, L.: Special issue on noisy text analytics. IJDAR 10(3-4), 127–128 (2007)
15. Levenshtein, V.I.: Binary codes capable of correcting deletions, insertions, and reversals. Technical Report 8 (1966)
16. Li, M., Zhu, M., Zhang, Y., Zhou, M.: Exploring distributional similarity based models for query spelling correction. In: ACL (2006)
17. Lopresti, D.P., Roy, S., Schulz, K., Subramaniam, L.V.: Special issue on noisy text analytics. IJDAR 12, 139–140 (2009)
18. Minock, M.: Where are the "killer applications" of restricted domain question answering? In: Proceedings of the IJCAI Workshop on Knowledge Reasoning in Question Answering, Edinburgh, Scotland, p. 4 (2005)
19. Needleman, S.B., Wunsch, C.D.: A general method applicable to the search for similarities in the amino acid sequence of two proteins. Journal of molecular biology 48(3), 443–453 (1970)
20. Oard, D.W., Hedin, B., Tomlinson, S., Baron, J.R.: Overview of the trec 2008 legal track. In: TREC (2008)
21. Venkata Subramaniam, L., Roy, S., Faruquie, T.A., Negi, S.: A survey of types of text noise and techniques to handle noisy text. In: AND 2009: Proceedings of The Third Workshop on Analytics for Noisy Unstructured Text Data, pp. 115–122. ACM, New York (2009)
22. Vila, K., Ferrández, A.: Developing an ontology for improving question answering in the agricultural domain. In: MTSR, pp. 245–256 (2009)
23. Vinciarelli, A.: Noisy text categorization. IEEE Trans. Pattern Anal. Mach. Intell. 27(12), 1882–1895 (2005)
24. Winkler, W.E.: Overview of record linkage and current research directions. Research report series, rrs, Statistical Research Division, U.S. Census Bureau, Washington, DC 20233 (2006)

Measuring Tree Similarity for Natural Language Processing Based Information Retrieval

Zhiwei Lin, Hui Wang, and Sally McClean

Faculty of Computing and Engineering
University of Ulster, Northern Ireland, United Kingdom
{z.lin,h.wang,si.mcclean}@ulster.ac.uk

Abstract. Natural language processing based information retrieval (NIR) aims to go beyond the conventional bag-of-words based information retrieval (KIR) by considering syntactic and even semantic information in documents. NIR is a conceptually appealing approach to IR, but is hard due to the need to measure distance/similarity between structures. We aim to move beyond the state of the art in measuring structure similarity for NIR.

In this paper, a novel tree similarity measurement *dtwAcs* is proposed in terms of a novel interpretation of trees as multi dimensional sequences. We calculate the distance between trees by the way of computing the distance between multi dimensional sequences, which is conducted by integrating the all common subsequences into the dynamic time warping method. Experimental result shows that *dtwAcs* outperforms the state of the art.

1 Introduction

Information retrieval (IR) is a branch of computing science, which is a key component of search engines. Most IR systems, both operational and experimental, have traditionally been keyword-based. Google is currently the most successful Internet search engine, at the center of which is a keyword-based IR system coupled with the famous PageRank algorithm. The techniques used in Google and many other search engines are successful in utilizing the information available to the keyword-based approach to IR. They are quite effective if our information need (described in queries) is sufficiently characterized by the joint occurrence of a set of keywords. However our information need is sometimes better described by the joint occurrence of keywords along with the *structure of keywords*.

For example, a keyword-based search engine cannot differentiate between the following two queries: 'measurement of models' and 'model of measurements'. To answer such queries sensibly and sufficiently, we need to analyze the grammatical (syntactic) structure of the queries and identify the grammatical roles of keywords in the structure. The keywords and their structure are then used as a basis to search through the document collection to find the relevant ones. This is a *natural language processing (NLP) based approach to IR* (NIR for short).

It has long been felt intuitively that NLP techniques would benefit IR significantly, and this intuition has been validated with the research of natural language processing [1–6]. The Natural Language Information Retrieval (NLIR) project at New York

C.J. Hopfe et al. (Eds.): NLDB 2010, LNCS 6177, pp. 13–23, 2010.

University was a collaborative effort between a number of leading institutions in USA and Europe [3, 7, 8]. This project was to demonstrate that robust NLP techniques can help to derive better representation of texts for indexing and search purposes than any simple word or string-based methods commonly used in statistical full-text retrieval. In question/answering area, studies have shown the NLP based approach to questions classification is better than the bag-of-words method [5, 6].

This paper aims to go beyond the conventional bag-of-words based information retrieval (KIR) and extend the current NLP based approach by considering syntactic and even semantic information in documents. To this end, each sentence in a corpus will be parsed into a constituent tree, in which the grammatical information and the relationship between words are preserved so that much more semantic information can be captured. Significantly different from bag-of-words method, we need to measure the similarity between these trees. However, this measurement is not trivial due to the inherent complexity of a tree. In order to efficiently measure the similarity between tree, a novel tree similarity measurement *dtwAcs* is proposed in terms of a novel interpretation of trees as multi dimensional sequences. We calculate the distance between trees by the way of computing the distance between multi dimensional sequences, which is conducted by integrating the all common subsequences into the dynamic time warping method. Experimental results show that *dtwAcs* outperforms the state of the art.

This paper is organized as follows. In Section 2, the notations used in the paper are summarized. Section 3 briefly review some existing methods of tree similarity measurement. In Section 4, we use multi dimensional sequence to interpret a tree, and then the distance between trees becomes the distance between multi-dimensional sequences, which is calculated by integrating dynamic time warping and all common subsequences. Finally, experiment was conducted to show the algorithm's performance. The paper is concluded with a summary in Section 6.

2   Preliminaries

In this section, key concepts and notations are formally defined and introduced.

A *tree* T is denoted by $T = (V, E)$, where V and E, respectively, are the set of nodes and the set of edges of tree T. If the tree has a *root* node, denoted by $root(T)$, the tree is called *rooted tree*.

Let Σ be a finite *alphabet*. If tree T is a *labeled tree* over Σ, let the label of the node $v \in T$ be $\sigma(v)$, then $\sigma(v) \in \Sigma$. An *ordered tree* T is a labeled tree, which has left-right order over the children of each node. In the rest of this paper, we assume trees are *rooted*, *labeled*, and *ordered* unless otherwise stated.

Let Σ^T be the set of rooted, labeled, and ordered trees over Σ. The *size* of $T = (V, E)$, denoted by $|T|$, is the number of nodes in T, i.e., $|T| = |V|$. An empty tree ϵ has a size of 0. Let v be a node of T, $|v|$ be the number of children of v, and $ch(v, i)$ be the i-th child of v. A *leaf* v is a node without any child, i.e., $|v| = 0$, and $\mathcal{L}(T)$ is the set of all leaves in tree T. For $v \in T$ and $u \in T(v)$, we use $\pi_T(u, v)$ to denote the acyclic *path* from u to v. Let u be the i-th leaf in tree T, we use $path(T, i)$ to denote the acyclic path starting from $root(T)$ to u, i.e., $path(T, i) = \pi_T(root(T), u)$. The *depth* of $v \in V$, $depth(v)$, is the number of nodes in the path between $root(T)$ and v.

The depth of tree $\mathcal{T}$, denoted by $depth(\mathcal{T})$, is the maximum depth of nodes in $\mathcal{T}$. A *subtree* rooted at v, denoted by $\mathcal{T}(v)$, is a tree that includes v and all descendants of v. Let $v_1, v_2, \cdots, v_n$ be the children of v, the *pre-order traversal* of $\mathcal{T}(v)$ first visits v and then $\mathcal{T}(v_k)$ $(k = 1, 2, \cdots, n)$.

Sequence is a special tree over Σ where each element in the sequence does not have any siblings. An k-long sequence s is a tuple $s =< s(1), \ldots, s(k) >$ and $s(i) \in \Sigma$ $(1 \le i \le k)$. The length of s is denoted by $|s|$, i.e., $k = |s|$. The *empty sequence*, which has a length of zero, is denoted by λ, i.e., $|\lambda| = 0$. A k-long sequence $y = y_1 \ldots y_k$ is a *subsequence* of s, denoted by $y \preceq s$, if there exist $k + 1$ sequences $x_1, \ldots, x_{k+1}$ such that $x_1 y_1 \ldots x_k y_k x_{k+1} = s$. Actually, a subsequence of s can be obtained by deleting 0 or more elements from s. We let Σ^s be the set of sequences over Σ.

A multi-dimensional sequence $\mathbb{S}$ with k dimensions is a tuple over Σ^s, i.e., each entry $s_i (1 \le i \le k)$ in $\mathbb{S}$ is a sequence, or $s_i \in \Sigma^s$. We use $|\mathbb{S}|$ to denote the dimensions of $\mathbb{S}$, i.e., $k = |\mathbb{S}|$.

In the rest of paper, we use notations $\mathcal{S}$ and $\mathcal{T}$ to denote trees. Notations $\mathbb{S}$ and $\mathbb{T}$ are used to denote multi-dimensional sequences and lower letters, e.g., s_i for $\mathbb{S}(1 \le i \le |\mathbb{S}|)$ and t_j for $\mathbb{T}(1 \le j \le |\mathbb{T}|)$ respectively, to denote the elements in multi-dimensional sequences.

3 Review of Tree Similarities

Measuring tree similarity or distance plays an important role in the analysis of tree structured data. The best known and widely used measures are tree edit distance [9–13] and tree kernel [14–20]. In this section, we review tree edit distance and tree kernel in order to provide a context for our work.

3.1 Tree Edit Distance

Tree edit distance is the minimum number of edit operations to transform one tree to another. Different edit operations can be used, but the most common edit operations are: (1) insert a node into a tree; (2)delete a node from a tree; (3)change label of a node [9–13]. However these operations, especially insert and delete, may be interpreted differently for different purposes. For example, insertion may be allowed only when the new node becomes a leaf, and deletion is strictly allowed at the leaves [10–12].

Let $\mathcal{S}$ and $\mathcal{T}$ be two trees, and $c(e)$ be the cost associated with edit operation e. The cost of a sequence of edit operations $e_1, e_2, \cdots, e_n$ is $\sum_{i=1}^n c(e_i)$. The edit distance between $\mathcal{S}$ and $\mathcal{T}$, denoted by $ed(\mathcal{S}, \mathcal{T})$, is the minimum cost of transforming $\mathcal{S}$ to $\mathcal{T}$ through a sequence of edit operations.

Dynamic programming is often used to calculate the edit distance and there are many algorithms for calculating edit distance without any constraints on insert and delete operations [9]. A commonly cited dynamic programming algorithm to compute edit distance between two ordered trees is that in [13], where the space requirement is $O(|\mathcal{S}| \times |\mathcal{T}|)$, and the time complexity is $O(|\mathcal{S}| \times |\mathcal{T}| \times \min(depth(\mathcal{S}), |\mathcal{L}(\mathcal{S})|) \times \min(depth(\mathcal{T}), |\mathcal{L}(\mathcal{T})|))$.

Without constraints on edit operations, these algorithms are time consuming, especially when they are used to compare large trees. A recent study in [10] extended earlier work in [12], where insertion and deletion are allowed only at leaf nodes and the notion of *edit operation graph* (EOG) is introduced so that the tree edit distance problem can be efficiently solved by the Dijkstra algorithm. The algorithm has a lower complexity of $O(|\mathcal{S}| \times |\mathcal{T}|)$, but is not guaranteed to produce optimal distance due to restrictions on edit operations. This kind of TED has been extensively studied and applied in XML clustering by structure [11].

3.2   Tree Kernel

Tree kernel, as a similarity measure, was first studied for tasks in natural language processing [15, 17]. It is an extension of the concept of *convolution kernel* [21] to tree structured data. The main idea behind tree convolution kernels [15, 17] is to map trees to an n-dimensional feature space $\chi = \{f_1, f_2, \cdots, f_n\}$, where feature f_i corresponds to a subtree. A tree $\mathcal{T}$ can be represented as a feature vector $< h_1(\mathcal{T}), h_2(\mathcal{T}), \cdots, h_n(\mathcal{T}) >$, where $h_i(\mathcal{T})$ is the frequency of subtree f_i in $\mathcal{T}$. The kernel of two trees $\mathcal{S}$ and $\mathcal{T}$ is defined as the inner product of their feature vectors, i.e., $K(\mathcal{S}, \mathcal{T}) = \sum_{1 \leq i \leq n} h_i(\mathcal{S}) h_i(\mathcal{T})$. For each $v \in \mathcal{T}$, let $I_i(v) = 1$ if the root of f_i has the same label as v; otherwise, $I_i(v) = 0$, then $K(\mathcal{S}, \mathcal{T})$ can be changed to $K(\mathcal{S}, \mathcal{T}) = \sum_{v_1 \in \mathcal{S}} \sum_{v_2 \in \mathcal{T}} C(v_1, v_2)$, where $C(v_1, v_2) = \sum_{1 \leq i \leq n} I_i(v_1) I_i(v_2)$. Thus $K(\mathcal{S}, \mathcal{T})$ can be calculated through $C(v_1, v_2)$.

$C(v_1, v_2)$ can be efficiently calculated through dynamic programming. In natural language processing, a parse tree has *pre-terminal* nodes, each having only one leaf child. With a pre-terminal constraint, an efficient algorithm was proposed to calculate $C(v_1, v_2)$ according to the following [15]:

$$
C(v_1, v_2) = \begin{cases} 0, & \text{if } label(v_1) \neq label(v_2); \\ 1, & \text{otherwise, if } v_1, v_2 \text{ are pre-terminals}; \\ \prod_{j=1}^{|v_1|} (1 + C(ch(v_1, j), ch(v_2, j))), & \\ & \text{otherwise, if } prod(v_1) = prod(v_2); \end{cases}
\tag{1}
$$

where $prod(v_1) = prod(v_2)$ means that $label(v_1) = label(v_2)$, $|v_1| = |v_2|$, and the labels of $ch(v_1, j)$ and $ch(v_2, j)$ are identical. Equation 1 implies an algorithm with $O(|\mathcal{S}| \times |\mathcal{T}|)$ time complexity. To efficiently compute Equation 1, a fast computation algorithm was proposed in [17], but it still has $O(|\mathcal{S}| \times |\mathcal{T}|)$ complexity in the worst case. To improve the performance of Equation 1, $C(v_1, v_2)$ is multiplied by a decay factor λ $(0 < \lambda \leq 1)$ [15, 17].

The pre-terminal condition was relaxed so that the resulting tree kernel can be applied to trees whose leaves may have siblings [6]. This tree kernel has the restriction that the root of a subtree has a matched node v in its source tree and the subtree contains all the descendants of v. Another type of subtree is *partial subtree*, which is meaningful for grammar rules [5], and the corresponding tree kernel can be calculated efficiently as in [17].

Another tree kernel, called GTK for short, was presented in [16] under a more general definition of subtree. A subtree could be any part of its source tree, but needs to keep parent-child relationship. A different recursive equation was proposed to calculate $C(v_1, v_2)$, for $1 \leq i \leq |v_1|, 1 \leq j \leq |v_2|$:

$$
\begin{cases}
C(v_1, v_2) = S[|v_1|, |v_2|] \\
S[i, j] = S[i - 1, j] + S[i, j - 1] \\
\quad + S[i - 1, j - 1] * (C(ch(v_1, i), ch(v_2, j)) - 1)
\end{cases}
\tag{2}
$$

where S is a $(|v_1| + 1) \times (|v_2| + 1)$ matrix and $S[0, 0], S[0, j], S[i, 0]$ are initiated with 0. Another extension in this paper [16] is the definition of *elastic subtree*, which breaks the parent-child relation and allows the subtree be ancestor-descendant relation in the original tree. However, the authors note that it does not outperform existing methods in terms of classification accuracy. A recent work, called *route kernel*, takes into account the information of relative position between subtrees but this information does not make sense in some situations [22].

Spectrum tree kernel advocates using the frequency of the subtree with a fixed size q in trees [19], which is conceptually different from the above mentioned tree convolution kernels. Spectrum tree kernel, however, may not be able to capture much common information between trees. Another problem is due to the fixed size q, which must be given. The experimental result in [19] has confirmed that when $q = 1$, the computational time increase dramatically.

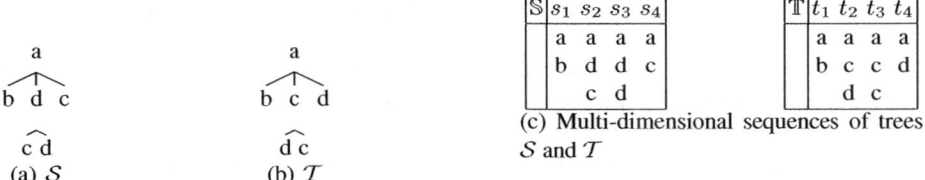

$\mathbb{S}$	s_1	s_2	s_3	s_4
	a	a	a	a
	b	d	d	c
		c	d	

$\mathbb{T}$	t_1	t_2	t_3	t_4
	a	a	a	a
	b	c	c	d
		d	c	

(a) $\mathcal{S}$ (b) $\mathcal{T}$

(c) Multi-dimensional sequences of trees $\mathcal{S}$ and $\mathcal{T}$

Fig. 1. Trees $\mathcal{S}$ and $\mathcal{T}$

4   Multi-dimensional Sequence Approach to Tree Similarity

This section introduces how we compare two trees by computing the distance between two multi dimensional sequences. In this method, we firstly present how to interpret a tree by a multi dimensional sequence (*MDS*). Then the distance between two trees equals to the distance between two *MDS*s, which is computed by the algorithm dynamic time warping.

4.1   Multi-dimensional Sequence of a Tree

Suppose tree $\mathcal{S}$ has l leaves and let $path(\mathcal{S} : i)(1 \leq i \leq l)$ be an acyclic path starting from $root(\mathcal{S})$ to the i-th leaf. We let $\mathbb{S}$ be a multi-dimensional sequence of $\mathcal{S}$ where each element $s_i(1 \leq i \leq l)$ in $\mathbb{S}$ is $path(\mathcal{S} : i)$.

Example 1. Fig. 1(c) shows the multi dimensional sequences of trees in Fig. 1, where $\mathbb{S}$ and $\mathbb{T}$, respectively, are multi dimensional sequences of trees $\mathcal{S}$ and $\mathcal{T}$.

4.2 Distance between Elements of Two *MDS*s

For any $\mathcal{S}, \mathcal{T}$, their multi dimensional sequences are interpreted by $\mathbb{S}$ and $\mathbb{T}$, respectively. Before introducing the distance between $\mathbb{S}$ and $\mathbb{T}$, we need to know, for the elements s_i in $\mathbb{S}(1 \leq i \leq |\mathbb{S}|)$ and t_j in $\mathbb{T}(1 \leq j \leq |\mathbb{T}|)$, how similar two elements s_i and t_j are. There are many methods can be adopted to compute distance or similarity between two s_i and t_j in the literature [23–27]. This paper adopts *all common subsequences* [23, 27].

All common subsequences. *All common subsequences* (ACS) was studied in [27] as a similarity measure, and as a computer science problem, ACS is concerned with the search for an efficient method to find all common subsequences of two or more sequences [23].

Consider two sequences $s, t \in \Sigma^s$, where $|s| = m$ and $|t| = n$. For $x \in \Sigma$, if there exists $k(1 \leq k \leq i)$, such that $x = s(k)$, let $\ell_s(i, x) = \max\{k|s(k) = x\}$; otherwise, let $\ell_s(i, x) = 0$. Then, for $1 \leq i \leq m, 1 \leq j \leq n$,

$$\phi[i,j] = \begin{cases} \phi[i-1,j], \text{ if } \ell_t(j, s(i)) = 0; \\ \phi[i-1,j] + \phi[i-1, \ell_t(j, s(i)) - 1], \\ \qquad \text{if } \ell_t(j, s(i)) > 0, \ell_s(i-1, s(i)) = 0; \\ \phi[i-1,j] + \phi[i-1, \ell_t(j, s(i)) - 1] - \\ \phi[\ell_s(i-1, s(i)) - 1, \ell_t(j, s(i)) - 1], \qquad \text{otherwise.} \end{cases} \quad (3)$$

where, $\phi[i,j]$ is an $(m+1) \times (n+1)$ matrix, and $\phi[i, 0] = 1, \phi[0, j] = 1$ and $\phi[0, 0] = 1$. Then the number of all common distinct subsequences between s and t, denoted by $acs(s, t)$, is $\phi[m, n]$.

For the sake of comparability, for any $s, t \in \Sigma^s$, it is necessary to normalize $acs(s, t)$. To this end, we define similarity function $\delta_{acs}(s, t)$ between s and t as

$$\delta_{acs}(s, t) = \frac{acs(s, t)}{\sqrt{acs(s, s) \times acs(t, t)}} \quad (4)$$

Because $\ell_s(i, x)$ and $acs(s, s)$ can be pre-processed before computing $\delta_{acs}(s, t)$, Equation 4 implies quadratic complexity.

4.3 Tree Distance with Dynamic Time Warping

In this section, we firstly introduces the *dynamic time warping* algorithm for handling sequences with various length. Then we propose a hybrid approach by integrating dynamic time warping algorithm with either all common subsequence or the largest common subsequence to measure the similarity between two trees.

Dynamic time warping (DTW) is a sequence alignment method for finding an optimal match between two sequences which may have different time or speed [28]. This method has been known and widely used in speech recognition to cope with different speaking speeds. DTW has also been used extensively in many database/data mining applications, e.g., time series analysis. The following equation shows how DTW calculate similarity between two sequences x and y^1 (with m and n elements, respectively) by a dynamic programming algorithm ($1 \leq i \leq m, 1 \leq j \leq n$) with cost value over their elements $x(i)$ and $y(j)$, denoted by $\theta(x(i), y(j))$:

$$\Phi[i,j] = \theta(x(i), y(j)) + \min\left\{\Phi[i, j-1], \Phi[i-1, j], \Phi[i-1, j-1]\right\} \qquad (5)$$

where, $\Phi[i,j]$ is a $(m+1) \times (n+1)$ matrix, $\Phi[0,j]$ and $\Phi[i,0]$ are initiated with ∞ and $\Phi[0,0] = 0$. Then the distance between x and y, denoted by $dtw(x, y)$ equals to $\Phi[m, n]$.

As can be seen from Equation 5, $dtw(x, y)$ highly relies on the cost value function $\theta(x(i), y(j))$ between two elements $x(i)$ and $y(j)$. In 1-dimensional sequence, for the case that the elements $x(i)$ and $y(j)$ are categorical, if $x(i) = y(j)$, then $\theta(x(i), y(j)) = 1$; otherwise, $\theta(x(i), y(j)) = 0$. If the elements $x(i)$ and $y(j)$ are numerical, then $\theta(x(i), y(j)) = |x(i) - y(j)|$.

However, in multi-dimensional sequences, given two such sequences $\mathbb{S}$ and $\mathbb{T}$, to calculate $dtw(\mathbb{S}, \mathbb{T})$, $\theta(x(i), y(j))$ must be computed in an additional way. Here, we present how to embed all common subsequences, and possibly other sequence methods, i.e., the *largest common subsequence*[24], *string edit distance* [26] or *sequence kernel* [25], into dynamic time warping. This approach works out $\theta(x(i), y(j))$ by computing the possibility of similarity with all common subsequences (i.e., $acs(x(i), y(j))$).

Consider two trees $\mathcal{S}$ and $\mathcal{T}$, which are interpreted as multi dimensional sequences $\mathbb{S}$ and $\mathbb{T}$, where obviously $|\mathbb{S}| = |\mathcal{L}(\mathcal{S})|$ and $|\mathbb{T}| = |\mathcal{L}(\mathcal{T})|$. To embed ACS into DTW, we define $\theta(x(i), y(j))$ as follows, for $1 \leq i \leq |\mathbb{S}|, 1 \leq j \leq |\mathbb{T}|$:

$$\theta(x(i), y(j)) = 1 - \delta_{acs}(x(i), y(j)) \qquad (6)$$

Equation 6 is denoted by *dtwAcs*. Then for *dtwAcs* , the distance between trees $\mathcal{S}$ and $\mathcal{T}$ equals to $\Phi[|\mathbb{S}|, |\mathbb{T}|]$.

We have shown the method to interpret a tree $\mathcal{S}$ as a multi-dimensional sequence $\mathbb{S}$, whose dimension equals to the number of the leaves of tree $\mathcal{S}$ (i.e., $|\mathbb{S}| = |\mathcal{L}(\mathcal{S})|$), and whose elements $x(i)$ are the paths from the root to the leaf i, which implies that $\theta(x(i), y(j))$ is $O(depth(\mathcal{S}) \times depth(\mathcal{T}))$. Together with the computation of $\Phi[i, j]$, for $1 \leq i \leq |\mathbb{S}|, 1 \leq j \leq |\mathbb{T}|$, the time complexity of *dtwAcs* is $O(|\mathcal{L}(\mathcal{S})| \times |\mathcal{L}(\mathcal{T})| \times depth(\mathcal{S}) \times depth(\mathcal{T}))$.

4.4   More on *dtwAcs*

The analysis shows that *dtwAcs* has high computational complexity. It is time consuming to calculate *dtwAcs* in this way. For example, consider $s_2 = adc$ and $s_3 = add$ in Fig. 1(c), s_2 and s_3 share two common prefixes "ad", which are ancestors of both c and d in original tree $\mathcal{S}$ in Fig. 1. If we want to compute $dtw(\mathbb{S}, \mathbb{T})$ via Equation 6, then the computation of $\delta_{acs}(s_2, t_j)$ ($1 \leq j \leq 4$) and $\delta_{acs}(s_3, t_j)$ ($1 \leq j \leq 4$) results

---

[1] We use x, y, instead of s, t, as the sequences in this section have distinct and more general concepts from that of the last section, e.g., the elements in sequences may be numerical or sequential.

in comparing the common subsequence "ad" with t_j more than once. In order to avoid the repeated calculation, we seek to preserving the value for "ad" after calculation over s_2 so that it can be used for calculating over s_3.

To this end, we need to use the pre-order traversal sequences of trees, and when calculate δ_{acs} for a new node, we return to its parent position instead of its previous position in pre-order traversal sequence. And if the node is a leaf node, we turn to calculate Φ. Therefore, the complexity of *dtwAcs* algorithm becomes $O(\mathcal{S} \times \mathcal{T})$.

5   Evaluation

We conducted an experimental evaluation to demonstrate the utility of *dtwAcs* in terms of classification accuracy by comparison with the commonly used tree kernels in [5] [19] and tree edit distance in [11].

5.1   Datasets

Two datasets are used in this paper. The first one is from predicate arguments, in which the trees for predicate arguments' task are generated from the corpus PropBank[2]. We consider arguments from Arg.0 to Arg.4 and argument Arg.M. Fig. 5.2 shows three different arguments (Arg.0, Arg.1, and Arg.M) of the predicate 'give' in the sentence. More details about predicate arguments can be found in [5]. The training dataset contains more than 174,100 trees and we leave 16,319 trees for test.

In the second experiment, we apply our method in the area of question/answering in information retrieval. We use the dataset from UIUC Cognitive Computation Group[3], which contains 5452 questions for training and 500 questions for testing. These questions are categorized into 6 coarse grained classes and 50 fine grained subclasses (More details can be found at their website[4]). This paper considers the problem of classification of questions into coarse gained classes. We use Stanford Parser[5] to parse these questions into constituent tree (An example of constituent tree can be found in Fig. 2(a)).

5.2   Performance Analysis

A kernel must be *positive semi-definite* (PSD for short), which is very important to kernel based classifier, e.g., support vector machine (SVM), otherwise it may take a long time for the classifier to optimize its parameters, or even worse, it can't converge at all. Unfortunately, it has been proved that similarity measurement derived from dynamic time warping (DTW) and tree edit distance (TED) can't hold the PSD property, which means that they are not suitable for kernel machines. Therefore, we turn to *kNN* classifier for DTW and TED which is widely used in comparisons of similarity measurements. The parameter k for *kNN* is set to 1, 3, 5, 7, 9, 11, 15, 17 and 19, and the parameters for tree kernels follow the preferred values in the literature [5] [19].

---

[2] http://verbs.colorado.edu/~mpalmer/projects/ace.html
[3] Available at http://l2r.cs.uiuc.edu/~cogcomp/Data/QA/QC/
[4] http://l2r.cs.uiuc.edu/~cogcomp/Data/QA/QC/definition.html
[5] Available at http://nlp.stanford.edu/software/lex-parser.shtml

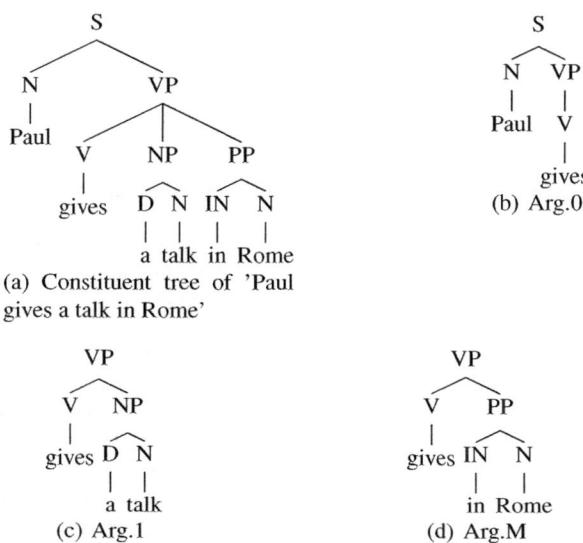

(a) Constituent tree of 'Paul gives a talk in Rome'

(b) Arg.0

(c) Arg.1

(d) Arg.M

Fig. 2. Constituent tree of 'Paul gives a talk in Rome' and its predicate arguments

Table 1. Classification accuracy (%)

	Predicate dataset	Questions classification
dtwAcs	82.74	78.20
Tree kernel in [5, 15]	78.42	75.20
Tree kernel in [19]	76.05	67.00
Tree edit distance [11]	51.09	68.60

Table 1 shows the classification accuracy of different algorithms on the dataset. In the table, we can find that *dtwAcs* is significantly better than the other measures.

6   Conclusion

Information retrieval (IR) is a branch of computing science, which is a key component of search engines. Most IR systems are based on bag-of-words, and hence lack of mechanism of understanding the meaning of what user needs.

A novel tree similarity measurement *dtwAcs* is proposed in terms of a novel interpretation of trees into multi dimensional sequences. We calculate the distance between trees by the way of computing the distance between multi dimensional sequences, which is conducted by integrating the all common subsequences into the dynamic time warping method. Experimental result shows that *dtwAcs* outperforms the state of the art. Future work includes an experiment in our IR system and test our hypothesis with TREC dataset.

Acknowledgements. Thank Professor Alessandro Moschitti for source his codes and datasets.

References

1. Mauldin, M.: Retrieval performance in FERRET: a conceptual information retrieval system. In: SIGIR 1991 (1991)
2. Strzalkowski, T. (ed.): Natural language Information Retrieval. Kluwer, New York (1999)
3. Carballo, J.P., Strzalkowski, T.: Natural language information retrieval: progress report. Information Processing Management 36(1), 155–178 (2000)
4. Mittendorfer, M., Winiwarter, W.: Exploiting syntactic analysis of queries for information retrieval. Journal of Data and Knowledge Engineering (2002)
5. Moschitti, A.: Efficient convolution kernels for dependency and constituent syntactic trees. In: Fürnkranz, J., Scheffer, T., Spiliopoulou, M. (eds.) ECML 2006. LNCS (LNAI), vol. 4212, pp. 318–329. Springer, Heidelberg (2006)
6. Zhang, D., Lee, W.S.: Question classification using support vector machines. In: SIGIR 2003: Proceedings of the 26th annual international ACM SIGIR conference on Research and development in informaion retrieval, pp. 26–32. ACM, New York (2003)
7. Strzalkowski, T.: Natural Language Information Retrieval Project Homepage, http://www.cs.albany.edu/tomek
8. Strzalkowski, T., Perez-Carballo, J., Karlgren, J., Hulth, A., Tapanainen, P., Lahtinen, T.: Natural language information retrieval: TREC-8 report. In: TREC 1999, pp. 381–390 (1999)
9. Bille, P.: A survey on tree edit distance and related problems. Theor. Comput. Sci. 337(1-3), 217–239 (2005)
10. Chawathe, S.S.: Comparing hierarchical data in external memory. In: VLDB 1999, Edinburgh, UK, pp. 90–101 (1999)
11. Dalamagas, T., Cheng, T., Winkel, K.J., Sellis, T.: A methodology for clustering xml documents by structure. Information System 31(3), 187–228 (2006)
12. Selkow, S.M.: The tree-to-tree editing problem. Information Processing Letters 6(6), 184–186 (1977)
13. Zhang, K., Shasha, D.: Simple fast algorithms for the editing distance between trees and related problems. SIAM Journal on Computing 18(6), 1245–1262 (1989)
14. Che, W., Zhang, M., Aw, A., Tan, C., Liu, T., Li, S.: Using a hybrid convolution tree kernel for semantic role labeling. ACM Transactions on Asian Language Information Processing (TALIP) 7(4), 1–23 (2008)
15. Collins, M., Duffy, N.: Convolution kernels for natural language. In: Advances in Neural Information Processing Systems 14, pp. 625–632. MIT Press, Cambridge (2001)
16. Kashima, H., Koyanagi, T.: Kernels for semi-structured data. In: ICML 2002: Proceedings of the Nineteenth International Conference on Machine Learning, pp. 291–298. Morgan Kaufmann Publishers Inc., San Francisco (2002)
17. Moschitti, A.: Making tree kernels practical for natural language learning. In: Proceedings of the Eleventh International Conference on European Association for Computational Linguistics, Trento, Italy (2006)
18. Moschitti, A., Pighin, D., Basili, R.: Tree kernels for semantic role labeling. Computational Linguistics 34(2), 193–224 (2008)
19. Tetsuji, K., Kouichi, H., Hisashi, K., Kiyoko, F.K., Hiroshi, Y.: A spectrum tree kernel. Transactions of the Japanese Society for Artificial Intelligence 22(2), 140–147 (2007)
20. Vishwanathan, S.V.N., Smola, A.: Fast kernels for string and tree matching. Advances in Neural Information Processing Systems 15 (2003)

21. Haussler, D.: Convolution kernels on discrete structures. Technical report, Department of Computer Science, University of California at Santa Cruz (1999)
22. Aiolli, F., Da San Martino, G., Sperduti, A.: Route kernels for trees. In: ICML 2009: Proceedings of the 26th Annual International Conference on Machine Learning, pp. 17–24. ACM, New York (2009)
23. Elzinga, C., Rahmann, S., Wang, H.: Algorithms for subsequence combinatorics. Theoretical Computer Science 409(3), 394–404 (2008)
24. Hirschberg, D.S.: A linear space algorithm for computing maximal common subsequences. Communications of the ACM 18(6), 341–343 (1975)
25. Leslie, C., Eskin, E., Cohen, A., Weston, J., Noble, W.S.: Mismatch string kernels for svm protein classification. Neural Information Processing Systems 15, 1441–1448 (2003)
26. Levenshtein, V.: Binary Codes Capable of Correcting Deletions, Insertions and Reversals. Soviet Physics Doklady 10, 707 (1966)
27. Wang, H.: All common subsequences. In: IJCAI 2007: Proceedings of the 20th international joint conference on Artifical intelligence, Hyderabad, India, pp. 635–640 (2007)
28. Sakoe, H.: Dynamic programming algorithm optimization for spoken word recognition. IEEE Transactions on Acoustics, Speech, and Signal Processing 26, 43–49 (1978)

Sense-Based Biomedical Indexing and Retrieval

Duy Dinh and Lynda Tamine

University of Toulouse,
118 route de Narbonne, 31062 Toulouse, France
{dinh,lechani}@irit.fr

Abstract. This paper tackles the problem of term ambiguity, especially for biomedical literature. We propose and evaluate two methods of Word Sense Disambiguation (WSD) for biomedical terms and integrate them to a sense-based document indexing and retrieval framework. Ambiguous biomedical terms in documents and queries are disambiguated using the Medical Subject Headings (MeSH) thesaurus and semantically indexed with their associated correct sense. The experimental evaluation carried out on the TREC9-FT 2000 collection shows that our approach of WSD and sense-based indexing and retrieval outperforms the baseline.

Keywords: Word Sense Disambiguation, Semantic Indexing, Term Weighting, Biomedical Information Retrieval.

1   Introduction

Nowadays, the volume of biomedical literature is constantly growing at an ever increasing rate. The exploitation of information from heterogeneous and diverse medical knowledge sources becomes a difficult task for any automated methods of data mining and searching. In general, biomedical texts are expressed in documents using natural human language, which causes the common problem of *ambiguity*. Indeed, human natural language is ambiguous by its nature. For example, the term "gold" may be a noun or adjective depending on the context where it appears. The noun may refer to a yellow metallic element in chemistry, while the adjective may refer to the characteristic of something having the color of this metal. Ambiguity is a common problem in the general text as well as in the specific domain such as biomedicine. For instance, polysemous words that are written and pronounced the same way as another, but which have different senses, frequently indicate both a gene and encoded protein, a disease and associated proteins ... Such polysemous words are called ambiguous. Meanings or senses of a given word or term are usually defined in a dictionary, thesaurus, ontology and so on. Recognizing and assigning the correct sense to these words within a given context are referred to as Word Sense Disambiguation (WSD).

Many investigations on WSD in general text have been done during the last years. Current approaches to resolving WSD can be subdivided into four categories: *Knowledge-based* [1, 2, 3], *Supervised* [4, 5], *Unsupervised* [6] and *Bootstrapping* methods [7]. Knowledge-based approaches use external resources such

C.J. Hopfe et al. (Eds.): NLDB 2010, LNCS 6177, pp. 24–35, 2010.

as Machine Readable Dictionaries (MRDs), thesauri, ontologies, etc. as lexical knowledge resources. Supervised machine learning (ML) methods (*Decision
Trees, Naive Bayes, Vector Space Model, Support Vector Machines, Maximum
Entropy, AdaBoost* ...) use manually annotated corpus for training classifiers.
Unsupervised classifiers are trained on unannotated corpora to extract several
groups (clusters) of similar texts. Bootstrapping (semi-supervised) approaches
rely on a small set of seed data to train the initial classifiers and a large amount
of unannotated data for further training. Most of the developed systems for WSD
of biomedical text are based on the supervised ML approach [5, 8, 9], which depends totally on the amounts and quality of training data. This constitutes the
principal drawback of those approaches: they require a lot of efforts in terms of
cost and time for human annotators.

This paper proposes a sense-based approach for semantically indexing and
retrieving biomedical information. Our approach of indexing and retrieval exploits the poly-hierarchical structure of the Medical Subject Headings (MeSH)
thesaurus for disambiguating medical terms in documents and queries. The remainder of this paper is organized as follows. Section 2 presents related work
on WSD in biomedical literature. In section 3, we detail our methods of WSD
for biomedical terms in documents. Section 4 deals with the document relevance
scoring. Experiments and results are presented in section 5. We conclude the
paper in section 6 and outline some perspectives for our future work.

2   Word Sense Disambiguation in Biomedical Text

Term ambiguity resolution in the biomedical domain becomes a hot topic during
the last years due to the amount of ambiguous terms and their various senses used
in biomedical text. Indeed, the UMLS[1] contains over 7,400 ambiguous terms [10].
For instance, biomedical terms such as *adjustment, association, cold, implanta-
tion, resistance, blood_pressure, etc.* have different meanings in different contexts.
In MeSH, a concept may be located in different hierarchies at various levels of
specificity, which reflects its ambiguity. As an illustration, figure 1 depicts the
concept "Pain", which belongs to four branches of three different hierarchies
whose the most generic concepts are: *Nervous System Disease* (C10); *Patho-
logical Conditions, Signs and Symptoms* (C23); *Psychological Phenomena and
Processes* (F02); *Musculoskeletal and Neural Physiological Phenomena* (G11).

Biomedical WSD has been recently the focus of several works [11, 5, 12, 8, 9,
13, 14, 15]. Indeed, related works can be subdivided into two main categories:
Knowledge-based and *Supervised-based* WSD. In the knowledge-based approach,
ambiguous terms are processed using knowledge sources such as MRDs, thesauri, ontologies, list of words or abbreviations together with their meanings,
etc. For example, UMLS contains information about biomedical and health related concepts, their various names, and the relationships among them, including
the Metathesaurus, the Semantic Network, and the SPECIALIST Lexicon and
Lexical Tools. Humphrey *et al.* [11] employed the Journal Descriptor Indexing

---

[1] Unified Medical Language System.

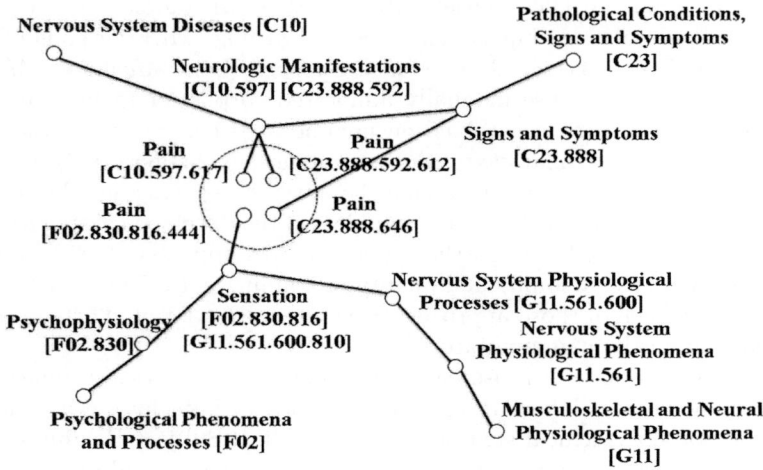

Fig. 1. Concept Pain in MeSH

(JDI) methodology to perform the disambiguation of terms in a large collection of MEDLINE citations when trying to map free text to the UMLS concepts. JDI-based WSD consists in selecting the best meaning that is correlated to UMLS semantic types assigned to ambiguous concepts in the Metathesaurus. More precisely, it is a statistical, corpus-based (training set of MEDLINE citations) method using pre-existing medical knowledge sources (a small set of journal descriptors in MeSH).

Most research works for WSD of biomedical text are based on the supervised ML approach [5, 12, 8, 9, 13, 14] and inspired by the Senseval[2] and BioCreative[3] challenges. The work in [13] used UMLS to disambiguate biomedical terms in documents using the information about term co-occurrence defined by using the naive Bayesian learning. Abbreviations in MEDLINE abstracts are resolved using SVM when trying to build a dictionary of abbreviations occurring with their long forms [12]. Briefly, works basically exploit features of general text then apply them to the biomedical text such as head word, part-of-speech, semantic relations and semantic types of words [8], unigrams, bigrams [9], surrounding words, distance, word collocations [5], lexical, syntactic features [14], etc. Most recently, the work in [15] proposed to disambiguate biomedical terms using the combination of linguistic features, which have been commonly used for WSD in general text, to the biomedical domain by augmenting it with additional domain-specific and domain-independent knowledge sources. The WSD method described in [15] combines the Concept Unique Identifiers (CUIs), which are automatically obtained from MetaMap [16] and MeSH terms, which are manually assigned to the abstracts to build feature vectors for training three WSD

---

[2] http://www.senseval.org/
[3] http://biocreative.sourceforge.net/

classifiers based on: Vector Space Model, Naive Bayes Network and Support Vector Machine. However, such methods become inflexible when always requiring a lot of efforts in terms of cost and time for human annotators.

Our contribution is outlined through the following key points:

1. We propose WSD methods that map free text to the MeSH concepts by assigning the most appropriate sense, indicated by its tree number, to each term or phrase in the local context of documents. Compared to other related WSD methods in the biomedical literature, our method has the following features: (1) do not require any training corpus, but only based on the local context of documents, and (2) exploit the MeSH semantic hierarchies to identify the correct sense for ambiguous concepts.
2. We then exploit our WSD algorithms as the basis of a sense-based indexing and retrieval model for biomedical text.

3 WSD Using MeSH Hierarchical Structure

Our objective here is to assign the appropriate *sense* related to a given *term* in the local context of the document mapped to the MeSH poly-hierarchy. Documents are at first tagged with Part-Of-Speech labels using a lexical tool such as TreeTagger [17]. A list of *concepts* is extracted in each document using the left to right maximum string matching algorithm. In what follows, we give some definitions and key notations and then detail our WSD methods.

3.1 Definitions and Notations

In MeSH, the preferred term, used for indexing, represents the name of the concept, also known as main heading. Otherwise, non-preferred ones, which are synonymous terms, are used for retrieval. In that poly-hierarchical structure, each concept is represented by a node belonging eventually to one (non-ambiguous) or multiple hierarchies (ambiguous), each of which corresponds to one of the sixteen MeSH domains: *A-Anatomy, B-Organisms, C-Diseases, ...* The following definitions are given based on the MeSH vocabulary.

- **Definition 1:** A **word** is an alphanumeric string delimited by spaces.
- **Definition 2:** A **term** is a group of one or more words comprising the basic unit of the vocabulary.
- **Definition 3:** A **concept** is the bearer of linguistic meaning consisting of synonymous term elements.
- **Definition 4:** The **sense** of a concept is represented by a tree node, indicated by the tree number in the poly-hierarchy. The set of **senses** of a concept c is denoted as $syn(c)$.
- **Definition 5:** The relationship **is-a** links concepts in the same hierarchy from various levels of specificity.

3.2   Left-To-Right Disambiguation

The first algorithm concerns the selection of the correct sense for each concept based on the following assumptions:

- The one-sense-per-discourse assumption [18], i.e., if a polysemous term appears two or more times in a discourse (sentence, paragraph, document), it is extremely likely that they will all share the same sense.
- The correlation between concepts in a local context expresses their semantic closeness.
- The priority of meanings is defined by the precedence of concepts: the leftmost concept impacts the overall meaning of the discourse, which inspires the semantic chain of the document from the beginning to the end.

Based on these hypotheses, we firstly compute the semantic similarity between the leftmost concept with its nearest neighbor. The third concept is disambiguated based on the meaning of the second and so on. Afterwards, their meanings will be propagated in all of their occurrences in the document. We visually illustrate such principle of WSD in Figure 2. Formally, given a sequence of n

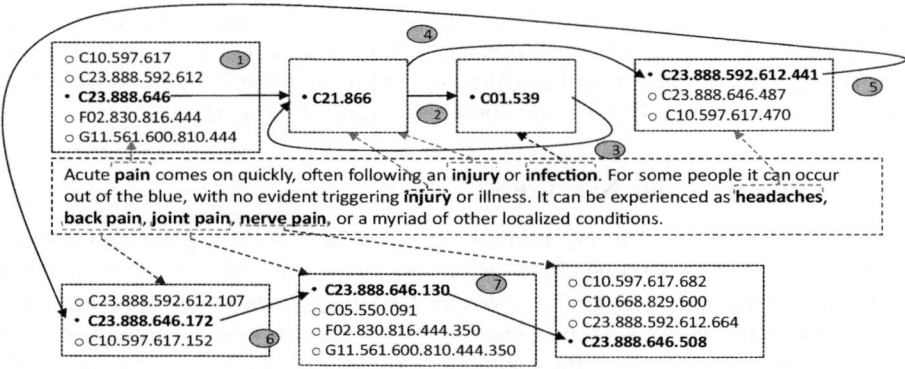

Fig. 2. Concept disambiguation

MeSH concepts, $L_n = \{c_1, c_2, ..., c_n\}$, we propose the following formula to identify the best sense for each concept c_k:

$$
\begin{cases}
(s_1, s_2) = \displaystyle\sum_{s_1 \in syn(c_1), s_2 \in syn(c_2)} sim(s_1, s_2) & \text{if } k \leq 2 \\
s_k = \arg\max(\displaystyle\sum_{s \in syn(c_k)} sim(s_{k-1}, s)) & \text{if } k > 2
\end{cases}
\tag{1}
$$

where s_k: the sense of the concept c_k,
$syn(c_k)$: the set of senses of c_k,
$sim(s_1, s_2)$: the similarity between s_1 and s_2.

The similarity between two senses referred to two concepts in a hierarchy is calculated using the graph-based similarity of related concepts [19]:

$$sim(s_1, s_2) = max \left\{ -\log \frac{length(s_1, s_2)}{2 * D} \right\} \qquad (2)$$

where $length(s_1, s_2)$ is the shortest path between s_1 and s_2, and D is the depth of the hierarchy.

3.3 Cluster-Based Disambiguation

Concept clusters have been mentioned in [20], but they are not employed for WSD. Inspired by this work, we exploit the concept clusters for biomedical WSD. In this approach, the one-sense-per-discourse [18] is applied and furthermore we assume that the sense of a concept depends on the sense of other hierarchical related concepts, whatever their location is within a document. Based on these assumptions, our algorithm functions as follows: related concepts in the document are grouped into clusters, each of which belongs to the same hierarchy. A concept can appear in one or more clusters since it can have multiple senses as defined in the thesaurus. Senses that maximize the similarity between concepts within a cluster will be assigned to the related concepts.

In our case, clusters are named according to the sixteen MeSH domains (A, B, C, ...), formally $K = \{k_1, k_2, ..., k_{16}\}$. Given a list of concepts in a document, $L_n = \{c_1, c_2, ..., c_n\}$, we assign to each concept c_i its correct sense based on the cluster-based similarity, defined as following:

$$s_i = \arg\max(\sum_{c_i, c_j \in k_u, i \neq j} \sum_{s_a \in syn(c_i), s_b \in syn(c_j)} sim(s_a, s_b)) \qquad (3)$$

The cluster-based WSD is similar to the Left-to-Right WSD while using the same similarity in formula 2. The difference between two WSD algorithms is that the considered context changes, i.e., the cluster of concepts for the former and the left-most concept and the previously disambiguated concept for the later.

4   Sense-Based Indexing and Retrieval

At this level, our objective is to compute the relevance score of documents with respect to each query. The principle of our proposed approach aims at representing both documents and queries using semantic descriptors and then matching them using a sense-based weighting scheme. Hence, each concept in the document (query) is tagged with the appropriate sense in the local context and indexed with its unique sense in the document. The retrieval makes use of the detected sense related to a given term in the query compared to documents where it appears. We formulate our sense-based indexing and retrieval process through the following steps:

Step 1: Build the document index. Let D_i be the initial document, then D_i contains both disambiguated concepts in the thesaurus and single words in the vocabulary. Formally:

$$D_i^s = \{d_{1i}^s, d_{2i}^s, ..., d_{mi}^s\}$$
$$D_i^w = \{d_{1i}^w, d_{2i}^w, ..., d_{ni}^w\} \tag{4}$$

where D_i^s, D_i^w are respectively the set of concepts and single words, m and n are respectively the number of concepts and words in D_i, d_{ji}^s is the j-th concept and d_{ji}^w is the j-th word in the document D_i.

Step 2: Build the query index. Queries are processed in the same way as documents. Thus, the original query Q can be formally represented as:

$$Q^s = \{q_1^s, q_2^s, ..., q_u^s\}$$
$$Q^w = \{q_1^w, q_2^w, ..., q_v^w\} \tag{5}$$

where Q^s, Q^w are respectively the set of concepts and single words, u and v are respectively the number of concepts and words in Q, q_k^s is the k-th concept and q_k^w is the k-th word in the query Q.

Step 3: Compute the document relevance score. The relevance score of the document D_i with respect to the query Q is given by:

$$RSV(Q, D_i) = RSV(Q^w, D_i^w) + RSV(Q^s, D_i^s) \tag{6}$$

where $RSV(Q^w, D_i^w)$ is the *TF-IDF* word-based relevance score and $RSV(Q^s, D_i^s)$ is the sense-based relevance score of the document w.r.t the query, computed as follows:

$$RSV(Q^w, D_i^w) = \sum_{q_k^w \in Q^w} TF_i(q_k^w) * IDF(q_k^w)$$
$$RSV(Q^s, D_i^s) = \sum_{q_k^s \in Q^s} \alpha_k * (1 + h(q_k^s)) * TF_i(q_k^s) * IDF(q_k^s) \tag{7}$$

where TF_i: the normalized term frequency of the word q_k^w or concept q_k^s in document D_i, IDF: the normalized inverse document frequency of q_k^w or q_k^s in the collection, α_k: the meaning rate of q_k^s between D_i and Q^s, $h(q_k^s)$: the specificity of q_k^s associated with its meaning in the query, calculated as follows:

$$h(q_k^s) = \frac{level(q_k^s)}{MaxDepth} \tag{8}$$

where $level(q_k^s)$: depth level of q_k^s, $MaxDepth$: maximum level of the hierarchy.

$$\alpha_k = \begin{cases} 1 & \text{if } sense(q_k^s, Q^s) = sense(q_k^s, D_i) \\ 1 - \beta & \text{otherwise} \end{cases} \tag{9}$$

where $sense(q_k^s, Q^s)$ (resp. $sense(q_k^s, D_i)$) indicates the sense of q_k^s in the query (resp. the document D_i) (see definition 4); β is an experimental parameter obtained the value in the interval $[0, 1]$. Indeed, we assume that information at a more fine-grained level of specificity is more relevant for search users. The specificity factor in formula 8 is integrated to favour documents containing concepts at a more fine-grained level of specificity. The meaning rate α_k is considered to alleviate the relevance score of any document D_i in which the sense of the concept q_j^s is different from the query.

5   Experimental Evaluation

In our experimental evaluation, we studied the effects of assigning the sense of concepts in documents during the process of biomedical IR. Hence, we performed a series of experiments to show the impact of sense tagging on the retrieval performance. We describe in what follows the experimental settings, then present and discuss the results.

5.1   Experimental Setup

- *Test collection*: We use the OHSUMED test collection, which consists of titles and/or abstracts from 270 medical journals published from 1987-1991 through the MEDLINE database [21]. A MEDLINE document contains six fields: *title (.T), abstract (.W), MeSH indexing terms (.M), author (.A), source (.S), and publication type (.P)*. To facilitate the evaluation, we converted the original OHSUMED collection into the TREC standard format. Some statistical characteristics of the collection are depicted in Table 1.
  We have selected 48 TREC standard topics, each one is provided with a set of relevant documents judged by a group of physicians in a clinical setting. The *title* field indicates *patient description* and the *description* field announces *information request*.

Table 1. Test collection statistics

Number of documents	293.856
Average document length	100
Number of queries	48
Average query length	6 (TITLE) 12 (TITLE+DESC)
Average number of concepts/query	1.50 (TITLE) 3.33 (TITLE+DESC)
Average number of relevant docs/query	50

- *Evaluation measures*: P@5, P@10 represent respectively the mean precision values at the top 5, 10 returned documents and MAP (Mean Average Precision) over the total of 48 testing queries. For each query, the first 1000 documents are returned by the search engine and average precisions (P@5, P@10, MAP) are computed for measuring the IR performance.

– *Medical Subject Headings:* MeSH is a medical domain knowledge resource that has been developed at the US National Library of Medicine (NLM) since 1960. The latest version of MeSH released in 2010 consists of 25,588 entries, each one represents a preferred concept for the indexing of publications included in the MEDLINE database.

5.2   Experimental Results

For evaluating the effectiveness of our WSD methods and the performance of our sense-based indexing approach, we carried out two sets of experiments: the first one is based on the classical index of titles and/or abstracts using Terrier standard configuration based on the state of the art weighting scheme OKAPI BM25 [22], used as the baseline, denoted *BM25*. The second set of experiments concerns our semantic indexing method and consists of three scenarios:

1. The first one is based on the naive selection of the first sense found in the hierarchy for each concept, denoted *WSD-0*,
2. The second one is based on the *Left-To-Right WSD*, denoted *WSD-1*,
3. The third one is based on the *Cluster-based WSD*, denoted *WSD-2*.

We use both terms representing MeSH entries and single words that do not match this thesaurus. In the classical approach, the documents were first indexed using the Terrier IR platform (http://ir.dcs.gla.ac.uk/terrier/). It consists in processing single terms occurring in the documents through a pipeline: removing stop words, identifying concepts in documents and stemming[4] of English words.

In our sense-based approaches that employ semantic information from WSD, documents and queries are firstly disambiguated and indexed with appropriate senses of concepts defined in the MeSH vocabulary. The semantic weighting scheme is then applied for each term in the query using Formula 7. In our experiments, the β parameter in formula 9 is set to 0.15 for the best results.

Table 2 depicts the IR performance based on the baseline and the proposed WSD methods both for querying with only *title* and with *title and description* together. Figure 3 shows the improvement rates of our methods over the baseline. We obtained the following results: both *WSD-1* and *WSD-2* methods outperform the *baseline* both for short queries (title) and long queries (title and description). The improving rates obtained are 5.61% and 4.77% for *WSD-1* and *WSD-2* respectively. This proves the interest of taking into account the document and query's semantics along with the specificity of the documents as well as of the queries in the IR process. Furthermore, the results show that randomly selecting the concept's sense (as in *WSD-0*) does not help, but correctly assigning the appropriate sense for each concept improves the IR performance. We see that *WSD-1* and *WSD-2* always give a better precision than *WSD-0*.

We have also tested the query evaluation with only *title* and *title and description* together in order to show the impact of the query length on the IR performance. We see that the semantic chain inspired the *WSD-1* gets a higher

---

[4] http://snowball.tartarus.org/

Table 2. Official results on the OHSUMED collection

(a) title

Measure	BM25	WSD-0	WSD-1	WSD-2
P@5	0.17500	0.1792	0.19170	0.17920
P@10	0.18540	0.1771	0.18750	0.18750
MAP	0.10270	0.1034	0.10800	0.10760

(b) title + description

Measure	BM25	WSD-0	WSD-1	WSD-2
P@5	0.50420	0.50830	0.52080	0.52080
P@10	0.45630	0.46040	0.47500	0.47500
MAP	0.24210	0.25450	0.26110	0.26110

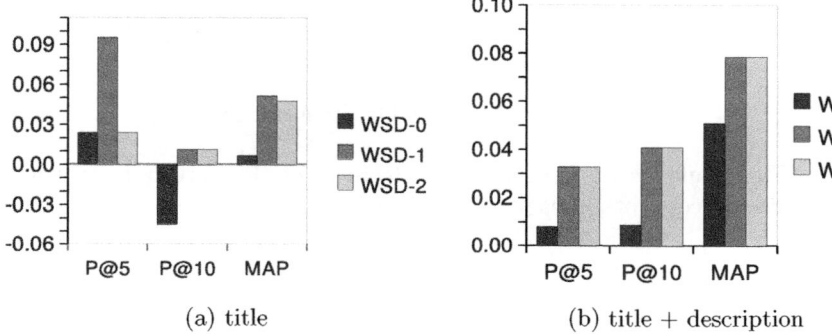

(a) title

(b) title + description

Fig. 3. Improving rate over the baseline

improving rate over the baseline (5.61%) compared to *WSD-2* (4.77%) for short queries. In addition, both of them get the same improving rate over the baseline (7.48%) for long queries. Indeed, for longer queries, both of our two methods identify better the sense of each concept and then induce its appropriate specificity level in the document. The only difference between the two methods is the selection of the context where a concept appears: the former from the left-side concepts and the later from clusters of concepts in the same hierarchy.

In a finer-grained analysis, we have reviewed the IR performance for each long query in the testing set to verify the impact of the concept's specificity with respect to the query length and the number of concepts in the query on the IR performance. For each query, we computed the average specificity of the query and we obtained a range from 2 to 6. For each group of queries having the specificity from 2 to 6, we compute the average query length, the average number of concepts and the average improving rates. Figure 4 shows the analysis of the results according to the query specificity. We notice that the more the query specificity, i.e., the average concept's specificity level in the query is, the less the number of concepts used in the query is. This could be explained by the fact that a few of more specific concepts are enough to express the user needs while many of generic ones are required to express better the user information need. In most of cases, our approach favours documents containing concepts at a higher level of specificity and shows a consistent improvement over the baseline. However, if the query is long but the number of concepts having a high specificity level in the query is less, our system tends to return first documents containing those

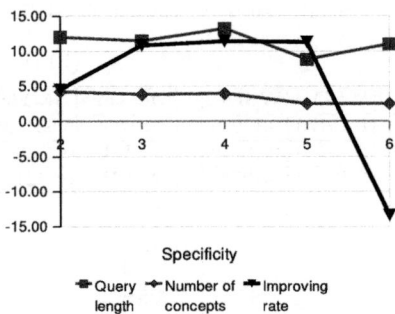

Fig. 4. Analysis of results according to query specificity

concepts. This could be the reason of the decrease of the performance when a few most specific concepts impact other terms in the query.

6 Conclusion

In this work, we have proposed and evaluated a sense-based approach of indexing and retrieving biomedical documents. Our approach relies on two WSD methods for identifying ambiguous MeSH concepts: *Left-To-Right WSD* and *Cluster-based WSD*. The evaluation of the indexing method on the standard OHSUMED corpus proves the evidence of integrating the sense of concepts in the IR process. Indeed, most of ongoing IR approaches match documents using term distribution without the sense of query terms along with the ones of the documents. The more the meaning of the query term matches the meaning of the document term, the more the retrieval performance is improved. The more the query terms are specific, the more the specificity of the returned documents is fine-grained. Future works will focus on automatically expanding the query using concepts extracted from the hierarchy indicated by the correct sense of concepts occurred in documents.

References

1. Lesk, M.: Automatic sense disambiguation using machine readable dictionaries: how to tell a pine cone from an ice cream cone. In: SIGDOC 1986, pp. 24–26 (1986)
2. Gale, W., Church, K., Yarowsky, D.: A method for disambiguating word senses in a large corpus. Computers and the Humanities, 415–439 (1993)
3. Mihalcea, R.: Unsupervised large-vocabulary word sense disambiguation with graph-based algorithms for sequence data labeling. In: HLT 2005, pp. 411–418 (2005)
4. Lee, Y.K., Ng, H.T., Chia, T.K.: Supervised word sense disambiguation with support vector machines and multiple knowledge sources. In: Senseval-3: Workshop on the Evaluation of Systems for the Semantic Analysis of Text, pp. 137–140 (2004)

5. Liu, H., Teller, V., Friedman, C.: A multi-aspect comparison study of supervised word sense disambiguation. J Am. Med. Inform. Assoc. 11(4), 320–331 (2004)
6. Yarowsky, D.: Unsupervised word sense disambiguation rivaling supervised methods. In: ACL 1995, pp. 189–196 (1995)
7. Abney, S.P.: Bootstrapping. In: ACL, pp. 360–367 (2002)
8. Leroy, G., et al.: Effects of information and machine learning algorithms on word sense disambiguation with small datasets. Medical Informatics, 573–585 (2005)
9. Joshi, M., Pedersen, T., Maclin, R.: A comparative study of support vector machines applied to the word sense disambiguation problem for the medical domain. In: IICAI 2005, pp. 3449–3468 (2005)
10. Weeber, M., Mork, J., Aronson, A.: Developing a test collection for biomedical word sense disambiguation. In: Proc. AMIA Symp., pp. 746–750 (2001)
11. Humphrey, S.M., Rogers, W.J., et al.: Word sense disambiguation by selecting the best semantic type based on journal descriptor indexing: Preliminary experiment. J. Am. Soc. Inf. Sci. Technol. 57(1), 96–113 (2006)
12. Gaudan, S., Kirsch, H., Rebholz-Schuhmann, D.: Resolving abbreviations to their senses in medline. Bioinformatics 21(18), 3658–3664 (2005)
13. Andreopoulos, B., Alexopoulou, D., Schroeder, M.: Word sense disambiguation in biomedical ontologies with term co-occurrence analysis and document clustering. IJDMB 2(3), 193–215 (2008)
14. Mohammad, S., Pedersen, T.: Combining lexical and syntactic features for supervised word sense disambiguation. In: CoNLL 2004, pp. 25–32 (2004)
15. Stevenson, M., Guo, Y., Gaizauskas, R., Martinez, D.: Knowledge sources for word sense disambiguation of biomedical text. In: BioNLP 2008, pp. 80–87 (2008)
16. Aronson, A.R.: Effective mapping of biomedical text to the UMLS Metathesaurus: the metamap program. In: Proceedings AMIA Symposium, pp. 17–21 (2001)
17. Schmid, H.: Part-of-speech tagging with neural networks. In: Proceedings of the 15th conference on Computational linguistics, pp. 172–176 (1994)
18. Gale, W.A., Church, K.W., Yarowsky, D.: One sense per discourse. In: HLT 1991: Proceedings of the workshop on Speech and natural Language, pp. 233–237 (1992)
19. Leacock, C., Chodorow, M.: Combining local context and wordnet similarity for word sense identification. An Electronic Lexical Database, 265–283 (1998)
20. Kang, B.Y., Kim, D.W., Lee, S.J.: Exploiting concept clusters for content-based information retrieval. Information Sciences - Informatics and Computer Science 170(2-4), 443–462 (2005)
21. Hersh, W., Buckley, C., Leone, T.J., Hickam, D.: Ohsumed: an interactive retrieval evaluation and new large test collection for research. In: SIGIR 1994, pp. 192–201 (1994)
22. Robertson, S.E., Walker, S., Hancock-Beaulieu, M.: Okapi at trec-7: Automatic ad hoc, filtering, vlc and interactive. In: TREC, pp. 199–210 (1998)

Semantic Content Access Using Domain-Independent NLP Ontologies

René Witte[1] and Ralf Krestel[2]

[1] Semantic Software Lab, Concordia University, Montréal, Canada
[2] L3S Research Center, University of Hannover, Germany

Abstract. We present a lightweight, user-centred approach for document navigation and analysis that is based on an ontology of text mining results. This allows us to bring the result of existing text mining pipelines directly to end users. Our approach is domain-independent and relies on existing NLP analysis tasks such as automatic multi-document summarization, clustering, question-answering, and opinion mining. Users can interactively trigger semantic processing services for tasks such as analyzing product reviews, daily news, or other document sets.

1 Introduction

Despite recent advances in semantic processing, users continue to be overloaded with information during everyday activities: Browsing product reviews on popular e-commerce websites, for example, can result in hundreds (or thousands) of user-generated reviews, which take considerable time to read and analyse for their relevance, informativeness, and opinion. Thanks to the success of the Web 2.0, the combined length of the user-generated reviews of a single, popular book can now exceed the length of the book itself—which defies the goal of reviews to save time when deciding whether to read the book itself or not. While some condensed views are often available (like a "star rating"), this in turn is usually too compressed to be used by itself.

Similar challenges are faced by everybody dealing with (natural language) content. Writers of reports or research papers need to survey extensive amounts of existing documents; email and other communication forms take up significant amounts of time, with no integrated semantic processing support available that could ease the analysis of questions or composition of answers while writing them. Both in research and industry, employees spend an ever increasing proportion of their time searching for the right information. Information overflow has become a serious threat to productivity.

In this paper, we focus on *user-driven NLP* for managing large amounts of textual information. That is, NLP analysis is explicitly requested by a user for a certain task at hand, not pre-computed on a server. To access the results of these analyses, we propose an NLP ontology that focuses on the *tasks* (like summarization or opinion analysis), rather than the domain (like news, biology, or software engineering).

C.J. Hopfe et al. (Eds.): NLDB 2010, LNCS 6177, pp. 36–47, 2010.

Background. In recent years, the fields of natural language processing and text mining have developed a number of robust analysis techniques that can help users in information analysis and content development: automatic summarization [1,2], question-answering [3] and opinion mining [4] can be directly applied to the scenarios outlined above.

Text analysis as described above is typically done within a component-based framework such as GATE [5] or UIMA [6]. However, these frameworks are targeted at language engineers, not end users. So far, none of the text mining methods described above have become available to a mass user base. To bring NLP directly to an end user's desktop, we previously developed *Semantic Assistants* [7], a service-oriented architecture that brokers the NLP pipelines through W3C Web services with WSDL descriptions directly to desktop applications (like word processors or email clients). After solving the technical integration, we can now focus on the *semantic integration* of NLP into end users' tasks.

Proposed Approach. In this work, we focus on semantic NLP services that can support users in common, yet time-consuming tasks, like browsing product reviews or analysing daily news. We argue that this can be achieved by building an ontology that integrates original content with the results of text mining pipelines. The following diagram illustrates our idea:

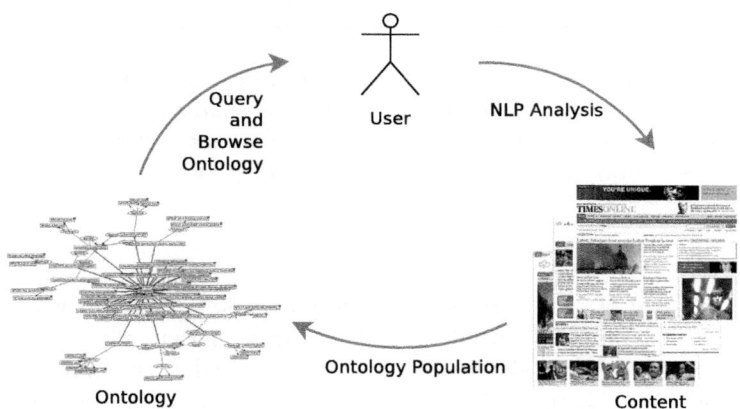

A user is faced with a large amount of natural language content—for example, hundreds of reviews for a single product on Amazon, or a cluster of thousands of news articles for a single event in Google News. Rather than dealing with this huge amount of text manually, the user triggers an NLP analysis of the document set. The results of the analysis is captured in a rich NLP ontology that now contains detected topics, summaries, contrastive information, answers to questions the user might have submitted to the NLP analysis, and also links back to the source documents. The user can now browse, query, or navigate the information through the highly structured populated ontology, and also follow links back to the source documents when needed. This approach empowers users by providing them with sophisticated text mining services, which both save time and deliver a richer information structure: for example, rather than reading

Table 1. Main concepts in the NLP ontology and their definition

Concept	Definition
Document	Set of source URIs containing information in natural language (e.g., news articles, product reviews, blog posts)
Content	Natural language text appearing either in a source document or generated as a result from text mining pipelines
DocContent	Natural language text appearing in a source document
Summary	NLP analysis artifact derived through applying specific algorithms to a set of input documents with optional contextual information
SingleSummary	An automatically generated summary of a single input document
ShortSumm	A keyphrase-like automatically generated summary indicating the major topics of a single document
Classical SingleSumm	An essay-like text of user-configurable length containing the most salient information of the source document
MultiSummary	An automatically generated summary of a set of input documents
ClassicalMultiSumm	An essay-like text of user-configurable length containing the most important (common) topics appearing in all source documents
FocusedSumm	An essay-like text of user-configurable length that addresses a specific user context (e.g., concrete questions the user needs to be addressed by the summary, or another reference document in order to find related content)
ConstrastiveSumm	Multi-document summarization method that generates (a) the commonalities (shared topics) across all input documents and (b) content specific to a single or subset of documents (contrasts)
Chain	Single- or cross-document coreference chain (NLP analysis artifact)
Chunk	Specific content fragments generated or manipulated by NLP analysis (e.g., noun phrases, verb groups, sentences)

just a few of hundreds of product reviews in order to reach a conclusion, the user can now see an automatically generated summary of the most important comments common to all reviews, and also directly see contrastive information (like disagreements), represented by specific ontology classes.

2   Design

The central goal of this work is to provide a semantically rich yet dense representation of large amounts of textual information that allows novel way of accessing content. Users should be able to see a highly summarized top-level view, but also be able to "drill-down" into specific aspects of the analyzed information. For example, a large number of product reviews could be summarized to a few sentences that contain information shared by the majority of individual entries. In addition, the major differences should also be analyzed and presented in a similar manner, allowing a user to detect the major diverging views without having to go through each individual review. Nevertheless, each of these summarized views should provide links to trace the analysis results back to their source statements. Essentially, we enrich content that is already available with automatically generated meta-information, thereby bringing the results of sophisticated text mining techniques to an end user, which is a significant improvement over current browsing methods, such as simple tag clouds.

Ontology Population. Such a semantically rich representation requires a suitable data model that can be queried, transformed, visualized, and exchanged

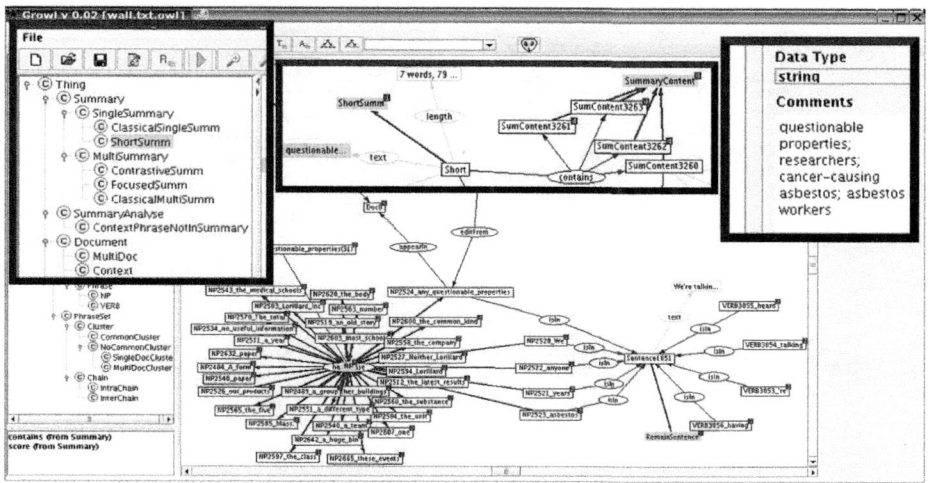

Fig. 1. An NLP-generated 10-word topic summary (right) of a single newspaper article represented in the NLP ontology

between multiple applications. The solution presented here is to provide a generic ontology for *natural language processing results* that can then be automatically populated with the concrete results (OWL individuals) of a particular analysis task (e.g., all the reviews for one product). This now allows executing the complete semantic toolchain on NLP analysis results, including (SPARQL) queries, OWL-DL reasoning, and visualization, which is a significant improvement compared to the static XML result formats typically employed in today's text mining frameworks.

NLP Ontology. To facilitate the outlined document analysis tasks, we developed an NLP ontology. The domain of discourse for this ontology comprises the artifacts involved in automatic document analysis—texts and their constituents (like sentences, noun phrases, words) and the results of specific analysis pipelines, like the various types of summaries outlined above.

Table 1 shows the main concepts of our NLP ontology, together with a brief definition. Its main goal is to facilitate *content* access. Content individuals are (useful) snippets of information, which a user can read while performing his tasks (e.g., analysing product reviews). This can be content that appears in a source text, like a Web page (**DocContent**), or some text has been generated by a summarization algorithm (**SummaryContent**). To provide support for drilling down into NLP analysis results, original documents that form the basis for analysis are modeled explicitly as well. The ontology further distinguishes between a single document, like a single Web page (**SingleDoc**), a document collection (**MultiDoc**) and contextual information (**Context**). The main NLP result artifact modeled in our ontology are *summaries*, including all the various types discussed in the introduction.

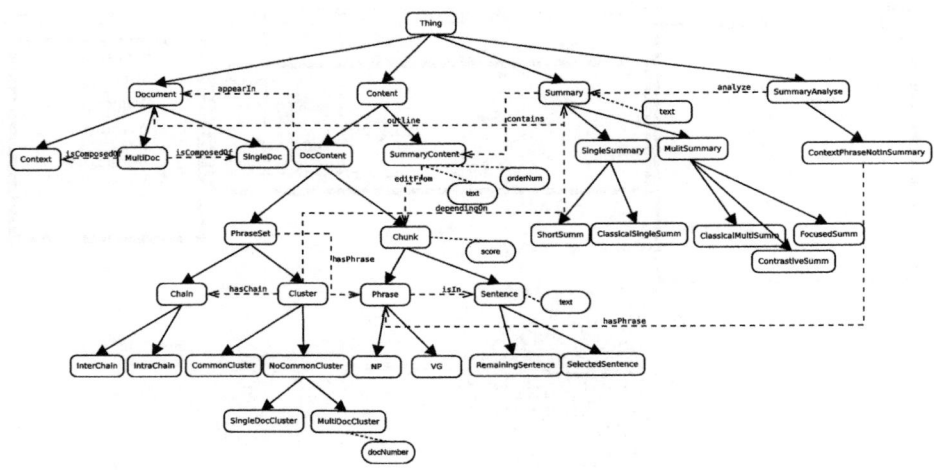

Fig. 2. Domain-Independent NLP Ontology Model

To allow further drill-down into the NLP analysis of a given document (set), we also model further important NLP concepts such as sentences, (phrase) chunks, coreference chains, and entity clusters. These provide for the navigational links between original documents and derived or generated content in the generated summaries—see Fig. 1 for an example of a populated ontology.

Fig. 2 shows an overview of the main concepts and relations in the resulting NLP ontology. For deployment, this ontology needs to be automatically populated with concrete analysis results (described in Section 3), and can then be used to facilitate content access across various domains and tasks (Section 4).

3   Implementation

Our implementation is based on the Semantic Assistants design [7], which allows bringing the results of NLP analysis to a connected desktop client (like a word processor or a Web browser) through semantic Web services. Natural language processing is performed within the GATE framework [5]. The concrete analysis pipelines deployed here, i.e., for summary generation and contrast analysis, are based on our previous work in automatic summarization [8]. Fig. 3 shows the processing pipeline with its various components. After some NLP processing (preprocessing, part-of-speech tagging, . . .) the task specific application is run – in this case a summarization component. The visualization component then populates the ontology described in Section 2 with the NLP results and exports the ontology using the OWL format.

Population of the ontology is achieved through a custom processing component in GATE, the OwlExporter [9], which combines the results from the analysis pipelines with the loaded documents, thereby enabling the combined view of original content with NLP-discovered analysis results, such as contrastive information. This ontology is rich enough to allow connected clients to present

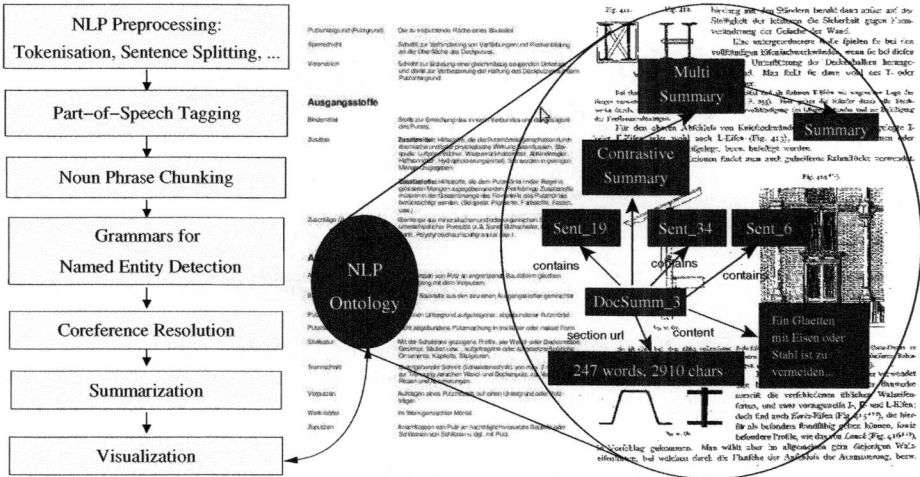

Fig. 3. Pipeline for populating the NLP ontology using text mining results

various views, depending on the output capabilities and the users' current need and context. For example, a Web browser plugin could directly annotate Web pages with the analysis results, while a small-screen device, such as a mobile phone, would need to present a more compressed view.

4 Applications

In this section, we evaluate the applicability of our ideas on three concrete scenarios: analyzing daily news, heritage documents, and product reviews. All three scenarios highlight the need for automated semantic support that so far has not been available to an end user.

Workflow. In all these scenarios, we employ the text mining pipelines discussed in Section 3. The documents for analysis, as well as relevant information about the user context—such as concrete questions or viewpoints the user needs to have addressed by the semantic services—is transmitted from the client to the NLP framework via the Semantic Assistants architecture described in [7]. The results of the analysis pipelines is captured in the OWL ontology described in Section 2 using a custom ontology population components as described in Section 3. This ontology is then transmitted back to the client.

Information Visualization. The product delivered to an end user is a populated OWL ontology that captures a number of NLP analysis results (topics, summaries, answers to explicit questions, contrastive information, etc.). How an end user interacts with this result ontology is an ongoing challenge. Within this work, we demonstrate the feasibility using standard ontology tools for browsing (SWOOP),[1]

---

[1] SWOOP Ontology Browser/Editor, `http://code.google.com/p/swoop/`

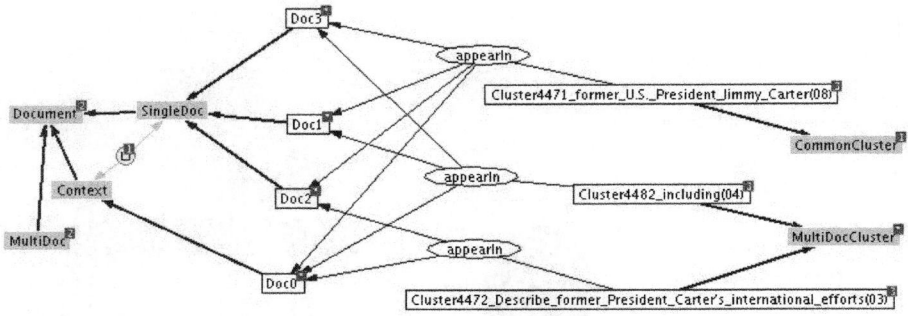

Fig. 4. Part of a populated NLP ontology showing topic clusters that are shared by a set of news documents (CommonCluster) and topic clusters that are specific to a subset of news articles (MultiDocCluster)

visualizing (GrOWL),[2] and querying (SPARQL)[3] the ontology. However, these require some level of understanding about OWL ontologies, which should not be presumed of an end user. Advanced query and visualization paradigms (e.g., graphical queries, NL queries) are a highly active area of research and the results from these efforts can be directly leveraged by our framework.

4.1   News Analysis

Despite the surge of user-generated content such as blogs and wikis, traditional newspaper and newswire texts continue to be an important source of daily news. News aggregators such as Google News[4] provide a condensed overview of current topics and the articles covering them. A number of research prototypes such as Newsblaster[5] or NewsExplorer[6] apply information extraction and summarization on the daily news to provide semantic query and navigation facilities. Compared with these server-side, precomputed content access options, our work provides semantic analysis "on demand" triggered by a user. This kind of analysis can be executed on a specific document set as selected by a user, and also be based on additional context information, e.g., a particular question a user needs to have addressed.

Topic analysis and summarization can augment a list of news articles with a brief list of topics (see Fig. 1, right side, for an example) and a summary of the main content of the article. This analysis aims to help a user in deciding whether he wants to read a specific article in full or not. However, our semantic analysis services can provide additional result ontologies when applied on a document set: multi-document summaries can extract the major common topics across a set of

---

[2] GrOWL ontology visualiser, http://ecoinformatics.uvm.edu/dmaps/growl
[3] SPARQL RDF query language, http://www.w3.org/TR/rdf-sparql-query/
[4] Google News, http://news.google.com
[5] Columbia Newsblaster, http://newsblaster.cs.columbia.edu/
[6] EMM NewsExplorer, http://press.jrc.it/NewsExplorer

documents; contrastive summaries can detect differences between documents in a set, and focused summaries can extract information related to a user's current context—e.g., a concrete question or another document he is working on, like an email or a report.

Example Use Case. A user is faced with a large set of news relating to a single event. In order to find the topics shared by all news, he initiates an NLP analysis for topic detection and common multi-document summarization, which provides him with a list of topics and a summary (of adjustable length). While it is often sufficient to summarize the commonalities, the user might also be interested in specific differences between the news: for example, when the same event is reported differently across North American, European, Asian, and Middle Eastern newspapers. Finding such differences is possible with contrastive summarization [8], which detects topic clusters that only appear in a single or subset of documents. Using the populated NLP ontology (see Fig. 4), the user can navigate directly to such topic clusters, and then also see an automatically generated summary containing these differences (see [8] for some examples).

4.2   Heritage Data Analysis

In addition to analyzing current news, blogs, or web pages, heritage data can provide a rich source of information. This data is usually not easily accessible on a semantic level. Outdated terms, different styles of writing, or just the huge amount of heritage data makes it impossible for experts to exploit this knowledge satisfactorily nowadays. In this context, contrasting the heritage data with current data and make these differences visible and browsable is a big asset.

Example User Case. We applied our contrast summarization framework to an old, German encyclopedia of architecture from the 19th century [10] and compared it to present-day Swiss construction regulations. Practitioners, like building historians or architects, can navigate through the heritage data based on extracted contrasts between the historic knowledge and the state-of-the-art building engineering regulations. Fig. 3 shows a page of the old encyclopedia together with a contemporary standards document.

As with the news analysis described above, the populated ontology enables the user to find contrasts for particular methods or materials and browse through the source text of the found information.

Not only the different styles and formats makes it difficult for architects/ historians to compare the heritage data with new data, but also the huge amount of available information. The populated ontology enables the user to find contrasts for particular engineering methods or building materials and browse through the source text of the found information.

4.3   Analyzing Product Reviews

E-commerce websites such as Amazon have long integrated user-contributed content in form of wikis and product reviews. These can provide important information to both buyer and vendor. However, when it takes longer to read all reviews for a book than the actual book itself, they are no longer a viable means for saving time.

In this application scenario, a user would delegate the analysis of a large set of reviews to an NLP service that analyses them for commonalities and differences. In the resulting ontology, the main topics in agreement will be detected and supplemented by a single summary. In addition, the contrastive cluster analysis will detect topics that only appear in a single or subset of reviews, allowing a semantic navigation of reviews based on content, rather then simple structure or surface feature as they are common today.

Example Use Case. In this example, a user obtained a number of product reviews for a computer science book in order to help in a purchase decision. Rather than reading all the reviews manually, the user invokes an NLP analysis of the obtained reviews and receives the populated ontology. Fig. 5 shows an excerpt of such a large, populated ontology summarizing multiple book reviews of a single book. As can be seen, contrastive statements can be easily identified and using the ontology a deeper analysis of the review content is possible. From this ontology, the user can navigate to the generated summary of the common opinions, as shown in Fig. 6. However, not all reviews share the positive outlook: some of diverging opinions can be found as OWL individuals in the ContrastiveSumm class, as shown in Fig. 5 (in particular the text in the middle of the figure). As before, each of the generated summaries and concepts can be traced back to sentences in the original reviews, in case a user needs to understand the results of the automated NLP analysis in their original context.

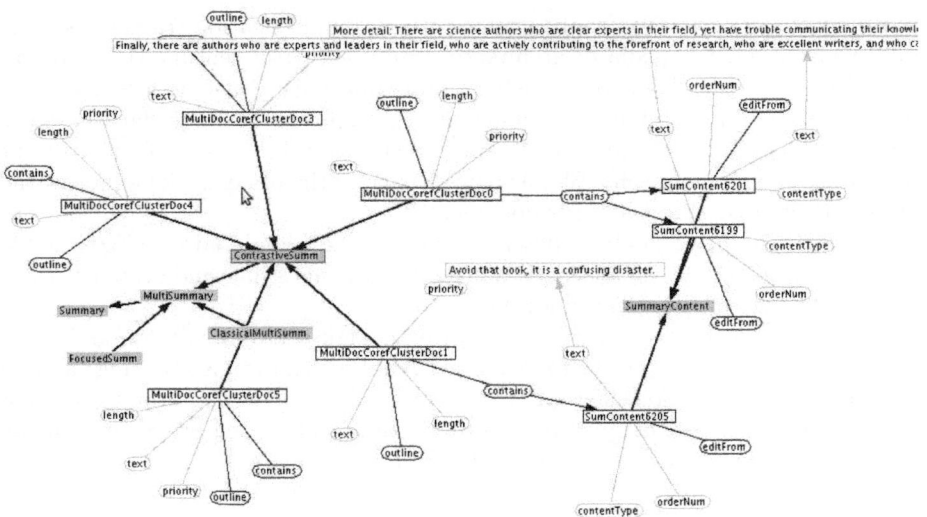

Fig. 5. Visualization of an Excerpt of an Ontology for Book Reviews

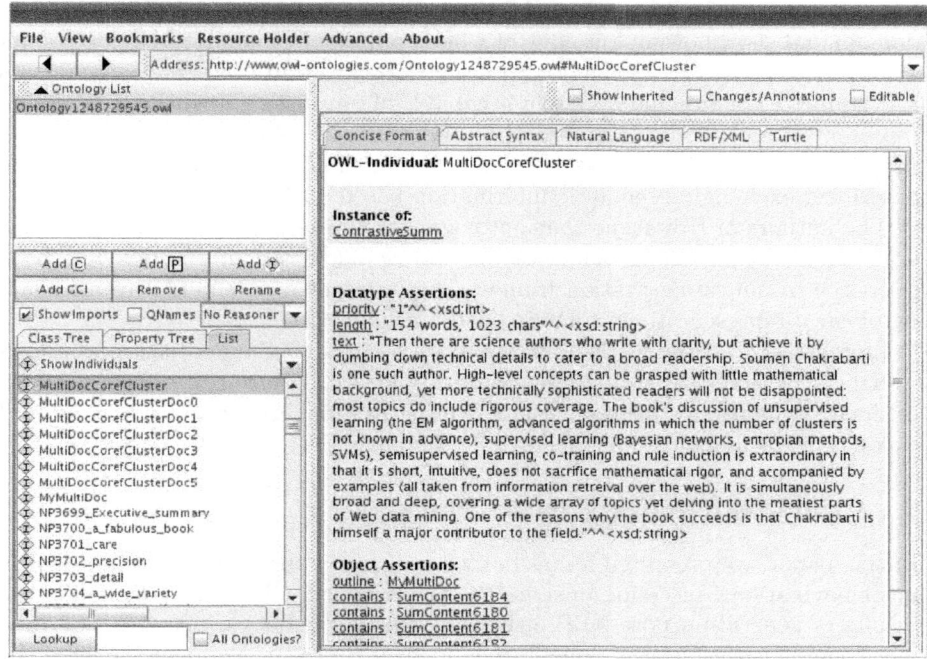

Fig. 6. An automatically generated summary of the commonalities of all the reviews of a book in the NLP ontology

5   Related Work

Using ontologies to facilitate the access to natural language documents was explored before. In [11], the authors develop an ontology for linguistic concepts to ease the sharing of annotated linguistic data and provides for searching and browsing of inhomogeneous corpora. The overall goal was to develop a way to conserve the rich linguistic concepts of (endangered) languages. The focus was therefore on creating an ontology that can deal with dynamic, changing data and with different source material.

The authors of [12] demonstrate the benefits of view-based search methods for RDF(S) repositories including semantic recommendations. The data is represented within an ontology and the user can refine his query by browsing through the views. This combines ontology-based search with multi-faceted search and enables the user to find information beyond keyword-based search results.

Ontology population from linguistic extractions is also the main focus of [13]. Knowledge acquisition rules are used to map concept tree nodes to ontology instances. The nodes are the result of extraction and annotation of documents. In particular the paper describes how to connect two components to work together in one framework: One for modeling domain knowledge using ontological concepts and one for linguistic extractions. They present their results in the legal domain and show how an ontology can be populated from annotated documents.

In [14] one way for a semantic representation of natural language documents is described. The presented system is capable of outputting its internal semantic representation of a document to the OWL format, allowing a semantic motivated browsing of the textual data. Also the possibilities of multiple agents are described, exchanging semantic information based on RDF or OWL.

The authors of [15] argue that ontologies need a linguistic grounding. RDFS or OWL offer not enough support for adding linguistic information like part-of-speech or subcategorization frames. They present a new model to associate ontological representations with linguistic information.

In contrast to these works, our approach focuses on the end-user and allows visualizing specific NLP analysis results. Also, we base our approach on user-initiated semantic annotation, whereas other systems use server-side, standard extractions to modeled documents using ontologies.

6   Conclusions and Future Work

In this paper, we presented a novel ontology-based approach for semantic document navigation. Accessing unstructured content is facilitated by modeling the results of general-purpose NLP algorithms, in particular various forms of automatic summarization, in a domain-independent ontology. We demonstrated the feasibility with a complete implementation including NLP analysis pipelines and applied it to a number of concrete application scenarios, including news analysis, cultural heritage data management, and product reviews. However, our approach is not limited to these examples; many tasks require a comparative study of a document set: applied to paper reviews in a conference system, contrastive summarization can help a conference chair find agreements and diverging views. In collaborative editing environments, like a wiki, a structured view that highlights the semantic differences between different versions can greatly facilitate the work of a maintainer—e.g., applied to Wikipedia articles where strong disagreements often lead to "edit wars."

A particular feature of our approach is that is user-driven: rather than relying on existing, pre-computed semantic annotations we envision a user that is supported by semantic analysis services in his tasks. NLP-driven analysis pipelines are executed on demand and the result ontology can be used for further document navigation, content access, or solving specific tasks.

Future work is specifically needed in two areas: first, more user-friendly ways of presenting the analysis results to the user that hide the OWL-specific implementation details. This can be achieved with client-specific plugins that provide user-friendly ontology browsing and querying facilities. And second, user studies that evaluate the effect of the provided semantic support on concrete tasks, comparing them with current approaches. While this will undoubtedly result in new requirements, we believe that empowering users by providing them with sophisticated semantic annotation support for existing content will be a significant improvement for accessing and processing content in the Web.

Acknowledgements. Ting Tang contributed to the OWL NLP ontology and its population GATE component.

References

1. Mani, I.: Automatic Summarization. John Benjamins, Amsterdam (2001)
2. NIST: DUC 2001. In: Proceedings of the ACM SIGIR Workshop on Text Summarization DUC 2001, New Orleans, Louisiana USA, NIST (2001)
3. Dang, H.: Overview of the tac 2008 opinion question answering and summarization tasks. In: Proceedings of the First Text Analysis Conference (TAC 2008), Gaithersburg, Maryland, USA, November 17-19, NIST (2008)
4. Pang, B., Lee, L.: Opinion mining and sentiment analysis. Foundation and Trends in Information Retrieval 2(1-2), 1–135 (2008)
5. Cunningham, H., Maynard, D., Bontcheva, K., Tablan, V.: GATE: A Framework and Graphical Development Environment for Robust NLP Tools and Applications. In: Proc. of the 40th Anniversary Meeting of the ACL (2002), http://gate.ac.uk
6. Ferrucci, D., Lally, A.: UIMA: An Architectural Approach to Unstructured Information Processing in the Corporate Research Environment. Natural Language Engineering 10(3-4), 327–348 (2004)
7. Witte, R., Gitzinger, T.: Semantic Assistants – User-Centric Natural Language Processing Services for Desktop Clients. In: Domingue, J., Anutariya, C. (eds.) ASWC 2008. LNCS, vol. 5367, pp. 360–374. Springer, Heidelberg (2008)
8. Witte, R., Bergler, S.: Next-Generation Summarization: Contrastive, Focused, and Update Summaries. In: International Conference on Recent Advances in Natural Language Processing (RANLP 2007), Borovets, Bulgaria, September 27–29 (2007)
9. Witte, R., Khamis, N., Rilling, J.: Flexible Ontology Population from Text: The OwlExporter. In: Int. Conf. on Language Resources and Evaluation, LREC (2010)
10. Witte, R., Gitzinger, T., Kappler, T., Krestel, R.: A Semantic Wiki Approach to Cultural Heritage Data Management. In: Language Technology for Cultural Heritage Data (LaTeCH 2008), Marrakech, Morocco June 1, (2008)
11. Farrar, S., Lewis, W.D., Langendoen, D.T.: A common ontology for linguistic concepts. In: Proceedings of the Knowledge Technologies Conference (2002)
12. Hyvönen, E., Saarela, S., Viljanen, K.: Application of ontology techniques to view-based semantic search and browsing. In: Bussler, C.J., Davies, J., Fensel, D., Studer, R. (eds.) ESWS 2004. LNCS, vol. 3053, pp. 92–106. Springer, Heidelberg (2004)
13. Amardeilh, F., Laublet, P., Minel, J.L.: Document annotation and ontology population from linguistic extractions. In: K-CAP 2005: Proceedings of the 3rd international conference on Knowledge capture, pp. 161–168. ACM, New York (2005)
14. Java, A., Nirenburg, S., McShane, M., Finin, T., English, J., Joshi, A.: Using a Natural Language Understanding System to Generate Semantic Web Content. International Journal on Semantic Web and Information Systems 3(4) (2007)
15. Buitelaar, P., Cimiano, P., Haase, P., Sintek, M.: Towards linguistically grounded ontologies. In: Aroyo, L., Traverso, P., Ciravegna, F., Cimiano, P., Heath, T., Hyvönen, E., Mizoguchi, R., Oren, E., Sabou, M., Simperl, E. (eds.) ESWC 2009. LNCS, vol. 5554, pp. 111–125. Springer, Heidelberg (2009)

Extracting Meronymy Relationships from Domain-Specific, Textual Corporate Databases

Ashwin Ittoo, Gosse Bouma, Laura Maruster, and Hans Wortmann

University of Groningen
9747 AE Groningen, The Netherlands
{r.a.ittoo,g.bouma,l.maruster,j.c.wortmann}@rug.nl

Abstract. Various techniques for learning meronymy relationships from open-domain corpora exist. However, extracting meronymy relationships from domain-specific, textual corporate databases has been overlooked, despite numerous application opportunities particularly in domains like product development and/or customer service. These domains also pose new scientific challenges, such as the absence of elaborate knowledge resources, compromising the performance of supervised meronymy-learning algorithms. Furthermore, the domain-specific terminology of corporate texts makes it difficult to select appropriate seeds for minimally-supervised meronymy-learning algorithms. To address these issues, we develop and present a principled approach to extract accurate meronymy relationships from textual databases of product development and/or customer service organizations by leveraging on reliable meronymy lexico-syntactic patterns harvested from an open-domain corpus. Evaluations on real-life corporate databases indicate that our technique extracts precise meronymy relationships that provide valuable operational insights on causes of product failures and customer dissatisfaction. Our results also reveal that the types of some of the domain-specific meronymy relationships, extracted from the corporate data, cannot be conclusively and unambiguously classified under well-known taxonomies of relationships.

Keywords: Meronymy, part-whole relations, natural language processing.

1 Introduction

Meronymy is an important semantic relationship that exists between a part and its corresponding whole [2]. Approaches exist for automatically learning meronymy relationships [1,2,3,10,12] from open-domain corpora (e.g. SemCor) to support traditional natural-language-processing (NLP) applications like question-answering. However, none of them targeted textual databases in corporate domains, despite the numerous application opportunities. Product development and/or customer service (PD-CS) are such corporate domains in which meronymy is of fundamental importance as a central structuring principle in artifact design [12]. Meronymy relationships harvested from PD-CS textual databases could support activities like product quality assurance [8], and generating domain ontologies and bills-of-materials from product descriptions.

C.J. Hopfe et al. (Eds.): NLDB 2010, LNCS 6177, pp. 48–59, 2010.
© Springer-Verlag Berlin Heidelberg 2010

Our primary motivation in learning meronymy relationships from textual PD-CS databases is that they encode valuable operational knowledge that PD-CS organizations can exploit to improve product quality and ensure customer satisfaction. Meronymy relationships in the PD-CS domains are useful for uncovering causes of customer dissatisfaction that are implicitly expressed in complaint texts. For example, in "...dots *appear on* the screen...", the meronymy pattern "appear-on" expresses a customer's dissatisfaction at dots being shown *as part of* the screen display. These types of customer dissatisfaction causes, which are lexically realized with subtle, meronymy patterns (e.g. "appear-on", "available-on"), are harder to detect than those which are unequivocally expressed by customers in their complaints, such as "screen does not work". Meronymy relationships mined from PD-CS data also enable service engineers to efficiently diagnose product failures and devise remedial measures. For example, the meronymy pattern "located-at" in "switch *located at* panel 1 is broken" helps engineers to precisely identify defective components in products. Other meronymy relationships, as in "calibration is *part of* upgrade", provide pertinent information about actions performed by engineers and about service/warranty packages to management of PD-CS organizations.

Our interest in mining meronymy relationships from textual databases in corporate domains is also attributed to the challenges they pose to extant approaches. A major challenge in many corporate environments is the absence of readily-usable knowledge resources (e.g. WordNet) to support supervised meronymy learning approaches [2,3,12]. Minimally-supervised algorithms [1,10] alleviate the need for elaborate knowledge resources. Instead, they rely on a small initial set of part-whole instance pairs (e.g. engine-car), known as seeds, and extract the meronymy relationships that connect co-occurring instances in a corpus. However, defining seeds over corporate texts is challenging. It requires proficiency in the domain-terminology to ensure that selected seeds are valid part-whole instance pairs, and to deal with terminological variations due to multiple corporate stakeholders (e.g. management, engineers and customers) using different terms to refer to a single concept. Seed selection from domain-specific corporate texts also requires prior knowledge of the textual contents to ensure that the selected part-whole instances co-occur in sentences so that the meronymy relationships instantiated by these co-occurring instances can be mined. This is in stark contrast to traditional open-domain corpora, which facilitate seed selection by offering an abundance of archetypal part-whole pairs that can reasonably be assumed to co-occur in sentences (e.g. engine-car, grape-wine). These challenges in learning meronymy relationships from textual corporate databases are compounded by the wide variety of lexical constructs that encode meronymy [4].

To address these issues, and support PD-CS organizations in creating better quality products, we develop and present in this paper a framework for automatically extracting accurate meronymy relationships from domain-specific, textual corporate databases. We realize our methodology in a prototype implemented as part of the DataFusion initiative[1]. DataFusion aims at facilitating product quality improvement by using relevant information extracted from PD-CS databases to align customers' expectations to products' specifications.

---

[1] The DataFusion or "Merging of Incoherent Field Feedback Data into Prioritized Design Information" initiative is a collaboration between academia and industry, sponsored by the Dutch Ministry of Economic Affairs under the IOP-IPCR program.

Our core contribution is a principled approach to extract accurate meronymy relationships from domain-specific textual corporate databases. Our approach starts by harvesting reliable meronymy patterns from a large, open-domain corpus to circumvent the difficulties posed by domain-specific texts to relationship extraction. Targeting such a corpus also enables the wide-variety of meronymy patterns [4] to be learnt. The acquired patterns are then used to extract meronymy relationship triples from the domain-specific textual databases. To overcome the drawbacks of traditional surface-pattern representations, we formalize the patterns harvested from the open-domain corpus by using sophisticated syntactic structures. Results of evaluations performed on real-life databases provided by our industrial partners indicate that our approach accurately uncovers valuable insights on causes of customer complaints and product failures that were implicitly encoded in meronymy constructs in the data. As an ancillary contribution, we also show that some of the domain-specific relationships identified from the corporate data are not conclusively classifiable by well-known taxonomies of meronymy relationships such as the taxonomy of Winston et al. [13].

This paper is organized as follows. Section 2 presents and compares related work. We present our approach in Section 3. Experiments are described in Section 4, before concluding and highlighting areas of future work in Section 5.

2 Related Work

Winston et al. [13] (Winston) developed a taxonomy of meronymy relationships based on psycholinguistics experiments on the linguistic usage of the term "part of". They mention six types of meronymy relationships: component-integral (e.g. engine-car), member-collection (e.g. soldier-army), portion-mass (e.g. metre-kilometre), stuff-object (e.g. grape-wine), place-area (e.g. Groningen-Netherlands), and feature-activity (e.g. chewing-eating).

Algorithms to automatically acquire meronymy relationships from texts are either (fully-)supervised or minimally-supervised. In the supervised approaches presented in [2,3], meronymy relationships connecting WordNet instances are manually extracted from 200,000 sentences of the SemCor and L.A. Times corpora. The relationships are used to train a decision-tree classifier, which achieves a precision of 80.95% and a recall of 75.91% in predicting whether previously unseen constructs encode meronymy. The supervised algorithm in [12] relies on 503 part-whole pairs, acquired from specialized thesauri, to extract 91 reliable part-whole patterns from web documents with a precision of 74%.

Minimally-supervised approaches [1,10] do not require external knowledge resources. The algorithm in [10] uses part-whole instance pairs (e.g. city-region) as initial seeds to extract meronymy surface-patterns from the Acquaint (5,951,432 words) and Brown (313,590 words) corpora. An iterative procedure bootstraps the patterns, and uses them to induce new part-whole instances and patterns. The precisions reported over the Acquaint and Brown corpora are respectively 80% and 60%. However, this approach requires large numbers of surface-patterns to be manually authored, and fails to detect long-range dependencies (relationships) between words in text. The minimally-supervised technique in [1] uses six "whole" instances (e.g. school, car) and infers their corresponding "parts" (e.g. room, engine) from the North American News Corpus (1 million words) with an accuracy of 55%.

Compared to the open-domain corpora (e.g. Acquaint) targeted by the above approaches, domain-specific corporate texts present new challenges yet to be addressed. The absence of knowledge resources (e.g. ontologies) in corporate environments compromise the performance of supervised algorithms. Selecting appropriate seeds from domain-specific texts to support minimally-supervised meronymy mining algorithms is also challenging. Furthermore, the types of meronymy relationships mined from domain-specific, corporate texts could be different from those mentioned in existing taxonomies [13]. We address these challenges by developing and presenting, in the next section, our novel framework to extract accurate meronymy relationships from domain-specific, textual corporate databases. Although we focus on the product development and customer service (PD-CS) domains, our approach can be considered generic enough to be applied in other corporat contexts.

3   Methodology for Meronymy Relationships Extraction

Our proposed methodology to learn meronymy relationships from domain-specific textual corporate databases consists of three major phases: Pattern Induction, Meronymy Pattern Selection and Meronymy Relationships Extraction, as depicted in Figure 1 (dotted objects represent inputs and outputs).

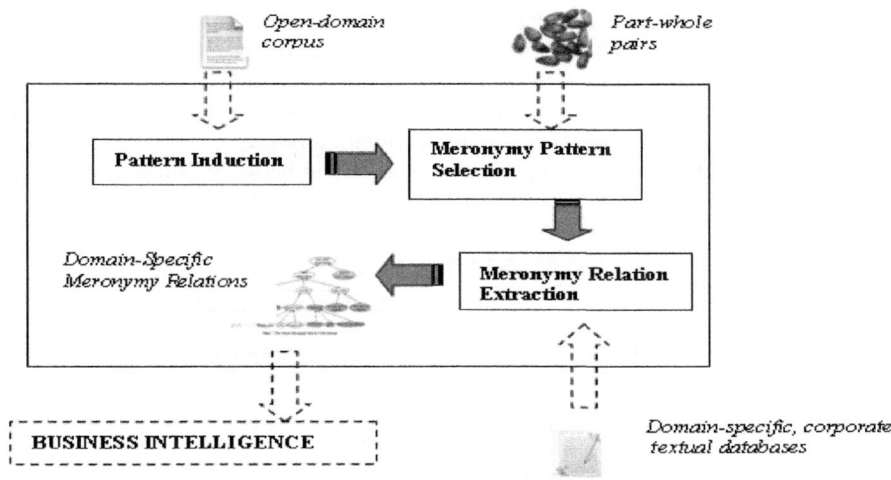

Fig. 1. Phases underlying the proposed methodology

The Pattern Induction phase (Section 3.1) induces lexico-syntactic patterns from a large, open-domain and broad-coverage corpus known as the "learning-corpus". Our rationale for initially targeting such a corpus is that it offers abundant typical and co-occurring part-whole instances (e.g. engine-car) that are suitable seeds for minimally-supervised meronymy learning algorithms. Thus, it circumvents the difficulties of seed selection over domain-specific corporate texts. Targeting a large corpus, following the predication that "it is useful to have more data than better data" [6], also captures the wide variety of patterns that encode meronymy. A minimally-supervised algorithm in the Meronymy Pattern Selection stage (Section 3.2) determines which of the induced patterns express meronymy. The Meronymy Relationships Extraction phase (Section 3.3) then uses these patterns to extract meronymy relationships from the domain-specific texts.

3.1 Pattern Induction

Pattern induction starts by syntactically parsing the sentences of the learning-corpus to derive their parse-trees. Co-occurring instances (entities) in the sentences are then detected based on their parts-of-speech (PoS) by term-recognition filters [7]. Instead of representing the relationships between co-occurring instances with traditional surface-patterns as in [10], we adopt linguistically-sophisticated dependency paths. A dependency path is the shortest sequence of lexico-syntactic elements, i.e. the shortest lexico-syntactic pattern, connecting instances in their parse-trees. Dependency paths abstract from surface texts, and alleviate the manual authoring of large numbers of surface-patterns. They formally characterize relationships by capturing long-range dependencies between words regardless of position and distance in surface texts [11].

The output of this phase is a set of lexico-syntactic patterns, the instances they sub-categorize (connect) and statistics about the occurrence/co-occurrence frequencies as harvested from the learning-corpus. Sample patterns, instance pairs they connect and the co-occurrence frequencies of the pairs and patterns are in the 4th, 3rd, 2nd and 1st columns of Figure 2 (N1 and N2 are generic markers to represent the actual instances). These patterns denote various types of relationships (including meronymy) between their instance pairs. The 1st pattern in Figure 2 depicts a "cause-of" relationship between instances "hiv" and "aids" in the learning-corpus, as in "hiv (is the) cause of aids". The 2nd pattern denotes a meronymy relationship between "stanza" and "poem", as in "poem consists of stanza".

```
11 |  hiv  | aids  | N1+nsubj < cause > prep+of+pobj+N2
5  | stanza | poem  | N1+pobj+of+prep < consists > nsubj+N2
```

Fig. 2. Sample lexico-syntactic patterns extracted, representing various semantic relationships

3.2 Meronymy Pattern Selection

The Meronymy Pattern Selection phase uses a minimally-supervised algorithm [10] that takes as input typical part-whole instance pairs (seeds), e.g. engine-car, to determine which of the previously acquired lexico-syntactic patterns express meronymy. Our algorithm considers a pattern to encode meronymy if it sub-categorizes any of the seeds in the learning-corpus. We rank the inferred meronymy patterns according to their reliability scores $r(p)$, computed by equation (1). It measures the reliability of a pattern p in expressing meronymy as its average strength of association with part-whole instance pairs i, weighted by the reliability $r(i)$ of these instance pairs. Initially, the reliability of the seeds is set to 1 (i.e. r(i) =1). In equation (1), pmi(i,p) is the point-wise mutual information between an instance pair i=x-y (e.g. i= engine-car) and a meronymy pattern p (e.g. "consists of"). It is calculated by equation (2), where |x,p,y| is the probability that p sub-categorizes x and y, and * represents any character.

$$r(p) = \frac{\sum_{i \in I} \left(\frac{pmi\ (i,p)}{\max\ _{pmi}} \times r(i) \right)}{|\ I\ |} \tag{1}$$

$$pmi(i,p) = \log \frac{|\ x,p,y\ |}{|\ x,*,y\ \|\ *,p,*\ |} \tag{2}$$

After calculating their reliability, the top-k most reliable meronymy patterns are bootstrapped, and used to induce new part-whole instance pairs from the learning-corpus. The reliability of these instance pairs, r(i), analogous to the patterns' reliability, is computed using equation (3), where |P| is the set of top-k meronymy patterns selected earlier.

$$r(i) = \frac{\sum_{p \in P} \left(\frac{pmi\ (i,p)}{\max\ _{pmi}} \times r(p) \right)}{|\ P\ |} \tag{3}$$

The top-m most reliable part-whole instance pairs are then bootstrapped for inferring other meronymy patterns. This recursive procedure of learning reliable meronymy patterns from reliable part-whole instance pairs is repeated until t patterns are extracted. The values of k, m and t are experimentally determined.

This phase yields a set of reliable meronymy-encoding lexico-syntactic patterns. Two example patterns, which read as "N2 released on N1" and "N1 includes N2" are shown in Figure 3. N1 and N2 are generic slot-fillers, respectively representing the "whole" (e.g. car) and the "part"(e.g. engine) instances.

```
N1+pobj+on+prep < release > nsubjpass+N2
N1+nsubj < include > dobj+N2
```

Fig. 3. Reliable meronymy-encoding lexico-syntactic patterns identified

3.3   Meronymy Relationships Extraction

The Meronymy Relationships Extraction stage extracts meronymy relationships as triples from the domain-specific, textual corporate databases. Each triple consists of an instance pair, in the corporate data, that is sub-categorized by any of the reliable meronymy patterns harvested earlier.

We applied standard linguistic pre-processing to the domain-specific texts by segmenting them into individual sentences, tokenizing the sentences, and determining the parts-of-speech (PoS) and lemmas of the word-tokens. The sentences are not syntactically parsed since they are short and ungrammatical, as in "image not available on console", and do not involve long-range dependencies between terms (e.g. "image", "console") and patterns (e.g. "available-on"). Furthermore, errors in the syntactic parse-trees of ungrammatical sentences can compromise our overall performance in mining meronymy relationships.

We identify instances (terms) from the domain-specific texts using the Textractor algorithm in [5], and we select the most frequent ones as domain-relevant instances. (Term identification is not discussed further in this paper).

Next, we re-write the previously acquired meronymy lexico-syntactic patterns (Section 3.2) into their equivalent surface-strings to facilitate their detection in the corporate texts, which were not syntactically parsed. Our automatic re-writing procedure starts at the patterns' subject, indicated by "nsubj" or "nsubjpass". It then collects the patterns' roots, enclosed in "<" and ">", and prepositional modifiers, indicated by "prep", to generate the corresponding surface-strings. Figure 4 shows a meronymy lexico-syntactic pattern and its equivalent surface-string. In this example, "V" is the PoS for verbs, and the regular-expression operator "?" makes the token "d" optional to cope with inflected verb forms (e.g. past-tense of regular verbs).

```
Lexico-syntactic form: N1+pobj+on+prep < release > nsubjpass+N2
Equivalent surface string: * release(d)?/V on/prep *
```

Fig. 4. Lexico-syntactic pattern and its surface-string equivalent

Finally, we extract occurrences of the meronymy patterns and the instance pairs that they connect in the domain-specific texts as meronymy relationship triples.

4   Experimental Evaluation

We conducted experiments to evaluate the performance of our approach in extracting meronymy relationships from real-life databases of our industrial partners. The data contained 143,255 textual narratives of customer complaints captured at helpdesks, and repair actions of service engineers.

4.1   Pattern Induction

We chose the English Wikipedia texts as our learning-corpus to infer reliable meronymy patterns that are subsequently used to extract meronymy relationship triples from the domain-specific databases. With two million articles in 18 Gb of texts, encompassing a broad spectrum of topics [9], the Wikipedia corpus satisfies the desiderata of being large, open-domain and broad-coverage. It offers abundant general part-whole instances (e.g. engine-car) that co-occur in its sentences, and thus, facilitates seeds selection for minimally-supervised meronymy mining algorithms. Its broad-coverage also ensures that it captures the wide variety of meronymy patterns.

We parsed a copy of the English Wikipedia corpus [14] (about 400 million words) using the Stanford parser [15], and extracted 2,018,587 distinct lexico-syntactic patterns, and 6,683,784 distinct instance pairs. Statistics about their occurrence and co-occurrence frequencies were also computed.

Figure 5 shows a Wikipedia sentence describing various relationships between instances (terms) "church", "Romanesque style", and "naves". The bottom row (last column) depicts the corresponding lexico-syntactic pattern that we derived to formalize the relationship between instances "church" and "naves". Lemmatized instances, and their co-occurrence frequency with the pattern are respectively in the 3rd, 2nd and 1st columns. N1 and N2 are generic slot-fillers for the "part" and "whole" instances. As can be seen, our lexico-syntactic patterns concisely encode the semantic relationships between instance pairs regardless of their distance and position in texts.

```
"..The church is build in a mostly Romanesque style and consists of three naves..."
5 | church | nave | N1+nsubj < consist > prep+of+pobj+N2
```

Fig. 5. Wikipedia sentence and corresponding dependency path extracted

4.2   Meronymy Pattern Selection

We implemented a minimally-supervised meronymy learning algorithm similar to [10], and defined a seed set of 143 typical part-whole instance pairs that are likely to co-occur in Wikipedia sentences. Seeds were equally distributed across the six types of meronymy relationships mentioned in [13]. Examples include engine-car, wine-grape, director-board, m-km, municipality-town, and paying-shopping. In each of its iteration, our algorithm bootstraps the top-k most reliable meronymy patterns to induce new part-whole instance pairs, and the top-m most reliable part-whole instance pairs to infer new meronymy patterns, until t patterns are extracted. In our experiments, we set k = |P| + 5 and m=|I|+20, where |P| and |I| are respectively the number of patterns and instance pairs from the previous iterations. The largest set of most reliable meronymy patterns (i.e. optimal precision and recall) was obtained in the 45th iteration, which yielded 162 patterns (i.e. t=162). Patterns introduced in subsequent iterations were noisy and irrelevant. Incrementing k with more than 5 patterns and m with more than 20 instance pairs in each iteration resulted in smaller sets of reliable patterns. Smaller values for the pattern and instance increments did not have significant effects on the performance. However, the largest set of most reliable meronymy patterns was then obtained in later iterations. Table 1 shows the five most reliable meronymy lexico-syntactic patterns inferred by our approach from Wikipedia, and their possible linguistic interpretations.

Table 1. Inferred meronymy lexico-syntactic patterns and their interpretations

Meronymy Pattern	Linguistic Interpretation
N1+nsubj < include > dobj+N2	Whole *includes* Part
N1+nsubj < contain > dobj+N2	Whole *contains* Part
N1+nsubj < consist > prep+of+pobj+N2	Whole *consists-of* Part
N1+pobj+on+prep <release> nsubjpass+N2	Part *released-on* Whole
N1+pobj+in+prep < find > nsubjpass+N2	Part *found-in* Whole

4.3   Meronymy Relationships Extraction

We pre-processed the domain-specific corporate texts to determine the parts-of-speech tags and lemmas of the word-tokens using the Stanford tagger and morphological analyzer [15]. The most frequently occurring terms were then identified as relevant domain instances using the Textractor algorithm in [5]. We also automatically transformed our meronymy lexico-syntactic patterns (Section 4.2) into their equivalent surface-strings, as described in Section 3.3, to facilitate their detection in the domain-specific texts, which were not syntactically parsed.

Out of our 162 distinct meronymy patterns, 63 were found to connect the domain-specific instance pairs in the corporate texts. The instance pairs-patterns combinations yielded 10,195 domain-specific meronymy triples that we extracted. Examples are in Table 2. Meronymy patterns are shown in the 2nd column. The 1st column indicates the patterns' occurrence frequency in the domain-specific texts. Meronymy triples extracted, of the form <*meronymy pattern, part-instance, whole-instance*>, are in column 3. Relationships that cannot be conclusively classified in existing taxonomies of meronymy relationships are marked with "*". Lexical manifestations of the patterns and part-whole instance pairs in the corporate texts are illustrated in the last column.

Table 2. Sample domain-specific meronymy triples extracted from corporate data

Freq (%)	Pattern	Triple	Example
	Available-on	<available-on, image , monitor>*	Image not *available on* monitor
71-75	Show-in	<show-in, artifact, image>*	Artifact *shown in* image
	Include	<include, calibration, corrective action >	Corrective action *includes* calibration
66-70	Perform-in	<perform-in, reboot, configuring>	Reboot *performed in* configuring
	Locate-in	<locate-in, adaptor board, pc>	Adaptor board *located in* PC
	Find-in	<find-in, blown fuse, settop box>	Blown fuse *found in* settop box
51-65			
	Come-from	<come-from, noise, generator>*	Noise *comes from* generator
1-30	Reach	<reach, c-arm, table base>	C-arm unable to *reach* table base
	Release-on	<release-on, software upgrade, processor>*	Software upgrade *released on* processor

The most frequently extracted domain-specific meronymy relationships involved patterns like "appear-on/in", "show-in" and "available-on". They accounted for 71-75% of our extracted triples. These types of relationships, between intangible parts (e.g. image) and their wholes, are not defined in Winston's taxonomy of meronymy relationships [13]. They are relevant to PD-CS organizations as they enable the identification of product malfunctioning that are implicitly expressed in customer complaint texts, for example in "Horizontal line *appears on* screen". The high frequency of such relationships can be attributed to the contents of the data we investigated, which pertained to video/imaging equipment. The next most frequent relationships that we identified were realized with patterns such as "include" and "perform-in". They occurred in 66-70% of our triples, and related activities (processes) to their constituent phases (steps). These relationships correspond to the "Feature-Activity" relationship type in Winston's taxonomy. They provide pertinent information about repair actions of engineers, and about service/warranty packages to management of PD-CS organizations, as in "Reboot *performed in* configuring". Mereotopological relationships (i.e. 3-D containment) [8] were also frequent in our data, constituting around 51-65% of the extracted triples. These relationships, which exist between parts and their containers/regions, were manifested with patterns like "find in/on/at", "contains", and "consist-of". They can be classified under the "Component-Integral" relationship type of Winston's taxonomy. Mereotopological relationships, such as "Blown fuse *found in* settop box", enable PD-CS engineers in precisely identifying product components that fail, and in efficiently devising corresponding remedial measures. Meronymy relationships involving patterns like "reach", "come-from", "release-on" and "incorporate-in" were rarer, accounting for at most 30% of our extracted triples. The meronymy pattern "reach" was found to relate discrete (physical) parts to their wholes. These relationships are classifiable under the "Component-Integral" type of Winston's taxonomy. They are suitable in PD-CS organizations for determining causes of product failures, as in "C-arm unable to *reach* table base". Meronymy relationships involving the pattern "come-from" related parts to their originating wholes. These relationships are not classifiable in Winston's taxonomy, and are useful in identifying sources of customer dissatisfaction as in "Noise *comes from* generator". Patterns like "incorporate-in" or "release-on" were found to relate intangible information artifacts (e.g. software) in the domain-specific texts. These relationships do not correspond to any of the types mentioned in Winston's taxonomy. They provide pertinent information on patches or upgrades released in existing software applications, and can be applied in software versioning in PD-CS organizations.

The remaining 99 (out of 162) open-domain meronymy patterns were not found in our domain-specific, corporate texts since they are unlikely to occur in narratives of customer complaints and repair actions of engineers in PD-CS databases. Examples of such patterns with zero frequency are "divide-into", "character-from", "publish-in", "member in/of", "add-to", "record in/on", and "collection-of".

We manually evaluated 2500 of the extracted meronymy relationship triples with the help of industrial domain experts since no gold-standard knowledge resources were available. Our evaluation sample consisted of relationships which were identified with a frequency of at least 30% from the domain-specific texts. The relationships were chosen such that they were equally distributed across the various frequency ranges of Table 2. Less frequent relationships, i.e. with frequency below than 30%, (e.g. <released-on, software upgrade, processor>) were not taken into account since they were very precise, and could positively bias our evaluation results. We calculated the precision of the meronymy triples according to equation (4), where *true_positive* is the number of valid domain-specific meronymy triples that our approach identified, and *false_positive* is the number of triples suggested by our approach, but deemed invalid by the domain experts.

$$\text{Precision} = \frac{\text{true\_positive}}{\text{true\_positive}+\text{false\_positive}} \tag{4}$$

Our manual evaluations identified 2023 *true_positives*, and 477 *false_positives*. The majority of *false_positives* involved the pattern "make-in", which did not always encode meronymy as in "monitor *made in* factory". The overall precision of our approach was thus 81%. This result compares favorably with the precisions reported by state-of-the-art techniques that mine meronymy relationships from open-domain corpora, such as Pantel's [10] 80%, Girju's [2] 81% and van Hage's [12] 74%. We did not compute the recall measure as the number of valid meronymy relationships in the corporate databases was unknown. However, we can expect a reasonably high recall score since the patterns used to extract the meronymy relationships from the domain-specific texts were harvested from a much larger corpus.

5   Conclusion and Future Work

We have described the design and implementation of a principled approach to learn precise meronymy relationships from domain-specific, textual, corporate databases. Our approach efficiently addresses the challenges of meronymy relationships extraction from domain-specific corporate texts by leveraging on linguistically sophisticated meronymy lexico-syntactic patterns harvested from a large, open-domain corpus. Evaluations on real-life, industrial databases indicate that our approach uncovers, with high precision, valuable insights on causes of customer complaints and product failures that are implicitly encoded in meronymy constructs. Our results also reveal that the types of some of the domain-specific meronymy relationships extracted from corporate data cannot be unambiguously classified in Winston's well-known taxonomy of meronymy relationships. Future work will involve learning ontologies from the extracted information to semantically integrate heterogeneous, but complementing, data sources to support business intelligence activities. We will also investigate the extraction of other semantic relationships that are relevant in the product development and/or customer service domains, such as "caused-by" to discover the "causes" of product failures. Our other research efforts will be dedicated towards a deeper examination of the identified domain-specific meronymy relationships that could not be classified in Winston's taxonomy.

Acknowledgements. This work is being carried out as part of the project "Merging of Incoherent Field Feedback Data into Prioritized Design Information (DataFusion)", sponsored by the Dutch Ministry of Economic Affairs under the IOP-IPCR program.

References

1. Berland, M., Charniak, E.: Finding parts in very large corpora. In: 37th Annual Meeting of the Association for Computational Linguistics, pp. 57–64. University of Maryland (1999)
2. Girju, R., Badulescu, A., Moldovan, D.: Automatic Discovery of Part-Whole Relations. Computational Linguistics 32, 83–135 (2006)

3. Girju, R., Badulescu, A., Moldovan, D.: Learning semantic constraints for the automatic discovery of part-whole relations. In: Conference of the NAACL on HLT, pp. 1–8. Association for Computational Linguistics, Morristown (2003)
4. Iris, M., Litowitz, B., Evens, M.: Problems with part-whole Relation. In: Evens, M.W. (ed.) Relational Models of the Lexicon: Representing Knowledge in Semantic Networks, pp. 261–288. Cambridge University Press, Cambridge (1988)
5. Ittoo, A., Maruster, L., Wortmann, H., Bouma, G.: Textractor: A Framework for Extracting Relevant Domain Concepts from Irregular Corporate Textual Datasets. In: Abramowicz, W., Tolksdorf, R. (eds.) BIS 2010. LNBIP, vol. 47, pp. 71–82. Springer, Heidelberg (2010)
6. Jijkoun, V., de Rijke, M., Mur, J.: Information extraction for question answering: improving recall through syntactic patterns. In: 20th Intl Conference on Computational Linguistics. Association for Computational Linguistics, Morristown, NJ, USA
7. Justeson, J., Katz, S.M.: Technical terminology: some linguistic properties and an algorithm for identification in text. Natural Language Engineering 1, 9–27 (1995)
8. Keet, C.M., Artale, A.: Representing and reasoning over a taxonomy of part-whole relations. Appl. Ontol. 3, 91–110 (2008)
9. Medelyan, O., Milne, D., Legg, C., Witten, I.H.: Mining meaning from Wikipedia. International Journal of Human-Computer Studies 67, 716–754 (2009)
10. Pantel, P., Pennacchiotti, M.: Espresso: leveraging generic patterns for automatically harvesting semantic relations. In: 21st Intl Conference on Computational Linguistics, Association for Computational Linguistics, Morristown, NJ, USA, pp. 113–220 (2006)
11. Shen, D., Kruijff, K.G.-J., Klakow, D.: Exploring syntactic relation patterns for Question-Answering. In: Dale, R., Wong, K.-F., Su, J., Kwong, O.Y. (eds.) IJCNLP 2005. LNCS (LNAI), vol. 3651, pp. 507–518. Springer, Heidelberg (2005)
12. van Hage, W.R., Kolb, H., Schreiber, G.: A method for learning part-whole relations. In: Cruz, I., Decker, S., Allemang, D., Preist, C., Schwabe, D., Mika, P., Uschold, M., Aroyo, L.M. (eds.) ISWC 2006. LNCS, vol. 4273, pp. 723–735. Springer, Heidelberg (2006)
13. Winston, M., Chaffin, R., Hermann, D.: A taxonomy of part-whole relations. Cognitive Science 11, 417–444 (1987)
14. English Wikipedia (2007-08-02), ISLA, University of Amsterdam,
    http://ilps.science.uva.nl/WikiXML/
15. Stanford Natural Language Processing Group,
    http://nlp.stanford.edu/software/index.shtml

Automatic Word Sense Disambiguation Using Cooccurrence and Hierarchical Information

David Fernandez-Amoros, Ruben Heradio Gil,
Jose Antonio Cerrada Somolinos, and Carlos Cerrada Somolinos

ETSI Informatica, Universidad Nacional de Educacion a Distancia
Madrid, Spain
david@lsi.uned.es,
{rheradio,jcerrada,ccerrada}@issi.uned.es

Abstract. We review in detail here a polished version of the systems with which we participated in the SENSEVAL-2 competition English tasks (all words and lexical sample). It is based on a combination of selectional preference measured over a large corpus and hierarchical information taken from WordNet, as well as some additional heuristics. We use that information to expand sense glosses of the senses in WordNet and compare the similarity between the contexts vectors and the word sense vectors in a way similar to that used by Yarowsky and Schuetze. A supervised extension of the system is also discussed. We provide new and previously unpublished evaluation over the SemCor collection, which is two orders of magnitude larger than SENSEVAL-2 collections as well as comparison with baselines. Our systems scored first among unsupervised systems in both tasks. We note that the method is very sensitive to the quality of the characterizations of word senses; glosses being much better than training examples.

1   Introduction

We advocate unsupervised techniques for Word Sense Disambiguation (WSD). Supervised techniques often offer better results but they need reliable training examples which are expensive in terms of human taggers. Furthermore, the problem is considerably more complex than others that have been successfully tackled with machine learning techniques (such as part-of-speech tagging) so it is unclear what amount of training examples will be enough to solve the problem to a reasonable extent, provided that it is a matter of quantity. In the next section we describe some related work. In section 3, the process of constructing the relevance matrix is resumed. In section 4, we present the particular heuristics used for the competing systems. We show the results in section 5. Finally, in section 6 we discuss the results and draw some conclusions.

2   Related Work

We are interested in performing in-depth measures of the disambiguation potential of different information sources. We have previously investigated the informational value of semantic distance measures in [4]. For SENSEVAL-2, we have combined word cooccurrence and hierarchical information as sources of disambiguation evidence [5].

C.J. Hopfe et al. (Eds.): NLDB 2010, LNCS 6177, pp. 60–67, 2010.
© Springer-Verlag Berlin Heidelberg 2010

Cooccurrence counting has played an important role in the area of WSD, starting with [7] whose algorithm, consisted in counting co-occurrences of words in the sense definitions of the words in context. To disambiguate a word, the sense with a higher number of co-occurrences with the sense definitions of the other words in the sentence was chosen. He did not make a rigorous evaluation of his algorithm but it has been widely used as a starting point in developing more complex techniques, such as those presented in [2]. They realized that the amount of noise could be reduced if the process of counting co-occurrences was limited to one sense of a word at a time. To solve the combinatorial explosion problem they employed the non-linear optimization technique known as *simulated annealing*. [9] makes the following claim about co-occurrences of words and the senses associated with them:

1. The probability of a relationship between two word-senses occurring in the same sentence is high enough to make it possible to extract useful information from statistics of co-occurrence.
2. The extent to which this probability is above the probability of chance cooccurrence provides an indicator of the strength of of the relationship.
3. If there are more and stronger relationships among the word-senses in one assignment of word-senses to words in a sentence than in another, then the first assignment is more likely to be correct.

Wilks et al. counted the co-occurrences of words in the LDOCE (Longman's Dictionary of Contemporary English) and used a combination of relatedness functions between words (using the co-occurrence information) and a set of similarity measures between sense-vectors and co-occurrence vectors. They used all this information to perform quite a number of experiments disambiguating the word *bank*.

[8] measured the co-occurrence in a Wall Street Journal corpus and expanded the disambiguation contexts by adding the words most related via co-occurrence with the words in context. They used this information to perform WSD applied to Information Retrieval (IR).

[10] computed the co-occurrences of words in the Grolier Enclyclopedia. He calculated $\frac{Pr(w|RCat)}{Pr(w)}$ that is, the probability of a word w appearing in the context of a Roget Category divided by its overall probability in the corpus.

3   The Relevance Matrix

Before building our systems we have developed a resource we have called the *relevance matrix*. The raw data used to build the matrix comes from the Project Gutenberg (PG)[1].

At the time of the creation of the matrix, the PG consisted of more than 3000 books of diverse genres. We have adapted these books for our purpose : First, discarding books not written in English; we applied a simple heuristic that uses the percentage of English stop words in the text. This method is considered acceptable in the case of large texts. We stripped off the disclaimers, then proceeded to tokenize, lemmatize, strip punctuation and stop words and detect numbers and proper nouns. The result is a collection of around 1.3GB of text.

---

[1] http://promo.net/pg

3.1 Cooccurrence Matrix

We have built a vocabulary of the 20000 most frequent words (or labels, as we have changed all the proper nouns detected to the label PROPER_NOUN and all numbers detected to NUMBER) in the text and a symmetric cooccurrence matrix between these words within a context of 61 words (we thought a broad context of radius 30 would be appropriate since we are trying to capture vague semantic relations).

3.2 Relevance Matrix

In a second step, we have built another symmetric matrix, which we have called *relevance matrix*, using a mutual information measure between the words (or labels), so that for two words a and b, the entry for them would be $\frac{P(a \cap b)}{P(b)P(a)}$, where $P(a)$ is the probability of finding the word a in a random context of a given size. $P(a \cap b)$ is the probability of finding both a and b in a random context of the fixed size. We approximated those probabilities with the frequencies in the corpus.

The mutual information measure is known to overestimate the relation between low frequency words or those with very dissimilar frequencies [3]. To avoid that problem we adopted a similar approach as [1] and ignored the entries where the frequency of the intersection was less than 50.

We have also introduced a threshold of 2 below which we set the entry to zero for practical purposes (we are only interested in strongly related pairs of words). We think that this is a valuable resource that could be of interest for many other applications other than WSD. Also, it will grow in quality as soon as we feed it with a larger amount of raw data. An example of the most relevant words for some of the words in the lexical sample task of the SENSEVAL-2 competition can be seen in table 1.

Table 1. Most relevant words for a sample of words

word	Relevant words
art	decorative pictorial thou proficient imitative hub archaeology whistler healing angling culinary sculpture corruptible photography handicraft adept
authority	vested abrogate municipal judiciary legislative usurped subordination marital parental guaranty federal papal centralized concurrent unquestioned ecclesiastical
bar	capstan shuttered barkeeper bartender transverse visor barmaid bolt iron cage padlock stanchion socket gridiron whitish croup
blind	deaf lame maimed buff unreasoning sightless paralytic leper dumb slat eyesight crustacean groping venetian necked wilfully
carry	stretcher loads portage suitcase satchel swag knapsack valise sedan litter petrol momentum connotation sling basket baggage
chair	bottomed sedan wicker swivel upholster cushioned washstand horsehair rocker seating rickety tilted mahogany plush vacate carven
church	congregational anglican methodist militant presbyterian romish lutheran episcopal steeple liturgy wesleyan catholic methodists spire baptists chancel
circuit	node relay integrated transmitter mac testing circuit billed stadium carbon installation tandem microphone platinum id generator

4   Cascade of Heuristics

We have developed a very simple language in order to systematize the experiments. This language allows the construction of WSD systems comprised of different heuristics that are applied in cascade so that each word to be disambiguated is presented to the first heuristic, and if it fails to disambiguate, then the word is passed on to the second heuristic and so on. We now present the heuristics considered to build the systems.

4.1   Monosemous Expressions

Monosemous expressions are simply unambiguous words in the case of the all words English task since we did not take advantage of the satellite features. In the case of the lexical sample English task, however, the annotations include multiword expressions. We have implemented a multiword term detector that considers the multiword terms from WordNet *index.sense* file and detects them in the test file using a multilevel backtracking algorithm that takes account of the inflected and base forms of the components of a particular multiword in order to maximize multiword detection. We tested this algorithm against the PG and found millions of these multiword terms.

We restricted ourselves to the multiwords already present in the training file since there are, apparently, multiword expressions that where overlooked during manual tagging (for instance the WordNet expression *the_good_old_days* is not hand-tagged as such in the test files) In the SemCor collection the multiwords are already detected and we just used them as they were.

4.2   Statistical Filter

WordNet comes with a file, *cntlist*, literally *a file listing number of times each tagged sense occurs in a semantic concordance* so we use this to compute the relative probability of a sense given a word (approximate in the case of collections other than SemCor). Using this information, we eliminated the senses that had a probability under 10% and if only one sense remains we choose it. Otherwise we go on to the next heuristic. In other words, we didn't apply complex techniques with words which are highly skewed in meaning.

Some people may argue that this is a supervised approach. In our opinion, the *cntlist* information does not make a system supervised *per se*, because it is standard information provided as part of the dictionary. In fact, we think that every dictionary should make that information available. Word senses are not just made up. An important criterion for incorporating a new sense is SFIP, that is Sufficiently Frequent and Insufficiently Predictable. If one wants to build a dictionary for a language it is customary to create meanings for the words in a large corpus and assign them. But in order to do so it is imperative that the frequency of distribution of the senses be kept. Besides, we don't use the examples to feed or train any procedure.

4.3   Relevance Filter

This heuristic makes use of the relevance matrix. In order to assign a score to a sense, we count the cooccurrences of words in the context of the word to be disambiguated with the words in the definition of the senses (the WordNet gloss tokenized, lemmatized and stripped out of stop words and punctuation signs) weighting each cooccurrence by the entry in the relevance matrix for the word to be disambiguated and the word whose cooccurrences are being counted. Also, not to favor senses with long glosses we divide by the number of terms in the gloss.

We use idf (inverse document frequency) a concept typically used in information retrieval to weight the terms in a way that favors specific terms over general ones. Finally, for each word in context we weight its contribution with a *distance* function. The distance function is a gaussian that favors terms coming from the immediate surroundings of the target word. So, if s is a sense of the word α whose definition is S and C is the context in which α is to be disambiguated, the score for s would be calculated by eq. 1.

$$\sum_{w \in C} R_{w\alpha} \text{freq}(w, C) \text{distance\_weight}(w, \alpha) \text{freq}(w, S) \text{idf}(w, \alpha) \qquad (1)$$

Where $\text{idf}(w, \alpha) = \log \frac{N}{d_w}$, with N being the number of senses for word α and d_w the number of sense glosses in which w appears. $\text{freq}(w, C)$ is the frequency of word w in the context C, $\text{freq}(w, S)$ is the frequency of w in the sense gloss S and distance_weight $(w, \alpha) = 0.1 + e^{-\text{distance}(w,\alpha)^2/2\sigma^2}$. σ has been assigned the value 2.17 in our experiments with SENSEVAL-1 data.

The idea is to prime the occurrences of words that are relevant to the word being disambiguated and give low credit (possibly none) to the words that are incidentally used in the context.

Also, in the all words task (where POS tags from the TreeBank are provided) we have considered only the context words that have a POS tag *compatible* with that of the word being disambiguated. For instance, adverbs are used to disambiguate verbs, but not to disambiguate nouns.

We also filtered out senses with low values in the *cntlist* file, and in any case we only considered at most the first six senses of a word. We finally did not use this heuristic. Instead, we used it as an starting point for the next ones.

4.4   Enriched Senses and Mixed Filter

The problem with the Relevance filter is that there is little overlapping between the definitions of the senses and the contexts in terms of cooccurrence (after removing stop words and computing idf) which means that the previous heuristic didn't disambiguate many words. To overcome this problem, we enrich the senses characteristic vectors taking for each word in the vector the words related to it via the relevance matrix weights. This corresponds to the algebraic notion of multiplying the matrix and the characteristic vector. In other words, if R is the relevance matrix and v our characteristic vector we would finally use Rv.

Since the matrix is so big, this product is very expensive computationally. Instead, given the fact that we are using it just to compute the score for a sense with equation 2, where, s is a sense of the word α to be disambiguated, whose definition is S and C is the context in which α is to be disambiguated, we only need to sum up a number of terms which is the product of the number of terms in C multiplied by the number of terms in S.

$$\sum_{i \in C} \sum_{j \in S} R_{ij}\text{freq}(i, C))\text{distance\_weight}(i, \alpha)\text{freq}(j, S) \tag{2}$$

One interesting effect of the relevance matrix being symmetric is that it can be easily proved that the effect of enriching the sense characteristic vectors is the same as enriching the contexts.

The *mixed filter* is a particular case of this one, when we also discard senses with low relative frequency in SemCor.

For those cases that could not be covered by other heuristics we employed the first sense heuristic. The difference is almost negligible since it is rarely used.

5   Systems and Results

The heuristics we used and the results evaluated for SENSEVAL-2 and SemCor for each of them are shown in Table 2. If the individual heuristics are used as standalone WSD systems we would obtain the results in Table 3.

We have also built a supervised variant of the previous systems. We have added the training examples to the definitions of the senses giving the same weight to the definition and to all the examples as a whole (i.e. definitions are given more credit than examples). The evaluation is only interesting for the lexical sample, the results are given in Table 4 and discussed in the next section.

Table 2. UNED Unsupervised heuristics

Heuristic	All Words task			Lexical Sample			SemCor		
	Coverage	Prec	Recall	Coverage	Prec	Recall	Coverage	Prec	Recall
Monosemous	18%	89%	16%	4%	58%	2%	20%	100%	20%
Statistical Filter	23%	68%	16%	25%	43%	11%	28%	83%	23%
Mixed Filter	34%	38%	13%	44%	34%	15%	33%	42%	13%
Enriched Senses	21%	50%	10%	23%	47%	11%	15%	46%	7%
First Sense	1%	59%	0%	0%	100%	0%	1%	61%	1%
Total	99%	57%	57%	99%	41%	40%	100%	67%	67%

Table 3. UNED Unsupervised vs baselines

System	All Words task			Lexical Sample			SemCor		
	Coverage	Prec	Recall	Coverage	Prec	Recall	Coverage	Prec	Recall
FIRST	99%	60%	59%	99%	43%	42%	100%	75%	75%
UNED	99%	57%	57%	99%	41%	40%	100%	67%	67%
Mixed Filter	76%	59%	46%	75%	38%	29%	82%	71%	58%
Enriched Senses	97%	46%	45%	98%	35%	34%	97%	50%	49%
RANDOM	99%	36%	35%	99%	18%	18%	100%	40%	40%
Statistical Filter	41%	77%	32%	30%	46%	13%	49%	90%	44%
Monosemous	18%	89%	16%	4%	58%	2%	20%	100%	20%

Table 4. UNED Trained heuristics & UNED Trained vs baselines

	Lexical Sample		
Heuristic	Coverage	Prec	Recall
Monosemous	4%	58%	2%
Statistical Filter	25%	43%	11%
Mixed Filter	44%	22%	10%
Enriched Senses	23%	24%	5%
First Sense	0%	0%	0%
Total	99%	30%	29%

	Lexical Sample		
System	Coverage	Prec	Recall
First Sense	99%	43%	42%
UNED	99%	30%	29%
Mixed Filter	75%	32%	24%
Enriched Senses	99%	15%	15%
Random	99%	18%	18%
Statistical Filter	30%	46%	13%
Monosemous	4%	58%	2%

It is worth mentioning the difference in the size of the collections: The all words task consisted of 2473 test cases, the lexical sample task had 4328 test cases and the SemCor collection, 192639. The SemCor evaluation, which is nearly two orders of magnitude larger than the SENSEVAL tasks, is perhaps the main contribution of this paper insofar as results are much more significant.

6   Discussion and Conclusions

The results obtained support Wilks' claim (as quoted in the related work section) in that co-occurrence information is an interesting source of evidence for WSD. If we look at table 3, we see that the Enriched Senses heuristic performs 27% better that random for the all words and 25% better for the SemCor collection. This relative improvement jumps to 94% in the case of the lexical sample. This is not surprising since the lexical sample contains no monosemous words at all, but actually rather ambiguous ones, which severely affects the random heuristic performance.

The results in table 4 are surprising, in the sense that using training examples to enrich the senses characteristic vectors actually harms the performance. So, while using just the sense glosses to enrich via the relevance matrix yielded good results, as we have just seen in the previous paragraph, adding the training examples makes the precision fall from 47% to 24% as a heuristic in the cascade, and from 35% to 15% used as a baseline on its own. It is pertinent to remind here that all the training examples for a word were given the same weight as the sense gloss, so using just training examples to enrich the sense would probably yield catastrophic results.

We think it would be worth experimenting with smoothing techniques for the matrix such as the one described by [6] since we have experienced the same kind of problems mixing the frequencies of dissimilarly occurring words.

We were very confident that the relevance filter would yield good results as we have already evaluated it against the SENSEVAL-1 and SemCor data. We felt however that we could improve the coverage of the heuristic enriching the definitions multiplying by the matrix. The quality of the information used to characterize word senses seems to be very important, since multiplying by the matrix gives good results with the glosses, however the precision degrades terribly if we multiply the matrix with the training examples plus the glosses.

As for the overall scores, the unsupervised lexical sample obtained the highest recall of the unsupervised systems in SENSEVAL-2, which proves that carefully implementing simple techniques still pays off. In the all words task we also obtained the highest recall among the unsupervised systems.

References

1. Church, K.W., Hanks, P.: Word association norms, mutual information and lexicography. In: 27th Annual Conference of the Association of Computational Linguistics, pp. 76–82 (1989)
2. Cowie, J., Guthrie, J., Guthrie, L.: Lexical disambiguation using simulated annealing. In: International conference in computational linguistics (COLING), Nantes, pp. 359–365 (1992)
3. Dunning, T.E.: Accurate methods for the statistics of surprise and coincidence. Computational Linguistics 19(1), 61–74 (1993)
4. Fernández-Amorós, D., Gonzalo, J., Verdejo, F.: The role of conceptual relations in word sense disambiguation. In: Applications of Natural Language to Information Systems (NLDB), Madrid, pp. 87–98 (2001)
5. Fernández-Amorós, D., Gonzalo, J., Verdejo, F.: The uned systems at senseval-2. In: Proceedings of the 2nd International Workshop on Evaluating Word Sense Disambiguation Systems (SENSEVAL), Toulouse (2001)
6. Gale, W.A., Church, K.W., Yarowsky, D.: A method for disambiguating word senses in a large corpus. Computers and the Humanities 26(5), 415–439 (1993)
7. Lesk, M.E.: Automatic sense disambiguation using machine readable dictionaries : How to tell a pine cone from an ice cream cone. In: Proceedings of SIGDOC (Special Interest Group for Documentation Conference), Toronto, Canada (1986)
8. Schuetze, H., Pedersen, J.: Information retrieval based on word senses. In: Proceedings of the 4th Annual Symposium on Document Analysis and Information Retrieval, Las Vegas, NV, pp. 161–175 (1995)
9. Wilks, Y., Fass, D., Guo, C., McDonald, J., Plate, T., Slator, B.: Providing machine tractable dictionary tools. Machine Translation 5(2), 99–151 (1990)
10. Yarowsky, D.: Word-sense disambiguation using statistical models of Roget's categories trained on large corpora. In: Proceedings of COLING 1992, Nantes, France, pp. 454–460 (1992)

Automatic Quality Assessment of Source Code Comments: The JavadocMiner

Ninus Khamis, René Witte, and Juergen Rilling

Department of Computer Science and Software Engineering
Concordia University, Montréal, Canada

Abstract. An important software engineering artefact used by developers and maintainers to assist in software comprehension and maintenance is source code documentation. It provides insights that help software engineers to effectively perform their tasks, and therefore ensuring the quality of the documentation is extremely important. Inline documentation is at the forefront of explaining a programmer's original intentions for a given implementation. Since this documentation is written in natural language, ensuring its quality needs to be performed manually. In this paper, we present an effective and automated approach for assessing the quality of inline documentation using a set of heuristics, targeting both *quality* of language and *consistency* between source code and its comments. We apply our tool to the different modules of two open source applications (ArgoUML and Eclipse), and correlate the results returned by the analysis with bug defects reported for the individual modules in order to determine connections between documentation and code quality.

1 Introduction

"Comments as well as the structure of the source code aid in program understanding and therefore reduce maintenance costs." – Elshoff and Marcotty (1982) [1]

Over the last decade, software engineering processes have constantly evolved to reflect cultural, social, technological, and organizational changes. Among these changes is a shift in development processes from document driven towards agile development, which focuses on software development rather than documentation. This ongoing paradigm shift leads to situations where source code and its comments often become the only available system documentation capturing program design and implementation decisions. Studies have shown that the effective use of comments "can significantly increase a program's comprehension" [2], yet the amount of research focused towards the quality assessment of in-line documentation is limited [3]. Recent advancements in the field of natural language processing (NLP) has enabled the implementation of a number of robust analysis techniques that can assist users in content analysis. In this work, we focus on using NLP services to support users in performing the time-consuming task of analysing the quality of in-line documentation. We focus mainly on in-line documentation due to its close proximity to source code. Additionally, we developed

C.J. Hopfe et al. (Eds.): NLDB 2010, LNCS 6177, pp. 68–79, 2010.

a set of heuristics based on new and existing metrics to assess the consistency between code and documentation, which often degrades due to changes in source code not being reflected in their comments. Results of the NLP analysis are exported into OWL ontologies, which allow to query, reason, and cross link them with other software engineering artefacts.

2   Background

In this section we discuss the background relevant for the presented work, in particular *Javadoc* and the impact of inline documentation on software maintenance.

2.1   Inline Documentation and Javadoc

Literate programming [4] was suggested in the early 1980s by Donald Knuth in order to combine the process of software documentation with software programming. Its basic principle is the definition of program fragments directly within software documentation. Literate programming tools can further extract and assemble the program fragments as well as format the documentation. The extraction tool is referred to as *tangle* while the documentation tool is called *weave*. In order to differentiate between source code and documentation, a specific documentation or programming syntax has to be used.

Single-source documentation also falls into the category of documents with inter-weaved representation. Contrary to the literate approach, documentation is added to source code in the form of comments that are ignored by compilers. Given that programmers typically lack the appropriate tools and processes to create and maintain documentation, it has been widely considered as an unfavourable and labour-intensive task within software projects [5]. Documentation generators currently developed are designed to lessen the efforts needed by developers when documenting software, and has therefore become widely accepted and used. Javadoc [6] is an automated tool that generates API documentation in HTML using Java source code and source code comments. In Fig. 1, we show a small section of an API document generated using Javadoc.

Javadoc comments added to source code are distinguishable from normal comments by a special comment syntax ($/**$). A generator (similar to the weave tool within literate programming) can extract these comments and transform the corresponding documentation into a variety of output formats, such as HTML,

Node	`lastNode()` Get the node with the largest offset
Node	`nextNode(Node node)` Get the first node that is relevant for this annotation set and which has the offset larger than the one of the node provided.
boolean	`remove(Object o)` Remove an element from this set.

Fig. 1. Part of an API Documentation generated using Javadoc

LATEX, or PDF. Most tools also provide specific tags within comments that influence the format of the documentation produced or the way documentation pages are linked. Both Doxygen [7] and Javadoc also provide an API to implement custom extraction or transformation routines [6]. Even during early stages of implementation, the Javadoc tool can process pure stubs (classes with no method bodies), enabling the comment within the stub to explain what future plans hold for the created identifiers.

Different types of comments are used to document the different types of identifiers. For example, a *Class* comment should provide insight on the high-level knowledge of a program, e.g., which services are provided by the class, and which other classes make use of these services [2]. A *Method* comment, on the other hand, should provide a low-level understanding of its implementation [2]. The default Javadoc doclet provides a limited amount of checks that are mainly syntactic in nature. However, more analyses can potentially be applied on Javadoc comments, measuring factors such as completeness, synchronization and readability.

2.2   Source Code Comments and Impact on Software Maintenance

With millions of lines of code written every day, the importance of good documentation cannot be overstated. Well-documented software components are easily comprehensible and therefore, maintainable and reusable. This becomes especially important in large software systems [8]. Since in-line documentation comes in contact with various stakeholders of a software project, it needs to effectively communicate the purpose of a given implementation to the reader. Currently, the only means of assessing the quality of in-line documentation is by performing time-consuming manual code checks. Any well-written computer program contains a sufficient number of comments to permit people to read it. Development programmers should prepare these comments when they are coding and update them as the programs change. There exist different types of guidelines for in-line documentation, often in the form of programming standards.[1] In general, each program module contains a description of the logic, the purpose, and the rationale for the module. Such comments may also include references to subroutines and descriptions of conditional processing. Specific comments for specific lines of code may also be necessary for unusual coding. For example, an algorithm (formula) for a calculation may be preceded by a comment explaining the source of the formula, the data required, and the result of the calculation and how the result is used by the program.

Writing in-line documentation is a painful and time-consuming task that often gets neglected due to release or launch deadlines. With such deadlines pressuring the development team it becomes necessary to prioritize. Since customers are mostly concerned with the functionality of an application, implementation and bug fixing tasks receive a higher priority compared to documentation tasks. Furthermore, finding a balance, describing all salient program features comprehensively and concisely is another challenge programmers face while writing in-line

---

[1] See, e.g., http://www.gnu.org/prep/standards/standards.html

documentation. Ensuring that development programmers use the facilities of the programming language to integrate comments into the code and to update those comments is an important aspect of software quality. Even though the impact of poor quality documentation is widely known, there are few research efforts focused towards the automatic assessment of in-line documentation [9].

3   Improving Software Quality through Automatic Comment Analysis: The JavadocMiner

The goal of our *JavadocMiner* tool is to enable users to 1) automatically assess the quality of source code comments, and 2) export the in-line documentation and the results returned from this analysis to an ontology.

3.1   Analysis Heuristics for Source Code Comments

In this section, we discuss the set of heuristics that were implemented to assess the quality of in-line documentation. The heuristics are grouped into two categories, *(i)* internal (NL quality only), and *(ii)* code/comment consistency.

Internal (NL Quality) Comment Analysis. We first describe the set of heuristics targeting the natural language quality of the in-line documentation itself.

Token, Noun and Verb Count Heuristic: These heuristics are the initial means of detecting the use of well-formed sentences within in-line documentation. The heuristic counts the number of tokens, nouns and verbs used within a Javadoc comment.

Words Per Javadoc Comment Heuristic (WPJC): This heuristic calculates the average number of words in a Javadoc comment [9]. It can be used to detect a Java Class that contains Fields, Constructors and Methods that are under- or over-documented.

Abbreviation Count Heuristic (ABB): According to "How to Write Doc Comments for the Javadoc Tool" [10] the use of abbreviations in Javadoc comments should be avoided; therefore, using "also known as" is preferred over "aka". This heuristic counts the number of abbreviations used within a Javadoc comment.

Readability Heuristics (FOG/FLESCH/KINCAID): In the early twentieth century linguists conducted a number of studies that used people to rank the readability of text [9]. Such studies require a lot of resources and are often infeasible for source code comments. A number of formulas were implemented that analyse the readability of text [11], for example:

The Fog Index: Developed by Robert Gunning, it indicates the number of years of formal education a reader would need to understand a block of text.

```
/** The following method parses the associations of a class diagram.
 * @return String          A String association is returned
 * @param  role            The AssociationEnd <em>text</em> describes.
 * @param  text            A String on the above format.
 * @throws ParseException  When is detects an error in the role string.
 *                         See also ParseError.getErrorOffset().
 */
protected String parseAssociationEnd(Object role, String text) throws ParseException
```

Fig. 2. An Example of a Javadoc Method Comment

Flesch Reading Ease Level: Rates text on a 100 point scale. The higher the scale the easier to read. An optimal score would range from 60 to 70.[2]

Flesch-Kincaid Grade Level Score: Translates the 100 point scale of the Flesch Reading Ease Level metric to a U.S. grade school level.

These readability formulas are also used by a number of U.S. government agencies such as the DoD and IRS to analyse the readability of their documents [9].

Code/Comment Consistency Analysis. The following heuristics analyse in-line documentation in relation to the source code being documented.

Documentable Item Ratio Heuristic (DIR): The DIR metric was originally proposed by [9]. For in-line documentation describing a method to be considered complete, it should document all its aspects. For example, for methods that have a *return type*, contain *parameters*, or *throw exceptions*, the *@return*, *@parameter* and *@throws* in-line tags must be used.

Any Javadoc Comment Heuristics (ANYJ): ANYJ computes the ratio of *identifiers with Javadoc comments* to the *total number of identifiers* [9]. This heuristic can be used to determine which classes provide the least amount of documentation.

SYNC Heuristics (RSYNC/PSYNC/ESYNC): The following heuristics detect methods that are documenting *return types, parameters* and *thrown exceptions* that are no longer valid (e.g., due to changes in the code):

RSYNC: When documenting the *return type* of a *method* the *@return in-line tag* must begin with the correct name of the *type* being returned followed by the *doc comment* explaining the *return type*.

PSYNC: When documenting the *parameter list* of a *method* the *@param in-line tag* should begin with the correct name of the *parameter* being documented followed by the *doc comment* explaining the *parameter*.

ESYNC: When documenting the *exceptions thrown* by a *method* the *@throws* or *@exception* in-line tags documentation must begin with the correct names of the *exceptions* being returned followed by the *doc comment* explaining the *exception* itself.

---

[2] A score between 90–100 would indicate that the block of text could be understood by an 11 year old and would therefore be overly simplified [11].

The `parseAssociationEnd` method in Fig. 2 is an example of a method that contains a *return type, parameter list,* and *exception* that is consistent with the in-line documentation used in the *@return, @param,* and *@throws in-line tags.*

4   Implementation

In this section, we discuss in detail the various parts that make up the JavadocMiner. We begin by covering the process of building a corpus from source code using a custom `doclet`.[3] We then explain the components used within our text mining application to process the corpus for the set of heuristics described above.

4.1   Corpus Generation from Javadoc

Javadoc's standard doclet generates API documentation using an HTML format. While this is convenient for human consumption, automated NLP analysis requires a more structured XML format. Generation of such an XML format is possible by developing a custom doclet using the Javadoc library, which provides access to the Abstract Syntax Tree (AST) generated from source code and source code comments. We implemented a custom doclet called the `SSLDoclet` [12],[4] which enables us to *(i)* control what information from the source code will be included in the corpus, and *(ii)* mark-up the information using a schema that our NLP application can process easily.

4.2   The JavadocMiner GATE Application

Our JavadocMiner is implemented as a GATE pipeline [13], which is assembled using individual *processing resources* (PRs) running on a corpus of documents.

Preprocessing Stage. Before running the JavadocMiner PR, we perform pre-processing tasks using components already provided by GATE, including tokenization, sentence splitting, and POS-tagging.

Abbreviation Detection. The abbreviations are detected using a gazetteering list that is part of ANNIE [13]. The list contains commonly used abbreviations within the English language.

JavadocMiner PR. We implemented a GATE processing resource component called the *JavadocMiner PR* that contains the set of heuristics described above to assess the quality of source code comments. Most of the heuristics were implemented by us; with the exception of the readability heuristics that makes use of an existing library [14].

---

[3] See
http://java.sun.com/j2se/1.4.2/docs/tooldocs/javadoc/overview.html

[4] SSLDoclet, see http://www.semanticsoftware.info/javadoclet

Table 1. Assessed Open Source Project Versions, Release Dates, Lines of Code (LOC), Number of Comments, Identifiers and Bugs

Project Version	Release Date	LOC	Num. of Comments	Num. of Identifiers	Num. of Bug Defects
ArgoUML v0.24	02/2007	250,000	6871	13,974	46
ArgoUML v0.26	09/2008	600,000	6875	14,262	54
ArgoUML v0.28.1	08/2009	800,000	7168	14,789	48
Eclipse v3.3.2	06/2007	7,000,000	32,172	158,009	176
Eclipse v3.4.2	06/2008	8,000,000	33,919	163,238	413
Eclipse v3.5.1	06/2009	8,000,000	34,360	165,945	153

OwlExporter PR. The OwlExporter [15] is the final step in our pipeline that is in charge of taking the annotations created by the text mining pipeline and exporting it to an ontology.

5   Application and Evaluation

In the section, we discuss how the JavadocMiner was applied on two open source projects for an analysis of comment quality and their consistency with the source code; we discuss the results gathered from our study, and finally show how we additionally correlated the NLP quality metrics with bug statistics.

5.1   Data

We conducted a case study where the JavadocMiner was used to assess the quality of in-line documentation found in three major releases of the UML modelling tool ArgoUML[5] and the IDE Eclipse.[6] In Table 1, we show the versions of the projects that were part of our quality assessment.

5.2   Experiments

We split the ArgoUML and Eclipse projects into their three major modules, for ArgoUML – *Top Level*, *View & Control*, and *Low Level*, and for Eclipse – *Plugin Development Environment (PDE)*, *Equinox*, and *Java Development Tools (JDT)*. The quality of the in-line documentation found in each module was assessed separately for a total of 43,025 identifiers and 20,914 comments from ArgoUML, and 487,192 identifiers and 100,451 comments from Eclipse. The complete quality assessment process for both open source projects took less than 3 hours. We continued by finding the amount of bug defects reported for each version of the modules using the open source project's issue tracker system. The Pearson product-moment correlation coefficient measure was then applied to the data gathered from the quality assessment and issue tracker systems to determine the varying degrees of correlation between the individual heuristics and bug defects.

---

[5] ArgoUML, http://www.ohloh.net/p/argouml/analyses/latest
[6] Eclipse, http://www.ohloh.net/p/eclipse/analyses/latest

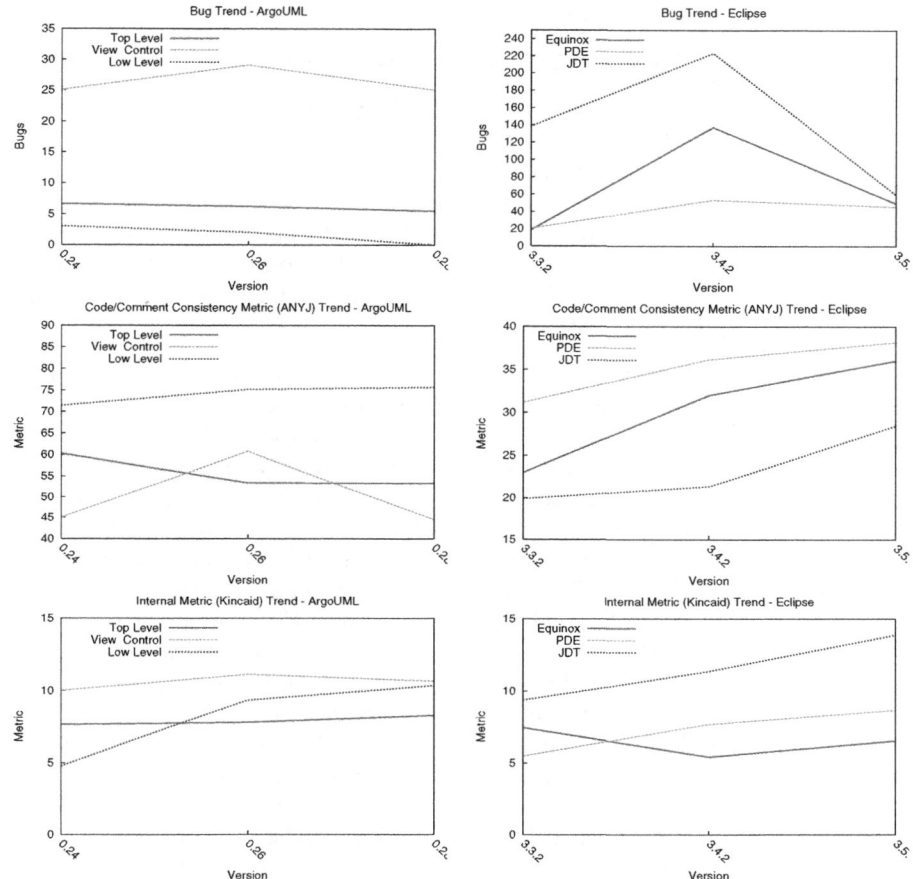

Fig. 3. Charts for Bug Trend, Code/Comment and Internal (NL Quality) Metrics

5.3   Results and Analysis

Results of our study indicated that the modules that performed best in our quality assessment were the *Low Level module* for ArgoUML (Fig. 3, left side) and the *PDE module* for Eclipse (Fig. 3, right side).

Quality Analysis. We believe that the reason for the Low Level module (ArgoUML) and the PDE module (Eclipse) outperforming the rest of the modules in every heuristic is that they are both the base libraries that every other module extends. For example, Eclipse is a framework that is extended using plug-ins that use the services provided by the PDE API module. The Eclipse project is separated into API and internal non-API packages, and part of the Eclipse policy states that all API packages must be properly documented [9].

Table 2. Pearson Correlation Coefficient Results for ArgoUML and Eclipse

Project	ANYJ	SYNC	ABB	FLESCH	FOG	KINCAID	TOKENS	WPJC	NOUNS	VERBS
ArgoUML	0.99	0.98	-0.94	0.32	0.80	0.79	0.89	0.91	0.98	0.87
Eclipse	0.97	0.89	-0.86	0.37	0.77	0.84	0.88	0.86	0.91	0.73

The readability results returned by the JavadocMiner indicate that the Low Level module contained in-line documentation that is less complicated compared to the View and Control module (Fig. 3, bottom left), yet more complicated than the in-line documentation found in the Top Level module. The PDE module in Eclipse also returned similar Kincaid results compared to the other two Eclipse modules (Fig. 3, bottom right).

Comment-Bug Correlation. As part of our efforts to correlate the results of our study with another software engineering artefact, we examined the amount of bug defects that were reported for each version of the modules for ArgoUML (Fig. 3, top left) and for Eclipse (Fig. 3, top right). We observed that the modules that performed best in our quality assessment also had the least amount of reported defects, and vice versa for the modules that performed poorly. In order to determine how closely each metric correlated with the number of reported bug defects, we applied the Pearson product-moment correlation coefficient on the data gathered from the quality assessment and the number of bug defects (Table 2). Fig. 3 also shows the quality assessments returned by ANYJ (Fig. 3, middle) and Kincaid (Fig. 3, bottom) for both ArgoUML and Eclipse.

The correlation and coefficient results showed that the ANYJ, SYNC, ABB, Tokens, WPJC and Nouns heuristics were strongly correlated to the number of bug defects, whereas the Flesch readability metric was amongst the least correlated. In Fig. 4 (top), we show the number bug defects reported at the level of quality assessments returned by the ANYJ and ABB metrics, and in Fig. 4 (bottom) for the readability and NLP metrics for ArgoUML. The same data is represented for Eclipse in Fig. 5.

Examining the charts showed a correlation between each metric and the number of bug defects; with the exception of the Flesch metric, which we previously determined as being the least correlated using the correlation statistic.

6   Related Work

There has been effort in the past that focused on analysing source code comments, for example in [16] human annotators were used to rate excerpts from *Jasper Reports*, *Hibernate* and *jFreeChart* as being either More Readable, Neutral or Less Readable. The authors developed a "Readability Model" that consists of a set of features such as the average and/or the maximum 1) line length in characters; 2) identifier length; 3) identifiers; and 4) comments represented using vectors. The heuristics used in the study were mostly quantitative in nature and based their readability scale on the length of the terms used, and not necessarily the complexity of the text as a whole. The authors also made no attempt to measure how up-to-date the comments were with the source code they were explaining.

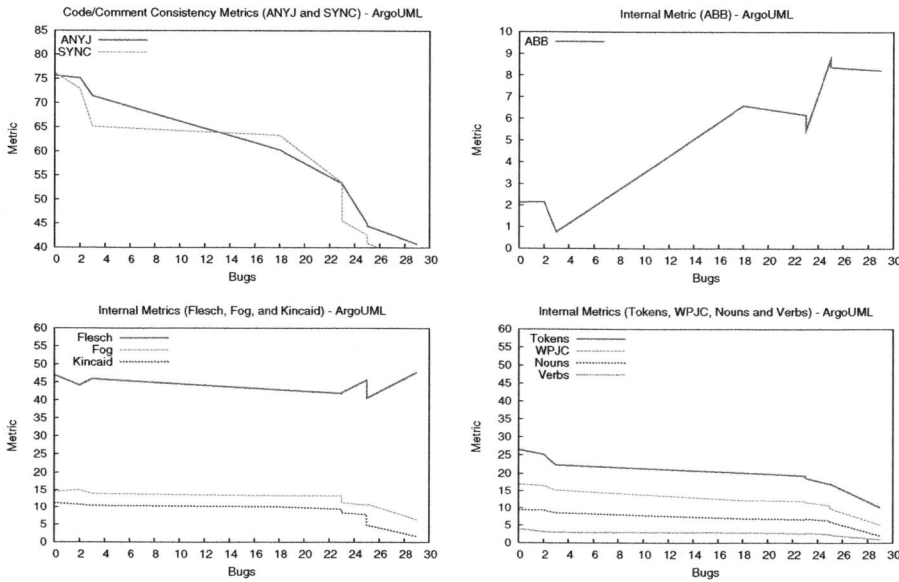

Fig. 4. Code/Comment Consistency and NL Quality Metrics vs. Bugs – ArgoUML

The authors of [3] manually studied approximately 1000 comments from the latest versions of Linux, FreeBSD and OpenSolaris. Part of their study was to see how comments can be used in developing a new breed of bug detecting tool, or how comments that use cross-referencing can be used by editors to increase a programmer's productivity by decreasing navigation time. The work attempts to answer questions such as 1) what is written in comments; 2) whom are the comments written for or written by; 3) where the comments are located; and 4) when the comments were written. Results from the study showed that 22.1% of the analysed comments clarify the usage and meaning of integers, 16.8% of the examined comments explain implementation, for example, which function is responsible for filling a specific variable, 5.6% of source code comments describe code evolution such as cloned code, deprecated code and TODOs. The purpose of the study was to classify the different types of in-line documentation found in software and not necessarily assess their quality. The authors also made no attempt to automate the process, nor was there any major correlations made with other software engineering artefacts.

The work described in [17] defines a Source Code Vocabulary (SV) as being the union of Class Name, Attribute Name, Function Name, Parameter Name and Comment Vocabularies. The work uses a combination of existing tools like diff to answer questions; such as how the vocabularies evolve over time, what type of relationships exist between the individual vocabularies, are new identifiers introducing new terms, and finally what do the most frequent terms refer to.

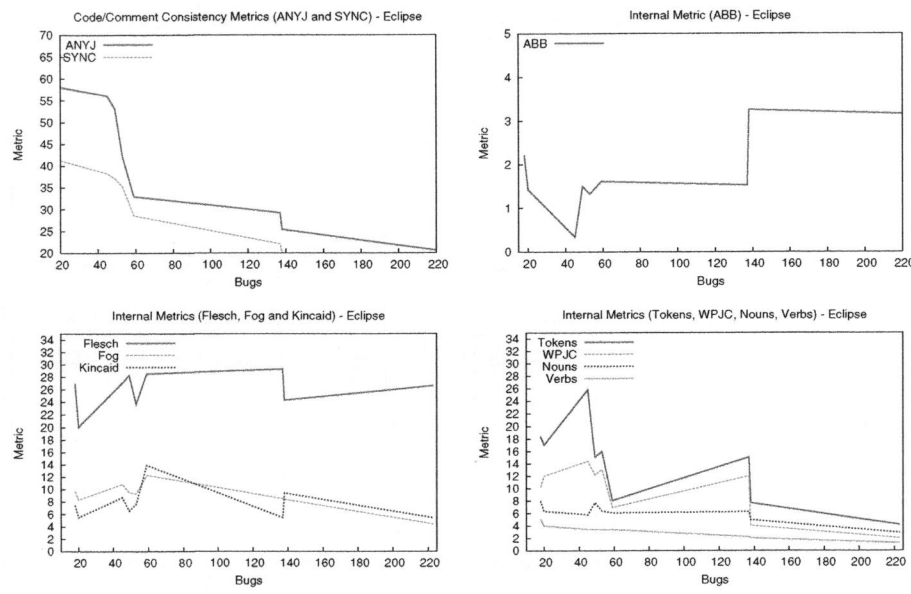

Fig. 5. Code/Comment Consistency and NL Quality Metrics vs. Bugs – Eclipse

The only work that we know of which focused on automatically analysing API documentation generated by Javadoc is [9]. The authors implemented a tool called QUASOLEDO that measures the quality of documentation with respect to its completeness, quantity and readability. Here, we extended the works of [9] by introducing new quality assessment metrics. We also analysed each module of a software project separately, allowing us to observe correlations between the quality of in-line documentation and bug defects. Both of the efforts mentioned above focus mostly on the evolution of in-line documentation and whether they co-change with source code, and not necessarily on the quality assessment of in-line documentation. None of the efforts mentioned in this section put nearly as much emphasis on correlating the quality of in-line documentation with reported bug defects as was done in our study.

7   Conclusions and Future Work

In this paper, we discussed the challenges facing the software engineering domain when attempting to manage the large amount of documentation written in natural language. We presented an approach to automatically assess the quality of documentation found in software. Regardless of the current trends in software engineering and the paradigm shift from documentation to development, we have shown how potential problem areas can be minimized by maintaining source code that is sufficiently documented using good quality up-to-date source code comments. Currently we are conducting a case study involving students and the quality assessment of 120 in-line documentation samples, in order to compare the results obtained from the JavadocMiner with the results of human intuition analysing the quality of in-line documentation.

Acknowledgements. This research was partially funded by DRDC Valcartier (Contract No. W7701-081745/001/QCV).

References

1. Fluri, B., Würsch, M., Gall, H.: Do Code and Comments Co-Evolve? On the Relation between Source Code and Comment Changes. In: WCRE, pp. 70–79 (2007)
2. Nurvitadhi, E., Leung, W.W., Cook, C.: Do class comments aid Java program understanding? Frontiers in Education (FIE) 1 (November 2003)
3. Padioleau, Y., Tan, L., Zhou, Y.: Listening to programmers Taxonomies and characteristics of comments in operating system code. In: ICSE 2009, Washington, DC, USA, pp. 331–341. IEEE Computer Society, Los Alamitos (2009)
4. Knuth, D.E.: Literate Programming. The Computer Journal 27(2), 97–111 (1984)
5. Brooks, R.E.: Towards a Theory of the Comprehension of Computer Programs. International Journal of Man-Machine Studies 18(6), 543–554 (1983)
6. Kramer, D.: API documentation from source code comments: a case study of Javadoc. In: SIGDOC 1999: Proceedings of the 17th annual international conference on Computer documentation, pp. 147–153. ACM, New York (1999)
7. van Heesch, D.: Doxygen (2010), http://www.stack.nl/~dimitri/doxygen/
8. Lehman, M.M., Belady, L.A. (eds.): Program evolution: processes of software change. Academic Press Professional, Inc., San Diego (1985)
9. Schreck, D., Dallmeier, V., Zimmermann, T.: How documentation evolves over time. In: IWPSE 2007: Ninth international workshop on Principles of software evolution, pp. 4–10. ACM, New York (2007)
10. Sun Microsystems: How to Write Doc Comments for the Javadoc Tool, http://java.sun.com/j2se/javadoc/writingdoccomments/
11. DuBay, W.H.: The Principles of Readability. Impact Information (2004)
12. Khamis, N., Witte, R., Rilling, J.: Generating an NLP Corpus from Java Source Code: The SSL Javadoc Doclet. In: New Challenges for NLP Frameworks (2010)
13. Cunningham, H., Maynard, D., Bontcheva, K., Tablan, V.: GATE: A framework and graphical development environment for robust NLP tools and applications. In: Proc. of the 40th Anniversary Meeting of the ACL (2002)
14. Ryan, K., Fast, G.: Java Fathom, http://www.representqueens.com/fathom/
15. Witte, R., Khamis, N., Rilling, J.: Flexible Ontology Population from Text: The OwlExporter. In: Int. Conf. on Language Resources and Evaluation, LREC (2010)
16. Buse, R.P.L., Weimer, W.R.: A metric for software readability. In: ISSTA 2008: Proceedings of the 2008 international symposium on Software testing and analysis, pp. 121–130. ACM, New York (2008)
17. Abebe, S.L., Haiduc, S., Marcus, A., Tonella, P., Antoniol, G.: Analyzing the Evolution of the Source Code Vocabulary. In: European Conference on Software Maintenance and Reengineering, pp. 189–198 (2009)

Towards Approximating COSMIC Functional Size from User Requirements in Agile Development Processes Using Text Mining

Ishrar Hussain, Leila Kosseim, and Olga Ormandjieva

Department of Computer Science and Software Engineering,
Concordia University, Montreal, Quebec, Canada
{h_hussa,kosseim,ormandj}@cse.concordia.ca

Abstract. Measurement of software size from user requirements is crucial for the estimation of the developmental time and effort. COSMIC, an ISO/IEC international standard for functional size measurement, provides an objective method of measuring the functional size of the software from user requirements. COSMIC requires the user requirements to be written at a level of granularity, where interactions between the internal and the external environments to the system are visible to the human measurer, in a form similar to use case descriptions. On the other hand, requirements during an agile software development iteration are written in a less formal way than use case descriptions — often in the form of user stories, for example, keeping with the goal of delivering a planned release as quickly as possible. Therefore, size measurement in agile processes uses methods (e.g. story-points, smart estimation) that strictly depend on the subjective judgment of the experts, and avoid using objective measurement methods like COSMIC. In this paper, we presented an innovative concept showing that using a supervised text mining approach, COSMIC functional size can be automatically approximated from informally written textual requirements, demonstrating its applicability in popular agile software development processes, such as Scrum.

1 Introduction

The agile development process breaks down the software development lifecycle into a number of consecutive iterations that increases communication and collaboration among stakeholders. It focuses on the rapid production of functioning software components along with providing the flexibility to adapt to emerging business realities [19]. In practice, agile processes have been extended to offer more techniques, e.g. describing the requirements with user stories [21]. Instead of a manager estimating developmental time and effort and assigning tasks based on conjecture, team members in agile approach use effort and degree of difficulty in terms of points, or size, to estimate of their own work, often with biased judgment [5]. Hence, an objective measurement of software size is crucial in planning and management of agile projects.

We know that effort is a function of size [22], and a precise estimation of software size right from the start of a project life cycle gives the project manager confidence

C.J. Hopfe et al. (Eds.): NLDB 2010, LNCS 6177, pp. 80–91, 2010.
© Springer-Verlag Berlin Heidelberg 2010

about future courses of action, since many of the decisions made during development depend on the initial estimations. Better estimation of size and effort allows managers to determine the comparative cost of a project, improve process monitoring, and negotiate contracts from a position of knowledge.

The above has led the industry to formulate several methods for functional size measurement (FSM) of software. Allan Albrecht first proposed the idea in his work on function point analysis (FPA) in 1979 [2], where he named the unit of functional size as function point (FP). His idea of effort estimation was then validated by many studies, like [3,17], and, thus, measuring the functional size of the software became an integral part of effort estimation. There have been many standards developed by different organizations on FSM methods, following the concepts presented in Albrecht's FPA method. Four of these standards have been accepted as ISO standards: they are IFPUG [14], Mark II [15], NESMA [16] and COSMIC [13].

There have been many studies in recent years [6,7,26], where researchers attempted to automate the process of different functional size measurement methods, but none of them, to our knowledge, addressed the problem where they would take the textual requirements as input to start the automatic measurement process. In addition, all these work depended on extracting manually the conceptual modeling artifacts first from the textual requirements, so that a precise functional size measurement can be performed. On the other hand, the work documented in this paper aims to develop a tool that would automatically perform a quicker approximation of COSMIC size without requiring the formalization of the requirements. This is in response to the high industrial demands of performing size estimation during agile development processes, where formalization of requirements are regarded as costly manipulation, and, thus, ignored during size estimation. Our methodology extends the idea presented in the Estimation by Analogy approach [25] and the Easy and Quick (E&Q) measurement approach, that was originated in the IFPUG standard [14]. The applicability of this approach in COSMIC was manually demonstrated by [24].

2 Background

2.1 COSMIC

For the purposes of this research, we have chosen to use the COSMIC FSM method developed by the Common Software Measurement International Consortium (COSMIC) and now adopted as an international standard [13]. We chose this method in particular, because it conforms to all ISO requirements [12] for FSM, focuses on the "user view" of functional requirements, and is applicable throughout the Agile development life cycle. Its potential of being applied accurately in the requirements specification phase compared to the other FSM methods is demonstrated by the study of [8]. Also, COSMIC does not rely on subjective decisions by the functional size measurer during the measurement process [13]. Thus, its measurements, taken from well-specified requirements, tend to be same among multiple measures. This is particularly important for validating the performance of the automatic size measurements that would be yielded by our solution.

In COSMIC, the size is measured in terms of the number of *Data-movements*, which accounts for the movement of one or more data-attributes belonging to a single *Data-group*. A data-group is an aggregated set of data-attributes. A *Functional Process*, in COSMIC, is an independently executable set of data-movements that is triggered by one or more *triggering events*. A triggering event is initiated by an "actor" (a functional user or an external component) that occurs outside the boundary of the software to be measured. Thus, a functional process holds the similar scope of a use case scenario, starting with the triggering event of a user-request and ending with the completion of the scenario.

The data-movements can be of four types: *Entry*, *Exit*, *Read* and *Write*. An *Entry* moves a data-group from a user across the boundary into the functional process, while an *Exit* moves a data group from a functional process across the boundary to the user requiring it. A *Write* moves a data group lying inside the functional process to persistent storage, and a *Read* moves a data group from persistent storage to the functional process.

COSMIC counts each of these data-movements as one CFP (COSMIC Function Point) of functional size, and measures the size of each of the functional processes separately. It then sums up sizes of all the functional processes to compute the total size of the system to be measured.

2.2   Size Measurement in Agile Development Processes

Agile development processes are driven by the motto of delivering releases as quickly as possible [19]. The size of every agile iteration is subjectively estimated by means of user requirements that are written less formally than use case descriptions. These textual requirements, which are mostly available in the form of smart use cases [1] or user-stories [21], although, do not provide detailed description of the scenarios like those found in use cases, they must hold "enough details" to perform the size estimation [21]. Size measurement methods in agile development processes include story-points [5] and smart estimation [1], and depend on the subjective judgment of human experts, and, therefore, are prone to biases and errors [5].

COSMIC offers an objective method of measuring functional size. It is built to be applied in the traditional processes of software development, where documentation of requirements using formalisms and templates is required. However, the IT industry recognized the traditional processes to cause many problems including delays and is now increasingly moving towards agile development processes, such as Scrum [29], an agile approach that does not impose documentation templates or formalisms on requirements. Lack of formalism in requirements restricts FSM methods, like COSMIC, to be applied for measuring the functional size of an iteration in an agile development process. For example, from the discussion in section 2.1, it can be understood that the number of data-groups, which is necessary to be known to carry out COSMIC FSM, cannot be identified by the measurer from a set of requirements statements alone unless he/she is supplied with a complete list of available data-groups that requires formalizing the requirements with conceptual model (e.g. a domain model).

Our work presents an alternative solution, which does not require the use of domain models; instead, it proposes an objective way of approximating the COSMIC functional size of a functional process (i.e. a use case) that is described by an informally written set of textual requirements, in forms likely to be used in agile size estimation.

3   Related Work

One of the leading work done in the area of automating COSMIC FSM is by Diab *et al* [7], where the authors developed a comprehensive system called, µcROSE, which accepts state charts as inputs to measure the functional size of real-time systems only. We find their work to be largely dependant on a set of hard-coded rules for mapping different objects of interest to different COSMIC components, and also require C++ code segments to be attached with the state transitions and supplied as input too, so that data-movements can be identified. They presented a very brief validation of their work by an expert, testing their system against only one case study, where it performed poorly in detecting the data groups, resulting in some erroneous measurement outputs.

Condori-Fernández *et al* [6] performed another study, where they presented step by step guidelines to first derive manually the UML modeling artifacts, e.g. the use case models and system sequence diagrams from the requirements, and then, apply their set of rules for measuring the COSMIC functional size of the system from the UML models. Their approach was validated on 33 different observations, showing reproducible results with 95% confidence.

4   Methodology

Most of the related work performed in this field were directed towards performing a precise measurement of COSMIC functional size. On the other hand, our goal is to develop an automated tool that would do quicker estimation of COSMIC size without requiring the formalization of the requirements. Our methodology requires the historical data of an organization to be stored for the purpose of generating a dataset for training/testing our application. The historical dataset needs to contain sets of textual user requirements written in any quality, where each set corresponds to a unique functional process, along with their respective functional size in COSMIC to be recorded by human measurers. We present our detailed methodology in the following sections.

4.1   CFP Measurement

Our first step is to build our historical dataset by manually measuring the COSMIC size of the functional processes in units of CFP (COSMIC Function Point). The available textual description of the user requirements corresponding to each functional process is used for this purpose. Here, for each requirements statement belonging to a functional process, the human measurer first identifies how many different types of data-movements are expressed by the statement, and then, how many data-groups participate in each of the types of data-movements present in the statement. Following COSMIC, the sum of number of data-groups for each type of data-movements indicates the total CFP size of one requirements statement. The measurer repeats this step for the rest of the requirements statements within the functional process and summing up their sizes results in the CFP count for the whole functional process. The measurer then again sums up CFP sizes for each of the functional processes to obtain the respective CFP count of the whole system. Table 1 illustrates the CFP counting process with a hypothetical example of a system consisting of two functional processes.

Table 1. A hypothetical example of precise CFP calculation

Functional processes	User requirements	Types of Data-movements expressed by the statement:	Number of Data-groups involved in a data-movement	Size in CFP
FPr:1	1.1 *User requests to view the detailed information of one item.*	Entry	2	2
		Read	1	1
		Size of statement 1.1 =		3
	1.2 *System displays detailed item information.*	Exit	1	1
		Size of statement 1.2 =		1
		Total size of **FPr:1** =		3+1 = **4**
FPr:2	2.1 *When user requests to add the item to the shopping cart, system adds it and displays the cart.*	Entry	2	2
		Write	1	1
		Exit	1	1
		Size of statement 2.1 =		4
		Total size of **FPr:2** =		4
		Total size of the whole system =		4 + 4 = 8

Our approach requires these measurement data to be saved in the historical database for the past completed projects. For this work, we will need the CFP count for each of the functional processes that have been measured, along with the textual requirements associated to a functional process.

4.2   Class Annotation of Functional Processes

Once we have gathered the historical dataset, we need to define classes of functional processes, based on their sizes in CFP, to be used later in the automatic classification task. To do this, we performed a box-plot analysis on the CFP size values from our historical dataset, to produce four different classes of functional processes, based on their sizes in CFP. Table 2 shows the defined ranges of these classes.

Table 2. Ranges of CFP values to define the classes

Classes	Ranges
Small	[0, Lower Quartile)
Medium	[Lower Quartile, Median)
Large	[Median, Upper Quartile)
Complex	[Upper Quartile, ∞)

Here, the lower quartile would cut off the lowest 25% of all the recorded CFP size data from the historical database. The median would divide the dataset by 50%, and the upper quartile cuts off the highest 25% of the dataset.

These four sets of ranges allow us to annotate the textual requirements of the functional processes automatically into four fuzzy size classes. In our class ranges, we keep the minimum and the maximum values as 0 and ∞, respectively, instead of the sample minimum or the sample maximum, like in an actual box-plot analysis; so that, if the new unseen sample is an outlier compared to the historical dataset, it would still get classified into a class.

After defining the class boundaries automatically, we then calculate the mean, the minimum and the maximum for each of the classes, to designate the range of the approximate size for each of the classes.

4.3 Text Mining

Our next step consists of extracting linguistic features from the textual requirements belonging to each of the functional processes from our training dataset, to train a text classification algorithm that can automatically classify a new set of textual requirements belonging to a functional process into one of the classes defined above (i.e. *Small*, *Medium*, *Large* or *Complex*). It will then simply approximate the size of the functional processes by outputting its size as the calculated mean value of the class it belongs to, along with the minimum and the maximum seen CFP value for that class to indicate possible variation in the approximation; and, thus, provide the quickest possible approximation of the COSMIC functional size from textual requirements that are not formalized and can be written in any quality.

5 Preliminary Study

As a proof of concept, we performed a small preliminary study with four different case studies: two industrial projects from SAP Labs, Canada, and two university projects. They are all completed projects and from different domains. Table 3 summarizes the case studies.

Table 3. Summary of the case studies

ID	Source	Title	Type of Application	Size of Requirements Document	Functional Processes extracted
C1	Industry (SAP)	*(undisclosed)*	Web (Internal)	11,371 words	12
C2	Industry (SAP)	*(undisclosed)*	Business	1,955 words	3
C3	University	Course Registration System	Business	3,072 words	14
C4	University	IEEE Montreal Website	Web (Public)	5,611 words	32
Total number of functional processes extracted =					61

We manually pre-processed these requirements to extract sets of requirements sentences each of which belongs to a distinct functional process. This mimics the available set of user requirements before an iteration starts in an agile development process. From all four requirements documents, we were able to extract 61 sets of textual requirements, each belonging to a distinct functional process.

We used five human measurers, all graduate students skilled to perform COSMIC FSM from requirements documents, to measure the CFP of these 61 functional processes, similarly to what is shown in Table 1. The CFP values and the textual requirements of the 61 functional processes built our historical dataset.

5.1 The Annotated Corpus

As mentioned in Section 3.2, we performed a box-plot analysis on the CFP values of our historical dataset that gave us — median = 6 CFP; lower-quartile = 5 CFP; and, upper-quartile = 8 CFP. Therefore, according to the ranges defined in Table 2 in section 3.2, the actual CFP ranges for the four size classes for our historical dataset are: *Small: [0,5)*; *Medium: [5,6)*; *Large: [6,8)*; and, *Complex: [8,∞)*. We then followed these ranges to automatically annotate the sets of textual requirements belonging to the 61 functional processes into the four size classes — where 9 (15%) functional processes were annotated as *Small*, 15 (25%) were *Medium*, 21 (34%) were *Large*, and 16 (26%) were annotated as *Complex*.

5.2 Syntactic Features

To perform the classification task, we considered a large pool of linguistic features that can be extracted by a syntactic parser. In this regards, we used the Stanford Parser [18] (equipped with Brill's POS tagger [4] and a morphological stemmer) to morphologically stem the words and extract many linguistic features, e.g. the frequency of words appearing in different parts-of-speech categories. As we have the actual CFP values in our historical dataset, we sorted the linguistic features based on the correlation between their values and the CFP values. The top ten highly correlated features are listed in Table 4.

Table 4. Ten linguistic features highly correlated with CFP

Features (*Frequency of..*)	Correlation with CFP
Noun Phrases	0.4640
Parentheses	0.4408
Active Verbs	0.4036
Tokens in Parentheses	0.4001
Conjunctions	0.3990
Pronouns	0.3697
Verb Phrases	0.3605
Words	0.3595
Sentences	0.3586
Uniques (*hapax legomena*)	0.3317

The correlation shows the ten syntactic features that influence COSMIC functional size the most. The reasons for some of them are explained below.

Noun Phrases. No matter how poorly a requirement is described, the involvement of a data-group in a particular data-movement is typically indicated by the presence of a noun phase. Therefore, if a functional process contains more noun phrases, the chances are that its data-movements involve more data-groups and its size is larger.

Parentheses & Number of tokens inside parentheses. When complex functional processes are often described in textual requirements, parentheses are used to provide brief explanations in a limited scope. Thus, a higher number of parentheses/Number of tokens inside parentheses can sometimes indicate a complex functional process.

Active Verbs & Verb Phrases. Verbs in active form define actions and are often used in larger numbers in textual requirements to explain data-movements, as data-movements result from actions carried out by the user or the system or an external system.

Pronouns. A longer description in textual requirements for a functional process often indicates its complexity, and requires the use of more pronouns within the functional process to maintain cohesion with references.

Words, Sentences and Uniques. They all account for lengthy description of the requirements in functional processes; and, as mentioned above, long description of requirements of a functional process may indicate its complexity.

Next, we looked at possible keyword features than can be extracted.

5.3 Keyword Features

A large body of work (e.g. [10,27]) has shown that using keywords grouped into particular parts-of-speech categories helps to obtain good results with Text Mining. For this study, we have, therefore, considered list of keywords, each list belonging to a given parts-of-speech category. We chose three part-of-speech groups for these keywords to be selected. They are: Noun-keywords (coded as: *NN_keyword*), Verb-keywords (coded as: *Verb_keyword*), and Adjective-keywords (coded as: *JJ_keyword*).

We generate finite lists of these keywords based on two different probabilistic measures, as described in [10], that takes into account how many more times the keywords occur in one class of the training set than the other class. A cutoff threshold is then used to keep the list significantly smaller. For example, the three lists that were automatically generated by this process from our training set during a fold of 10-fold-crossvalidation is shown in Table 5.

Table 5. Some of the keywords of POS group: Noun, Verb and Adjective

NN_keyword	VB_keyword	JJ_keyword
user	ensure	supplied
category	get	current
quota	choose	previous
content	start	available
default	restart	
chart	fill	
...	...	

These three lists constituted three additional features for our classification task. Thus, when we extract the features, we counted one of the keyword feature, for example, as how many times words from its keyword-list appears in the set of requirements of a functional process, and appearing in the same part-of-speech class.

5.4   Feature Extraction and Classification

To classify the sets of textual requirements belonging to different functional processes, we developed a Java-based feature extraction program that uses the Stanford Parser [18] to extract the values of all the syntactic and keyword features mentioned above, and uses Weka [28] to train and test the C4.5 decision tree learning algorithm [23]. We used the implementation of the C4.5 (revision 8) that comes with Weka (as J48), setting its parameter for the minimum number of instances allowed in a leaf to 6 to counter possible chances of over-fitting. The results are discussed in the next section. We also trained/tested with a Naïve Bayes classifier [9], and a logistic classifier [20]. The C4.5 decision tree-based classifier performed the best in comparison to the other classifiers with more consistent results during 10-fold-cross-validation.

6   Results and Analysis

The results of the classification were very moderate when using the whole dataset for training and testing. Since the dataset was not very large, we could not use a separate dataset for testing, and we could only use cross-validation, which can be very harsh on the performance, when the number of instances is very low. Yet, the classifier results did not drop significantly. Table 6 shows a summary of the results.

Table 6. Summary of the results

	Scheme	Correctly Classified Sentences	Incorrectly Classified Sentences	Kappa	Comment
Corpus Size = 61 (sets of textual requirements, each set representing a functional process)	Training + Testing on same set	45 (73.77%)	16 (26.23%)	0.6414	Tree is of desirable characteristics, not sparse, and also not flat. None of the branches are wrongly directed.
	Cross-validation (10 Folds)	41 (67.21%)	20 (32.79%)	0.5485	

The resultant decision tree after training on the complete dataset also came out well-formed. The tree is shown in Figure 1.

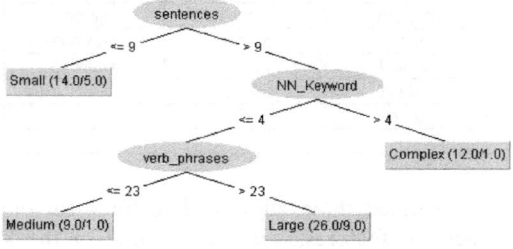

Fig. 1. The resultant C4.5 decision tree after training with the complete dataset

Detailed results of using 10-fold-crossvalidation with the final confusion matrix is shown in Table 7.

Table 7. Confusion matrix when using 10-fold-crossvalidation

	Classified as			
	Small	**Medium**	**Large**	**Complex**
Small	7	0	1	1
Medium	1	7	7	0
Large	2	1	16	2
Complex	2	0	3	11

The precision, recall and f-measure results for each of the classes in 10-fold-cross-validation are shown in Table 8.

Table 8. Precision, Recall and F-Measure, when using 10-fold-crossvalidation

Size Class	Precision	Recall	F-Measure
Small	0.583	0.778	0.667
Medium	0.875	0.467	0.609
Large	0.593	0.762	0.667
Complex	0.786	0.688	0.733
Mean	0.709	0.673	0.669

Although the kappa results of Table 6 shows stable and moderate results in terms of performance with the 10-fold-crossvalidation, the results of the confusion matrix in Table 7 and the results in Table 8 show that the classifier is struggling to attain a good recall with the fuzzy class *Medium*. The reason for this is that classifying an instance into the size classes that fall in the middle (e.g. functional processes that are not small and not large, but of medium size) is hard when we do not have a larger number of instances that can allow the learning algorithm to find the threshold values for the other features and thus, utilize them in making fine-grained distinction.

We can also demonstrate by showing that if we had less number of classes, i.e. two or three size classes, the available number of instances would have been enough for a more realistic classification task. To show that, we developed both a two-class size classifier (classifying functional processes into *Small* and *Large* classes), and a three-class size classifier (classifying functional processes into *Small*, *Medium* and *Large* classes) using the same principles and the same sets of features described earlier in this paper. The results were significantly better, attaining mean f-measures of 0.802 and 0.746 for the 2-class and the 3-class classifiers respectively.

7 Conclusions and Future Work

In this paper, we have shown that classification of textual requirements in terms of their functional size is plausible. Since our work uses a supervised text mining approach, where we needed experts to build our historical database by manually measuring the COSMIC functional size from textual requirements, we could not train and test our system with a large number of samples. Yet, the results that we were able to gather by crossvalidating on such small number of samples show a promising behavior of the classifier in terms of its performance. We have been able to identify automatically a set of highly discriminating features that can effectively help together with a classifier.

It should be mentioned that we have not yet tested our approach as to be used with requirements written in variable level of quality. Therefore, we believe that this approach would be organization-specific, where textual requirements saved in the historical dataset should all be written in the same format or writing style having similar quality. This would allow our classifier to pick the best set of features and set the best thresholds that would classify new requirements written in similar style and quality.

We are currently in the process of building larger datasets for training and testing our system. Our future work includes implementing a full-fledged prototype to demonstrate its use and a complete integration to the READ-COSMIC project [11], which is our umbrella project on software development effort estimation from textual requirements. We are also working on predicting the impact of non-functional requirements on the functional size for better precision in software effort estimation.

Acknowledgements. The authors would like to thank the anonymous reviewers for their valuable comments on an earlier version of the paper.

References

1. Accelerated Delivery Platform (2009) Smart use cases. (Retrieved on February 14, 2009) http://www.smartusecase.com/(X(1)S(hp3vxp242ym1mg45faqtegbg))/Default.aspx?Page=SmartUseCase
2. Albrecht, A.J.: Measuring Application Development Productivity. In: Proceedings of IBM Application Development Symp., pp. 83–92. Press I.B.M., Monterey (1979)
3. Albrecht, A.J., Gaffney, J.E.: Software function, source lines of code, and development effort prediction: A software science validation. IEEE Transactions on Software Engineering 9, 639–648 (1983)
4. Brill, E.: A Simple Rule-Based Part of Speech Tagger. In: Proceedings of the third conference on Applied natural language processing, pp. 152–155. Association for Computational Linguistics, Trento (1992)
5. Cohn, M.: Agile Estimating and Planning. Prentice Hall, Upper Saddle River, NJ (2005)
6. Condori-Fernández, N., Abrahão, S., Pastor, O.: On the estimation of the functional size of software from requirements specifications. Journal of Computer Science and Technology 22(3), 358–370 (2007)
7. Diab, H., Koukane, F., Frappier, M., St-Denis, R.: μcROSE: Automated Measurement of COSMIC-FFP for Rational Rose Real Time. Information and Software Technology 47(3), 151–166 (2005)

8. Gencel, C., Demirors, O., Yuceer, E.: Utilizing Functional Size Measurement Methods for Real Time Software System. In: 11th IEEE International Software Metrics Symposium (METRICS 2005). IEEE Press, Los Alamitos (2005)
9. John, G.H., Langley, P.: Estimating Continuous Distributions in Bayesian Classifiers. In: Proceedings of the Eleventh Conference on Uncertainty in Artificial Intelligence, San Mateo, pp. 338–345 (1995)
10. Hussain, I., Kosseim, L., Ormandjieva, O.: Using Linguistic Knowledge to Classify Non-functional Requirements in SRS documents. In: Kapetanios, E., Sugumaran, V., Spiliopoulou, M. (eds.) NLDB 2008. LNCS, vol. 5039, pp. 287–298. Springer, Heidelberg (2008)
11. Hussain, I., Ormandjieva, O., Kosseim, L.: Mining and Clustering Textual Requirements to Measure Functional Size of Software with COSMIC. In: Proceedings of the International Conference on Software Engineering Research and Practice (SERP 2009), pp. 599–605. CSREA Press (2009)
12. ISO/IEC 14143-1, Functional Size Measurement - Definition of Concepts. International Organization for Standardization (1998)
13. ISO/IEC 19761, COSMIC Full Function Points Measurement Manual v.2.2. International Organization for Standardization (2003)
14. ISO/IEC 20926, Software Engineering – IFPUG 4.1 Unadjusted functional size measurement method – Counting Practices Manual. International Organization for Standardization (2003)
15. ISO/IEC 20968, Software Engineering - Mk II Function Point Analysis - Counting Practices Manual. International Organization for Standardization (2002)
16. ISO/IEC 24570, Software Engineering – NESMA functional size measurement method version 2.1 – Definitions and counting guidelines for the application of Function Points Analysis. International Organization for Standardization (2005)
17. Kitchenham, B.A., Taylor, N.R.: Software cost models. ICL Technical Journal 4, 73–102 (1984)
18. Klein, D., Manning, C.D.: Accurate unlexicalized parsing. In: Proceedings of the 41st Meeting of the Association for Computational Linguistics, pp. 423–430. Association for Computational Linguistics (2003)
19. Larman, C.: Agile & Iterative Development: a Manager's Guide. Pearson Education, Boston (2003)
20. le Cessie, S., van Houwelingen, J.C.: Ridge Estimators in Logistic Regression. Applied Statistics 41(1), 191–201 (1992)
21. Martin, R.C.: Agile Software Development: Principles, Patterns and Practices. Prentice Hall, Upper Saddle River (2003)
22. Pfleeger, S.L., Wu, F., Lewis, R.: Software Cost Estimation and Sizing Methods. Issues and Guidelines. RAND Corporation (2005)
23. Quinlan, J.R.: C4.5: Programs for machine learning. Morgan Kaufmann, San Mateo (1993)
24. Santillo, L., Conte, M., Meli, R.: E&Q: An Early & Quick Approach to Functional Size. In: IEEE International Symposium on Software Metrics, p. 41. IEEE Computer Society, Los Alamitos (2005)
25. Shepperd, M., Cartwright, M.: Predicting with sparse data. IEEE Transactions on Software Engineering 27, 987–998 (2001)
26. Sneed, H.M.: Extraction of function points from source-code. In: 10th International Workshop, Proceedings of New Approaches in Software Measurement, IWSM, pp. 135–146. Springer, Berlin (2001)
27. Wiebe, J., Wilson, T., Bruce, R., Bell, M., Martin, M.: Learning Subjective Language. Computational Linguistics 30(3), 277–308 (2004)
28. Witten, I.H., Frank, E.: Data mining: Practical machine learning tools and techniques, 2nd edn. Morgan Kaufman, San Francisco (2005)

Semantic Enriching of Natural Language Texts with Automatic Thematic Role Annotation

Sven J. Körner and Mathias Landhäußer

Karlsruhe Institute of Technology (KIT)
sven.koerner@kit.edu, lama@ipd.uni-karlsruhe.de
http://svn.ipd.uni-karlsruhe.de/trac/mx/

Abstract. This paper proposes an approach which utilizes natural language processing (NLP) and ontology knowledge to automatically denote the implicit semantics of textual requirements. Requirements documents include the syntax of natural language but not the semantics. Semantics are usually interpreted by the human user. In earlier work Gelhausen and Tichy showed that SAL$_E$ мx automatically creates UML domain models from (semantically) annotated textual specifications [1]. This manual annotation process is very time consuming and can only be carried out by annotation experts. We automate semantic annotation so that SAL$_E$ мx can be completely automated. With our approach, the analyst receives the domain model of a requirements specification in a very fast and easy manner. Using these concepts is the first step into farther automation of requirements engineering and software development.

1 Introduction

Requirements engineering (RE) starts with the elicitation of the stakeholders' requirements, includes the management of the various user viewpoints and later leads to requirements analysis. Requirements analysis is often done by building domain models to visualize the processes. Domain models are used for *Model Driven Architecture* (MDA) [2], [1]. The requirements analyst uses these domain models to verify and rectify the stakeholders' input. Usually highly trained analysts build domain models manually. This process is as vital as time consuming in software development. So far, the analyst has little tool support [1] for his usecases. A recent survey [3] shows that many practitioners yearn for improvements.

In 2007, Gelhausen and Tichy [1] showed how UML domain models can be created automatically from text that is enriched with semantic information. The models that are created in the SAL$_E$ мx [4] process are complete and exhaustive, especially compared to the average quality of a human modeler. Their work shows that there is a direct connection of natural language and its corresponding UML model representations. The automatic model creation uses the implicit semantics of a phrase. The semantics is denoted manually via textual annotations. This

---

[1] Many tools support the later stages of the software development process, e.g. CASE tools.

C.J. Hopfe et al. (Eds.): NLDB 2010, LNCS 6177, pp. 92–99, 2010.
© Springer-Verlag Berlin Heidelberg 2010

makes semantic information computer processable. The problem is that semantic annotation is very time-consuming. The idea is to accelerate RE with automatic semantic annotation. The enivisioned toolkit enables the average requirements analyst to create domain models rapidly.

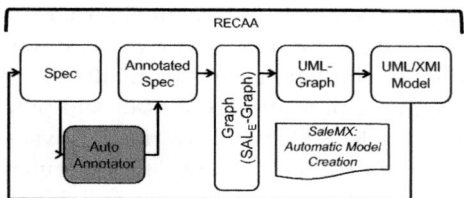

Fig. 1. AutoAnnotator Supports the Automatic Model Creation from Requirements

Our tool AUTOANNOTATOR (see Fig. 1) offers exactly this possibility and helps to put RE where it needs to be: next to the software development processes that have been proven for many years. To decrease the high error rates, RE needs to be defined more clearly and to be less dependent on the human factor. Our solution integrates into a larger scope software solution RECAA [4] that covers requirements engineering from requirements elicitation to implementation and back. The analyst profits from automatic model creation from natural language specifications while documenting his understanding in cooperation with the stakeholder. RECAA maintains the connection between textual specifications and their model representation in both directions, and AUTOANNOTATOR is an important part of this software solution.

The outline of this paper is as follows: Section 2 covers related work and the idea of automatic model creation. Section 3 describes our processing. Section 4 shows how we use semantic tools like ontologies to determine more complex structures and to verify the findings from NLP against world knowledge. Section 5 wraps up with a summary of our findings and the outlook to future work.

2   Related Work

Natural language is the main type of information in RE [3]. Its automatic processing is therefore especially interesting. Natural language was and will remain the main form of requirements documentation [5]. After elicitation, requirements are transformed into models that give a more formal representation of the described software system. These models are usually not intended for use with the client but with the software architects and the programming team. The average client cannot understand these models. As a result, the analyst usually maintains two models: one (semi) formal model for the development team, and one informal description in natural language for the client. These models have to be kept in sync during requirements evolution. The client signs a contract based on the model he understands. Dawson and Swatman argue in [6], that the mapping between informal and formal models is ad hoc and often results in divergent models. This strongly suggests to fill the gap between textual specifications and its models.

But requirements analysts cannot be replaced by NLP software, as Ryan argues in [7]. Most NLP tools are based on statistical approaches and have even greater error margins then human analysts. He highlights that NLP does not enable a machine to *understand* text, but it allows for a text's systematic transformation.

In 1997, Moreno[8] set the foundation for model extraction. Juzgado[9] et al. explain that a systematic procedure to dissect and process natural language information is strongly needed. They hint to the disadvantages of manual tasks which dominate the RE process until today. They postulate that this procedure must be independent from the analyst and his individual skills. In 2000, Harmain[10] developed CM-Builder, a NLP tool which generates an object oriented model from textual specifications. Additionally, Gildea and Jurafsky[11] showed in 2002 that statistical models can tag a sentence's semantics with a precision of 65% and a recall of 61%. Montes et al. describe in [12] a method of generating an object-oriented conceptual model (like UML class diagrams) from natural language text. Hasegawa[13] et al. describe a tool that extracts requirements models (abstract models of the system) from natural language texts. Instead of using only NLP, they perform text mining tasks on multiple documents to extract relevant words (nouns, verbs, adjectives etc.), assuming that important and correct concepts of the domain are contained in multiple distributed documents. Kof [14] reports that using NLP approaches is indeed feasible and worthwhile for larger documents (i.e. 80 pages and more). In [15] he shows that NLP is mature enough to be used in RE.

3   Combining NLP Tools - The Processing Pipeline

In this section, we explain how AUTOANNOTATOR derives semantic information from syntactic sentence structures automatically. First we need to describe how SALE ᴍx [1] extracts UML domain models from natural language text using annotations.

3.1   How Salₑ ᴍx Works

Thematic roles [1] can be used to extract domain models from natural language text. As an example we use the sentences `Chillies are very hot vegetables.` `Mike Tyson likes green chillies. Last week, he ate five of them.` Using the syntax of SALE [2], we need to tag the elements [3] with thematic roles. In SALE, elements containing more than one word are connected using a _ and obsolete elements are omitted using #. Furthermore, we prefix multiplicities with * and attributes with $. SALE contains 67 thematic roles [4] based on the works

---

[2] Semantic Annotation Language for English [1].

[3] Elements can be atomic or combined parts of a sentence that represent a semantic entity and can therefore represent thematic roles.

Table 1. Linguistic Structures of SAL$_E$ (excerpt)

Linguistic Structure	Explanation
AG *agens*	An acting person or thing executing an action
PAT *patiens*	Person or thing affected by an action
ACT (+AG +PAT) *actus*	An action, executed by AG on PAT
STAT (+AG +PAT) *status*	A relation between AG and PAT
FIN (+FIC) *fingens* and *fictum*	The FIN plays the role of/acts like/is a FIC
TEMP (+ACT) *tempus*	A time specification TEMP for an ACT

of Fillmore and others [16],[17],[18]. For our example, the roles shown in Tab. 1 are sufficient. Manually annotating the text with SAL$_E$ results in the following:

```
1  [ Chillies|FIN #are $very $hot vegetables|FIC ].
2  [ Mike_Tyson|AG likes|STAT $green chillies|PAT ].
3  [ $Last week|TEMP, he|AG ate|ACT *five #of them|PAT ].
4  [ @he|EQD @Mike_Tyson|EQK ].  [ @them|EQD @Chillies|EQK ].
5  [ @chillies|EQD @Chillies|EQK ].
```

The thematic role *fingens* (FIN) is used to denote a person or thing that is playing a role; vice versa, *fictum* (FIC) is the role played by somebody or something. The word are is encoded in the *fingens/fictum* relationship and thus can be omitted. very and hot are attributes – the former attributing hot, the latter attributing vegetables.

In the second sentence, the role *agens* (AG) tells the system that the according element is the active (not in a grammatical sense!) entity of the phrase. The *agens* in our case is Mike_Tyson which has been created by concatenating Mike and Tyson since it is an element consisting of two words. The role *actus* (ACT) is used for actions like *walk from A to B* while *status* (STAT) is used for general statements or relations like *A owns B*. Since like is a general statement, it is a *status*. Last but not least, we have chillies which is "the thing affected" by the *status* of Mike; therefore it is the *patiens* (PAT).

TEMP is a time, a date, or a "period". It modifies the roles it is used with in conjunction. Here week modifies *actus* and last is an attribute of week. he is the *agens* in the third phrase, performing the action ate. them is the thing being affected by the action of the *agens*, therefore it is the *patiens*. five is a multiplicity, determining the number of them. of is omitted.

Knowing that he refers to Mike_Tyson and them refers to chillies, the analyst includes the assertions listed in line 4. SAL$_E$ ᴍx replaces the element tagged with EQD [4] with a reference to the element tagged with EQK [5]. To preserve the same case, we replace chillies with Chillies in line 5 [6]. With this input, SAL$_E$ ᴍx generates an UML class diagram as shown in Fig. 2(a).

---

[4] EQD is an acronym for "equal drop". The element is marked for replacement.
[5] EQK is an acronym for "equal keep". The element replaces one or more EQD elements.
[6] Normalization could render assertions like this unnecessary. The model extraction of SAL$_E$ ᴍx is not yet capable of using this additional information.

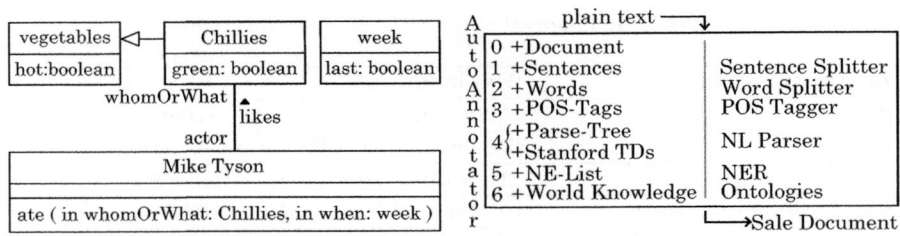

(a) The UML class diagram generated with SAL$_E$ мx.

(b) The Processing Pipeline of AUTOANNOTATOR.

Fig. 2.

3.2 Automating Annotation

To achieve a streamlined process and a holistic information extraction, we combine several NLP tools and check the results against "digital common sense", i.e. world knowledge from an ontology (see Sect. 4). Our process is outlined in Fig. 2(b). It starts with the plain text. Every stage of the pipeline adds or verifies some information.

First, the text[7] is converted into an internal data structure (0). It contains the plain text aside the additional information gathered during the conversion into a graph structure.

After loading the text and splitting it into chunks (steps 0, 1, 2), it is processed with a part-of-speech (POS) tagger (3), a statistical parser (4), and a named entity recognizer (5). All tools used are from the Stanford NLP Group [19]. Afterwards, the document contains the following information:

(PENN-like) POS Tags. as described in [20]:
```
Chillies/NNS are/VBP very/RB hot/JJ vegetables/NNS ./.
Mike/NNP Tyson/NNP likes/VBZ green/JJ chillies/NNS ./.
Last/JJ week/NN ,/, he/PRP ate/VBD five/CD of/IN them/PRP ./.
```

Stanford Typed Dependencies (SD). as described in [21]:
```
nsubj(vegetables-5, Chillies-1), cop(vegetables-5, are-2),
advmod(hot-4, very-3), amod(vegetables-5, hot-4), ...
```

Named Entities. The list of named entities contains only Mike Tyson.

Using the POS Tags and The Stanford Dependencies. One can derive, that

– Mike and Tyson should be concatenated because they are in the same noun phrase (NNP), that consists only of them. On top of that, there is a named entity 'Mike Tyson'.

---

[7] The text should comply to some rules: Since the described process is text-only, it should not contain images or rely on information given in images or illustrations. At the moment we cannot compute enumerations. They should be replaced beforehand.

- Likes is a verb having the subject Mike Tyson and the (direct, non-passive) object chillies. Since we do not know, if like is an action or a state, we can only tag it with METHODROLE[8]. Mike Tyson will be tagged with *agens* and chillies with *patiens*.
- Green modifies chillies, and is not a number; thus it will be marked as attribute.

Similar deductions can be made for the first and the third sentence:

1   <u>Chillies</u>|FIN #<u>are</u> $<u>very</u> $<u>hot</u> <u>vegetables</u>|FIC.
2   <u>Mike_Tyson</u>|AG <u>likes</u>|METHODROLE $<u>green</u> <u>chillies</u>|PAT.
3   $<u>Last</u> <u>week</u>|TEMPROLE, he|AG <u>ate</u>|METHODROLE *<u>five</u> #<u>of</u> <u>them</u>|PAT.

Comparing the AUTOANNOTATOR output with the manual annotation in Sect. 3.1, we realize that we need additional information to make the rest of the annotation decisions (6).

4   Semantic Information Enriching with Ontologies

We use two different knowlegde bases to gain the missing information: Word-Net [22] and Cyc [23]. These ontologies are built upon concepts (of a domain) and relationships between these concepts and can be used to answer queries.

First we determine a word's base form with WordNet. Only containing open-class words (nouns, verbs, adjectives and adverbs), WordNet has simply four POS tags. The POS tags we discovered in step (3) restrict the search space when querying WordNet, as e.g. ate is not only included as verb with the base form eat but also as noun Ate[9]. The PENN tags allow us to parametrize the query. Since we do know, that ate has the PENN tag VBD and therefore is a (past tense) verb, WordNet does not produce the goddess as a result to our query.

The Cyc ontology is one of the most exhaustive and compelling collections of structured computable world knowledge. Cyc offers a vast collection of assertions between the ontological representation of many real world objects. Cyc delivers additional semantics to the problems discovered in Sect. 3. Let's revisit the second sentence of our example: We have found that the verb like is some kind of action or relationship between two entities. When asked about like, Cyc answers:

```
Predicate:   likesRoleInEventType
     isa:    FirstOrderCollectionPredicate, TernaryPredicate
Collection: TernaryPredicate
   genls:    Predicate, TernaryRelation
```

Cyc shows that like is predicate type of word (a *TernaryPredicate* to be precise). The collection of all *TernaryPredicates* itself has a generalization *Predicate*. Predicates are modeled as relations [24] and therefore the thematic role that has to be assigned is *status*.

---

[8] Our role system allows inheritance. METHODROLE is the (abstract) parent of *actus* and *status*. Using METHODROLE, we (internally) mark the element to be processed later.

[9] Ate is recorded as the Greek *goddess of criminal rashness and its punishment*.

5 Summary

Using sentence grammar structures to determine the correct semantics of a sentence seems feasible with our approach. We use popular NLP tools for the preprocessing of natural language texts. Even though AUTOANNOTATOR is still work in progress, we have run a small qualitative case study using the technical specification of the WHOIS Protocol (IETF RFC 3912). The results suggest that the proposed approach is indeed capable of deriving the semantic tags of SAL$_E$ мх. Still there are some difficulties, which have to be addressed in future development.

First of all, subphrases are not yet handled correctly leading to confusing results. Errors of the pipelined NLP tools are not yet addressed adequately. Assigning a confidence value to each tool could improve results when information conflicts. On top of these future improvements, we plan to extend AUTOANNOTATOR with an interactive dialog tool. This allows the analyst to steer the analysis process. We expect this interactive component to be used to resolve obvious mistakes the algorithms make as part of a feedback loop in the annotation process. Together with an instant UML diagram building process, the analyst could identify and correct the derived semantics on the fly.

Eventually, our process improves the annotation process with a speedup which we are currently evaluating. Only if the analyst is faster and receives the same quality models than in the manual process, automatic model creation can help support and improve the software development process.

References

1. Gelhausen, T., Tichy, W.F.: Thematic role based generation of UML models from real world requirements. In: First IEEE International Conference on Semantic Computing (ICSC 2007), Irvine, CA, USA, pp. 282–289. IEEE Computer Society, Los Alamitos (2007)
2. Miller, J., Mukerji, J.: MDA Guide Version 1.0.1 (June 2003)
3. Mich, L., Franch, M., Inverardi, P.N.: Market research for requirements analysis using linguistic tools, vol. 9, pp. 40–56. Springer, London (February 2004)
4. Körner, S.J., Derre, B., Gelhausen, T., Landhäußer, M.: RECAA – the Requirements Engineering Complete Automation Approach, https://svn.ipd.uni-karlsruhe.de/trac/mx (2010.01.11)
5. Cheng, B.H.C., Atlee, J.M.: Research directions in requirements engineering. In: Proc. Future of Software Engineering FOSE 2007, May 2007, pp. 285–303 (2007)
6. Dawson, L., Swatman, P.A.: The use of object-oriented models in requirements engineering: a field study. In: ICIS, pp. 260–273 (1999)
7. Ryan, K.: The role of natural language in requirements engineering. In: Proceedings of IEEE International Symposium on Requirements Engineering, pp. 240–242. IEEE, Los Alamitos (1993)
8. Moreno, A.M., van de Riet, R.: Justification of the equivalence between linguistic and conceptual patterns for the object model (1997)
9. Juzgado, N.J., Moreno, A.M., López, M.: How to use linguistic instruments for object-oriented analysis, vol. 17 (2000)

10. Harmain, H.M., Gaizauskas, R.J.: CM-Builder: An automated NL-based CASE tool. In: ASE, pp. 45–54 (2000)
11. Gildea, D., Jurafsky, D.: Automatic labeling of semantic roles, vol. 28, pp. 245–288. MIT Press, Cambridge (2002)
12. Montes, A., Pacheco, H., Estrada, H., Pastor, O.: Conceptual model generation from requirements model: A natural language processing approach. In: Kapetanios, E., Sugumaran, V., Spiliopoulou, M. (eds.) NLDB 2008. LNCS, vol. 5039, pp. 325–326. Springer, Heidelberg (2008)
13. Hasegawa, R., Kitamura, M., Kaiya, H., Saeki, M.: Extracting conceptual graphs from Japanese documents for software requirements modeling. In: Kirchberg, M., Link, S. (eds.) APCCM. CRPIT, vol. 96, pp. 87–96. Australian Computer Society (2009)
14. Kof, L.: Natural language procesing for requirements engineering: Applicability to large requirements documents. In: Russo, A., Garcez, A., Menzies, T. (eds.) Proceedings of the Workshops, Automated Software Engineering, Linz, Austria (September 2004); In conjunction with the 19th IEEE Internationl Conference on Automated Software Engineering
15. Kof, L.: Natural language processing: Mature enough for requirements documents analysis? In: Montoyo, A., Muñoz, R., Métais, E. (eds.) NLDB 2005. LNCS, vol. 3513, pp. 91–102. Springer, Heidelberg (2005)
16. Fillmore, C.J.: Toward a modern theory of case. In: Reibel, D.A., Schane, S.A. (eds.) Modern Studies in English, pp. 361–375. Prentice-Hall, Englewood Cliffs (1969)
17. Krifka, M.: Thematische Rollen (June 2005)
18. Rauh, G.: Tiefenkasus, thematische Relationen und Thetarollen. Gunter Narr Verlag, Tübingen (1988)
19. Manning, C., Jurafsky, D.: The Stanford Natural Language Processing Group, http://nlp.stanford.edu (2009.11.07)
20. Santorini, B.: Part-of-speech tagging guidelines for the Penn Treebank Project (3rd revision). Technical Report MS-CIS-90-47, University of Pennsylvania Department of Computer and Information Science (1990)
21. de Marneffe, M.C., Manning, C.D.: The Stanford typed dependencies representation. In: COLING Workshop on Cross-framework and Cross-domain Parser Evaluation, pp. 1–8 (2008)
22. Miller, G.A.: WordNet: A lexical database for English, vol. 38, pp. 39–41. ACM Press, New York (1995)
23. Cycorp Inc.: ResearchCyc, http://research.cyc.com/ [checked 2010-02-15]
24. Körner, S.J., Gelhausen, T.: Improving automatic model creation using ontologies. In: Knowledge Systems Institute: Proceedings of the Twentieth International Conference on Software Engineering & Knowledge Engineering, July 2008, pp. 691–696 (2008)

Adaptive Topic Modeling with Probabilistic Pseudo Feedback in Online Topic Detection

Guoyu Tang and Yunqing Xia

Department of Computer Science and Technology, Tsinghua University,
Beijing 100084, China
sweetyuer@gmail.com, yqxia@tsinghua.edu.cn

Abstract. Online topic detection (OTD) system seeks to analyze sequential stories in a real-time manner so as to detect new topics or to associate stories with certain existing topics. To handle new stories more precisely, an adaptive topic modeling method that incorporates probabilistic pseudo feedback is proposed in this paper to tune every topic model with a changed environment. Differently, this method considers every incoming story as pseudo feedback with certain probability, which is the similarity between the story and the topic. Experiment results show that probabilistic pseudo feedback brings promising improvement to online topic detection.

Keywords: Adaptive topic modeling, probabilistic pseudo feedback, online topic detection, topic detection and tracking.

1 Introduction

Online topic detection (OTD) task is to makes decisions whether the input story is relevant to certain topic in a real-time manner.

In OTD system, every story is detected relevant to certain topic according to similarity between the story and the topic [1]. During this process, topic representation and clustering should be considered. The popular topic representation model is vector space model (VSM). Topic clustering in OTD system needs to handle sequential stories one at a time. Walls et al. (1999) proposed threshold model and time-based selection model in an incremental K-means algorithm [1].

Once a new story is assigned one topic, the topic modeling module should be executed again to adapt the topic model. For the topic modeling module, the new story is usually referred to as pseudo feedback. Pseudo feedback is a widely used technique in information retrieval. It assumes the system decisions correct and learns from them to improve the quality. For example, in UMASS system [7] and Dragon system [8], a story that is newly assigned one topic is in turn considered as training data to adjust the topic model.

However, system decisions on relevance are probabilistic, namely, wrong decisions are inevitable. So much uncertainty might occur if the wrong decisions were used as good training data. To alleviate the adverse effects, LIMSI system [9] adopts a bigger threshold to model the topics. However, the negative consequences are that too many

C.J. Hopfe et al. (Eds.): NLDB 2010, LNCS 6177, pp. 100–108, 2010.

stories are rejected by all topics; therefore too many topics are created. Rather differently, we consider similarity between story and topic as the probability for the pseudo feedback and apply the probability to adjust contribution of the pseudo feedback in topic modeling. Preliminary experimental result on test data show that the probabilistic pseudo feedback does bring promising repay in online topic detection.

The rest of this paper is organized as follows. Section 2 gives a re-observation on online topic detection and addresses how the probabilistic pseudo feedback is incorporated. Section 3 presents details of topic modeling. In Section 4, experiments as well as discussions are presented. We draw some conclusions and address future works in Section 5.

2   Online Topic Detection: A Re-observation

Architecture of typical online topic detection systems is given in Fig.1.

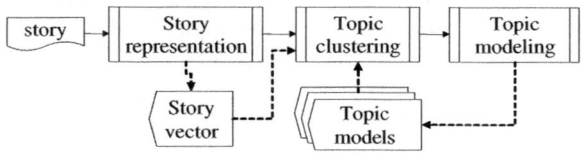

Fig. 1. Architecture of typical online topic detection (OTD) system

VSM is used in story representation. Based on the standard TF-IDF formulae, Allan et al. (1998) proposed the following enhanced equations to calculate term weights [6].

$$w_i^S = \alpha_i \times tf_i' \times idf_i', tf_i' = \frac{tf_i}{tf_i + 0.5 + 1.5 \times (dl / avg\_dl)}, idf_i' = \frac{\log\left(\frac{N + 0.5}{df_i}\right)}{\log(N + 1)} \quad (1)$$

α_i is empirically defined to reflect contribution of named entity and position that a word appears in a story. In OTD system, both tf and df are calculated online.

To measure similarity between a story and a topic, every story should be represented using feature terms that are specific to the counterpart topic. As topic is modeled adaptively, story representation is actually influenced by pseudo feedback incorporated in topic modeling.

Topic model is usually defined as centroid of the stories relevant to the topic. In practice, OTD systems incrementally adjust the weights once a new story is assigned.

Pseudo feedback is reflected when the income story is used to calculate the M+1-th topic model. Influence of the income story is uncertain. So a refined equation incorporating topic-story similarity will be discussed in Section 3.

Topic clustering aims to associate input stories to relevant topics. We use the popular cosine distance to compute similarity. Then the Threshold Model [1] is applied to determine which topic the story belongs to or whether a new topic should be created.

3 Adaptive Topic Modeling

3.1 Term Re-weighting

The adaptation happens on term weight w_i^S for feature term t_i that appears in story S. Thus two efforts were made in this work to improve term re-weighting: (1) incorporating pseudo feedback in calculating w_i^S and (2) assigning probability to adjust effect of w_i^S on w_i^T.

Incorporating Pseudo Feedback in Calculating w_i^S
The first effort was made on calculation of w_i^S in Equation (1). Two observations were made. Firstly, within one topic, we find a feature term tends to be more important if it appears in more stories. To be general, we define story dispersion degree (SDD) for every feature term within one topic.

$$SDD(t_i, T) = \frac{df_i^T}{M},$$
(2)

where df_i^T represents number of stories in the topic that contains term t_i. Incorporating SDD, we get Equation (3) as follows.

$$idf_i'' = \frac{\log\left(\frac{N \times df_i^T}{df_i \times (M+1)}\right)}{\log(N \times (df_i^T + 1))}$$
(3)

Secondly, we find a feature term tends to be more important if it appears in fewer topics. To be general, we define topic dispersion degree (TDD) for every feature term.

$$TDD(t_i) = \frac{TT_i}{L}$$
(4)

where TT_i represent number of topics that contain feature term t_i, and L total number of topics in the system. Incorporating topic dispersion degree, we get Equation (5) as follows.

$$idf_i'' = \frac{\log\left(\frac{N * L}{df_i * (TT_i + 1)}\right)}{\log(N * (L+1))}$$
(5)

We finally incorporate story dispersion degree and topic dispersion degree into one equation and obtain Equation (6).

$$idf_i'' = \frac{\log\left(\dfrac{N * df_i^T * L}{df_i * (M+1)*(TT_i+1)}\right)}{\log\left(N*(df_i^T+1)*(L+1)\right)} \tag{6}$$

Assigning Probability to Adjust Effect of w_i^S on w_i^T

The second effort was made on adjusting effect of w_i^S on w_i^T. It is proved by Zheng et al. (2007) that the similarity reflects how confidently the topic is assigned to the story [10]. We further propose to incorporate the similarity as the follow Equation (7).

$$w_i^{T,M+1} = \frac{1}{M+1}\left(M \times \sum_{k=1}^{M} w_i^{T,M} + Sim(S,T) \times w_i^S\right), \tag{7}$$

where $Sim(S,T)$ represents similarity between story S and topic T.

After term re-weighting, some terms may be excluded from the feature list while some others join in.

3.2 Relevance Threshold Estimation

Relevance threshold is adopted in this work to reject stories with smaller similarity values. Equation (8) is proposed by Stokes et al. (2000) to estimate dynamic threshold for every topic [11].

$$TH(T) = Sim(S,T) \times CCS, CCS \in (0,1), \tag{8}$$

where $TH(T)$ represents relevance threshold for topic T, and CCS cluster centroid similarity value. When a topic is created, an initial relevance similarity $InitSIM$ is used to replace $Sim(S,T)$.

According to Equation (8), as $Sim(S,T)$ is always smaller than 1, the topic tends to accept more stories. This is obviously problematic. We propose to incorporate story number to handle this problem. We believe that, with more stories joining one topic, feature weighting for this topic should be more precise and the threshold should be stricter. So the relevance threshold should be increased by incorporating number of stories in the topic. Equation (8) is revised as follows.

$$TH(T) = Sim(S,T) \times (CCS + \delta \times \log M), \tag{9}$$

where δ is a predefined parameter, which is empirically set 0.01 in this work.

4 Evaluation

4.1 Setup

Test data: Sogou dataset2,where 951 finely chronologically ordered selected Chinese stories are annotated by human annotators with 108 topics was used in our evaluation.

Evaluation methods: Two evaluation methods were adopted in our experiments.

First story detection (FSD): The FSD-based evaluation method was proposed in TDT evaluation task, which seeks to evaluate how accurate the first story which refers to the report that first address a topic is detected for each topic.

Topic detection (TD): The evaluation method was proposed by Steinbach et al. (2000) [13], which seeks to evaluate how accurate the stories are detected being relevant to certain topics. Let C_i represent set of stories that are contained in a system-generated cluster i, C_j set of stories that are contained in a human-generated cluster j. F measure of the system-generated cluster i is calculated as follows.

$$p_{i,j} = \frac{|C_i \cap C_j|}{|C_j|}, r_{i,j} = \frac{|C_i \cap C_j|}{|C_i|}, f_{i,j} = \frac{2 \times p_{i,j} \times r_{i,j}}{p_{i,j} + r_{i,j}} \qquad (10)$$

$$p_i = \max_j \{p_{i,j}\}, r_i = \max_j \{r_{i,j}\}, f_i = \max_j \{f_{i,j}\} \qquad (11)$$

where $p_{i,j}$, $r_{i,j}$ and $f_{i,j}$ represent precision, recall and F measure of cluster i in comparison with the human-generated cluster j, respectively. The general F measure of a system is sum of all F measures of the system-generated clusters. And so are precision and recall value.

4.2   Experiment I: Parameter Selection

In this experiment we intended to observe how the parameters i.e. *InitSIM* and *CCS* influence system performance. We altered *InitSIM* and *CCS* from 0.1 to 0.9, respectively, and ran the system on the test data. Experimental results in FSD task and TD task are presented in Fig.2 and Fig.3, respectively.

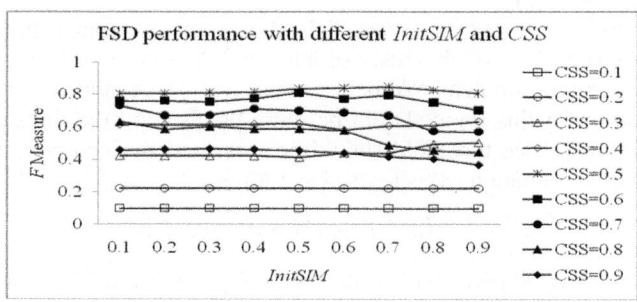

Fig. 2. FSD performance of TH-OTD system with different *InitSIM* and *CCS* values

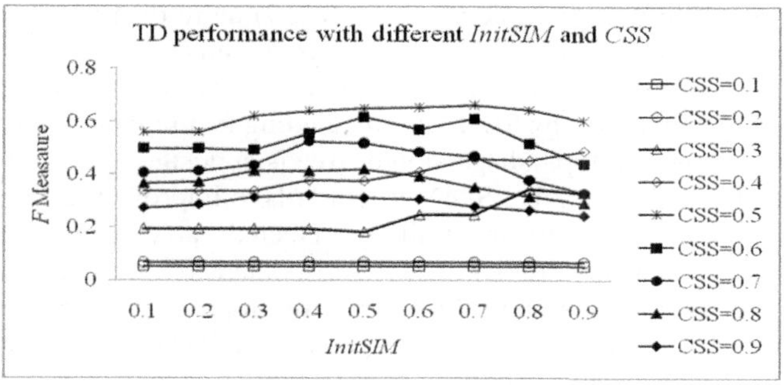

Fig. 3. TD performance of TH-OTD system with different *InitSIM* and *CCS* values

It is shown in Fig.2 and Fig.3 that TH-OTD system achieves the best performance in FSD and TD tasks when *CCS* is set 0.5 and *InitSIM* set 0.7. According to their definitions, *CCS* indicates similarity between story and topic centroid, and *InitSIM* specifies a starting point for the relevance threshold. Increase of either value alone leads to a higher relevance threshold according to Equation (9). As a consequence, topic detection becomes more and more strict. This is why system performance climbs until *CCS* is 0.5 and *InitSIM* is 0.7. In the following experiments, we set *CCS*=0.5 and *InitSIM*=0.7.

4.3   Experiment II: Term Re-weighting

To evaluate contribution of story dispersion degree and topic dispersion degree to term re-weighting, four systems were developed in this experiment: (1) BL1: The baseline system that adopts Equations (1). (2) SDD: The improvement system that incorporates SDD using Equation (3). (3) TDD: The improvement system that incorporates TDD using Equation (5). (4) TH-OTD: our system that incorporates both SDD and TDD using Equation (6). Experimental results are presented in Table 1.

Table 1. Experimental results of BL1, SDD, TDD and TH-OTD in FSD and TD task

System	FSD task			TD task		
	Precision	Recall	*F* Measure	Precision	Recall	*F* Measure
BL1	0.571	0.556	0.563	0.836	0.713	0.680
SDD	0.589	0.611	0.600	0.821	0.697	0.651
TDD	0.630	0.583	0.606	0.850	0.774	0.739
TH-OTD	0.614	0.722	0.664	0.883	0.864	0.846

On FSD task, two findings are disclosed in Table 1. Firstly, both SDD system and TDD system outperform the baseline system by around 0.04 on *F* measure in FSD task. We also find that TH-OTD system further improves the baseline system by 0.101 on *F* measure. This proves that either story dispersion degree or topic dispersion degree can improve term re-weighting alone but the two degrees work well with each other can further improve term re-weighting. Secondly, SDD system improves recall (i.e., by 0.055) more than precision (i.e., by 0.018) over the baseline. But TDD system improves precision (i.e., by 0.059) more than recall (i.e., by 0.027). Investigations disclose that story dispersion degree tends to assign bigger weights to terms that appear in more stories. As a consequence, the input stories are more likely detected carrying new topics by SDD system, leading to a higher gain on recall in first story detection. However, topic dispersion degree is opposite.

In TD task, Table 2 shows that SDD system fails to improve the baseline system in TD task. The input stories are more likely detected carrying new topics by SDD system, resulting that many stories are wrongly detected carrying new topics. As a comparison, we find TDD system that adopts the topic dispersion degree performs better. Furthermore, by adopting both degrees, TH-OTD system achieves further improvement. This is because, when the two degrees are adopted at the same time, the negative effect of story dispersion degree on topic detection is eliminated to greatest extent. Meanwhile, feature terms selected by TH-OTD system are made more precise. Contribution of the two degrees is thus vividly proved.

4.4   Experiment III: Adaptive Topic Modeling

In this experiment, we intended to evaluate contribution of the probabilistic pseudo feedback in adaptive topic modeling. Based on TH-OTD system, we created other three OTD systems: (1) BL2: the baseline system that incorporates pseudo feedback without the probability. (2) TRW: the system that incorporates probabilistic pseudo feedback merely in term-reweighting using Equation (7). (3) RTE: the system that incorporates probabilistic pseudo feedback merely in relevance threshold estimation using Equation (9). (4) TH-OTD: our system that incorporates probabilistic pseudo feedback in both term-reweighting and relevance threshold estimation. Experimental results are presented in Table 2.

Table 2. Experimental results of BL2, TRW, RTE and TH-OTD in FSD task

System	FSD task			TD task		
	Precision	Recall	F Measure	Precision	Recall	F Measure
BL2	0.622	0.685	0.652	0.858	0.866	0.822
TRW	0.641	0.694	0.667	0.850	0.879	0.829
RTE	0.713	0.597	0.650	0.805	0.872	0.790
TH-OTD	0.614	0.722	0.664	0.883	0.864	0.846

It is shown in Table 2 that TRW system performs best on F measure in FSD task. This proves that probabilistic pseudo feedback can improve term re-weighting. We however, find that RTE system performs worse than the baseline. After incorporating probabilistic pseudo feedback in term re-weighting, TH-OTD system is still slightly inferior to TRW system. Investigations disclose that much deviation occurs in relevance threshold estimation when wrong decisions are made by RTE system. This is because, when Equation (9) is used, the topic thresholds are increased compared to BL2 system. That may leads to the succeeding stories are rejected. Fortunately, the deviation can be eliminated to certain extent because features of topics can be refined when probabilistic pseudo feedback is incorporated in term re-weighting. As shown in Table 2, performance of TH-OTD is close to TRW.

Though proven incapable of improving first story detection, relevance threshold estimation is found helpful to improve topic detection according to Table 2. Similar to results in FSD task, RTE system fails to produce a performance better than the baseline in TD task. But when probabilistic pseudo feedback is incorporated in term re-weighting, TH-OTD system yields a best performance. This proves that relevance threshold estimation makes contribution to topic detection task.

5 Conclusion

Adaptive topic modeling is crucial to online topic detection because when a sequentially input story is assigned certain topic, the topic model must be adjusted accordingly to the changed environment. It is widely accepted that the story can be considered as pseudo feedback to facilitate the adaptation. But we find uncertainty of the feedback sometimes brings performance drop. In this work, probabilistic pseudo feedback is proposed to refine term re-weighting and relevance threshold estimation. Two conclusions can be drawn according to experimental results. Firstly, pseudo feedback is indeed capable of improving online topic detection if it is used appropriately. Secondly, probabilistic pseudo feedback is effective to improve online topic detection as uncertainty of the feedback is appropriately addressed in the new equations.

Acknowledgement

This work is partially supported by NSFC (No. 60703051) and MOST (2009DFA12970). We thank the reviewers for the valuable comments.

References

1. Walls, F., Jin, H., Sista, S., Schwartz, R.: Topic Detection in Broadcast News. In: Proc. of the DARPA Broadcast News Workshop (1999)
2. Chen, Y.J., Chen, H.H.: NLP and IR approaches to monolingual and multilingual link detection. In: Proc. of COLING 2002, pp. 1–7 (2002)
3. Clifton, C., Cooley, R.: Topcat: Data Mining for Topic Identification in a Text Corpus. In: Żytkow, J.M., Rauch, J. (eds.) PKDD 1999. LNCS (LNAI), vol. 1704, pp. 174–183. Springer, Heidelberg (1999)
4. Yang, Y., Carbonell, J., Jin, C.: Topic-Conditioned Novelty Detection. In: Proc. of KDD 2002, pp. 668–693 (2002)
5. Steinberger, R., Bruno, P., Camelia, I.: NewsExplorer: multilingual news analysis with cross-lingual linking. In: Information Technology Interfaces (2005)
6. Allan, J., Papka, R., Lavrenko, V.: On-line New Event Detection and Tracking. In: Proc. of SIGIR 1998, pp. 37–45 (1998)
7. Allan, J., Lavrenko, V., Frey, D., Khandelwal, V.: Proc. of Topic Detection and Tracking workshop, UMass at TDT 2000, pp. 109–115 (2000)

8. Yamron, J.P., Knecht, S., Mulbrget, P.V.: Dragon's Tracking and Detection Systems for the TDT 2000 Evaluation. In: Proc. of Topic Detection and Tracking workshop, 2000, pp. 75–79 (2000)

9. Lo, Y., Gauvain, J.L.: The LIMSI Topic Tracking system For TDT 2001. In: Proc. of Topic Detection and Tracking workshop (2001)

10. Zheng, W., Zhang, Y., Zou, B., Hong, Y., Liu, T.: Research of Chinese Topic Tracking Based on Relevance Model. In: Proc of 9th Joint Symposium on Computational Linguistics (2007) (in Chinese)

11. Stokes, N., Hatch, P., Carthy, J.: Topic Detection, a new application for lexical chaining? In: Proc. of the 22nd BCS IRSG Colloquium, pp. 94–103 (2000)

12. Steinbach, M., Kapypis, G., Kumar, V.: A Comparison of Document Clustering Techniques. In: KDD Workshop on Text Mining, pp. 109–111 (2000)

An Approach to Indexing and Clustering News Stories Using Continuous Language Models

Richard Bache[1] and Fabio Crestani[2]

[1] University of Glasgow, Glasgow, Scotland
[2] University of Lugano, Lugano, Switzerland

Abstract. Within the vocabulary used in a set of news stories a minority of terms will be topic-specific in that they occur largely or solely within those stories belonging to a common event. When applying unsupervised learning techniques such as clustering it is useful to determine which words are event-specific and which topic they relate to. Continuous language models are used to model the generation of news stories over time and from these models two measures are derived: bendiness which indicates whether a word is event specific and shape distance which indicates whether two terms are likely to relate to the same topic. These are used to construct a new clustering technique which identifies and characterises the underlying events within the news stream.

1 Introduction

Topic Detection and Tracking (TDT) traditionally uses a bag of words to index each of the news stories in a given stream of data. Common sense tells us that only a small subset of these words will be peculiar to a particular news topic or event and most of vocabulary will occur across topics and events. Here we distinguish between a topic e.g. industrial relations and an event, e.g. a particular strike [1] and focus on the latter. Since the event-specific terms are of particular interest, this has motivated the practice of labelling named entities within the text in the hope that these will provide an improved indexing vocabulary. Such named entities can be identified either by deep linguistic analysis [5] or by hand in which case it is a labour-intensive activity. In the case of tracking, which is an example of supervised learning, the training data can be used to identify the event specifc terms by techniques such as (discrete) Language Modelling. Detection, however, is unsupervised and thus we cannot know in advance which terms are event specific and which are not. When attempting to cluster documents into their respective events the event-specific terms will tend to draw same-event stories together whereas the other terms will introduce noise. Clearly it would be advantageous to be able to identify the event-specific terms in advance of clustering since this could improve the detection task.

We propose here an automatic technique for identifying the event-specific words and relating these words to the specific event from which they hail based on Continuous Language Modelling. Such a technique has shown demonstratable improvements in topic detection but for reasons of space the empirical study is

C.J. Hopfe et al. (Eds.): NLDB 2010, LNCS 6177, pp. 109–116, 2010.

given in outline only. We also argue that this approach also has applications for describing an event by identifying the key terms which characterise it.

The rest of the paper is organised as follows. In Section 2 we introduce continuous language models and extend their use by means of a polynomial (quadratic) function. Section 3 defines two measures derived from the continuous language model: bendiness and shape distance. In Section 4 we propose a model which incorporates both bendiness and shape distance to characterise events. Section 5 summarises the results of a small emprical study. In Section 6 we provide some conclusions.

2   Continuous Language Models

We use the retronym *discrete language models* to distinguish the language models normally used in Information Retrieval (IR) from the ones proposed here. A discrete language model represents a stochastic process which generates documents (or more specifically their indexes) by randomly emitting words, where for each word there is a fixed probability associated with its being emitted. In contrast, *continuous language models* have some scalar variable as a parameter and it is assumed that the probability with which a term will be emitted is some (continuous) function of that parameter.

2.1   Discrete versus Continuous Models

For discrete language modelling there are assumed to be two or more models which might have generated a given document. In the case of document retrieval [12], the generated document is the query and each language model is assocated with a document in the collection. For document classification [3,10,11], the generated document is an unclassified document and the language models correspond to each category. In each case, by using Bayes' Theorem, the likelihood function and some prior we calculate the probability that each of the models was the document generator. Continuous language models differ in that there is assumed to be one model generating all the documents. As its parameter varies some documents become more likely to be generated and others less so. We could alternatively view a continuous model as a continuum of discrete models each differing infinitessimally from the last.

2.2   Defining a Continuous Model

We assume that there is some real-valued parameter and the probability of each term the is a function which determines the probability of that term being generated. Training data is used to estimate each function. Such data comprises a set of documents and, for each document, a value of the scalar quantity of interest. Continous models have previously been applied to the analysis of crime descriptions to estimate a non-categoric attribute such as age of offender [2]. As age increased some words became more common in the descriptions and others less so.

For TDT, the model parameter will be time. We wish to develop more general continuous language models where the probability of a term may both increase and decrease over the time interval under consideration. A news event will typically last a period of time and words associated with that event with rise in frequency and then fall again. Words which do not relate to a specific event will show a different pattern. It is this key different we use to distinguish the event-specific words from the others. The idea of a model where probabilities of words being emitted changes of time has also been pursued by Kleinberg [7] but his model assumes that the probability rises and falls over time in discontinuous jumps rather than as a continuous function. The advantage of our approach is that by having an explicitly defined continuous function, it can be subjected to analysis.

We assume that there is a function for each word w_i which gives the probability of being emitted at a time t.

$$P(w_i|t) = f_i(t) \tag{1}$$

Two types of language model are used in IR: multinomial and multiple Bernoulli and there are studies [8,9] which determine their relative performance. However, here we are restricted to using multiple Bernoulli models since, unlike multinomial models, there is no constraint that the probabilities of each term being emitted add to 1. Thus we can estimate the functions for the probability of each term being emitted individually and treat each term in the vocabulary separately.

Since t can, in theory, take any value, we need a function which will always yield a value between 0 and 1. If we transform the probability to its log odds then it may have any real value and this means that we can equate the log odds probability to a polynomial. Thus we write:

$$\text{logit}(P(w_i|t)) = Q_i(t) \tag{2}$$

where Q_i is a polynomial. Although in principle any degree of polynomial could be used, the quadratic form was chosen. A linear formulation would yield to a monotonic function which therefore could not both rise and fall. A cubic function (and any higher odd degree polynomial) would rise, fall then rise again (or vice versa). Polynomials with degrees above three led to the logistic regression frequently failing to converge. Writing the probability functions gives us:

$$P(w_i|t) = f_i(t) = \frac{e^{a_i t^2 + b_i t + c_i}}{1 + e^{a_i t^2 + b_i t + c_i}} \tag{3}$$

2.3 Estimating the Model

We assume a sequence of N stories $D_1, \ldots, D_N$ with timestamps $t_1, \ldots, t_N$. At this stage we do not know which stories belong to which event. It is therefore not possible to define a separate language model for each event. Instead, we assume a single model is generating the entire collection and then use measures based on this model to separate the terms specific to each event. Using multivariate logistic regression we estimate the coefficients for each polynomial Q_i. This requires an approximation algorithm.

3   Measures Defined from the Continuous Model

By plotting the probability function against time for the interval between the first and last story it is possible to identify by hand the topic-specific terms. Figure 1 gives an example from transcripted news broadcasts where one event was an allegation that President Clinton had an inappropriate relationship with White House intern, Monica Lewinsky. The curve corresponding to the word 'president' is downward sloping; 'lewinsky' is above 'intern' and both have a global maximum in the graph at about the same point on the left side. The word 'court-martial' also has a rising-and-falling pattern further to the right. Words which show this particular pattern are more likely to be topic-specific terms. Indeed no other events contained these words. The word 'president' is not event specific since there are other events which use this word. The term 'court-martial' relates to an entirely different event. Whether or not a term is event specific is very largely dependent on the stream of stories under analysis. In another stream of stories the word 'president' was found to be event specific.

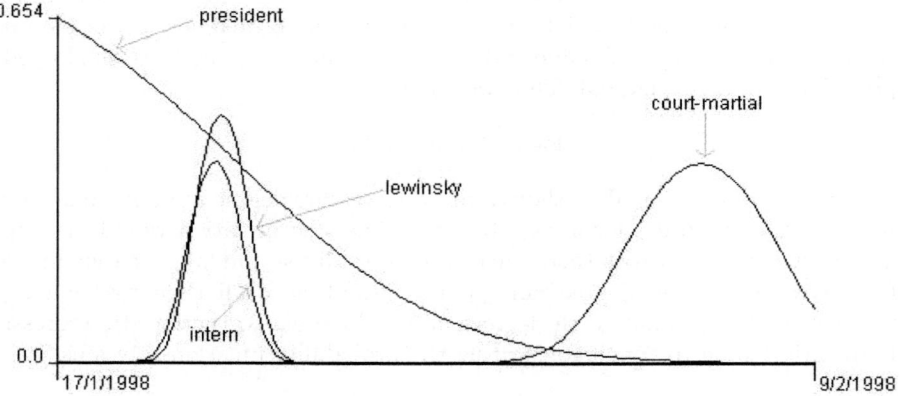

Fig. 1. Graph of Probability Function over Time for 3 Terms

Although these graphs could provide a useful form of visualisation to aid manual identification of key terms, our wish here is to automate the process. Thus we define two new measures: *bendiness* and *shape distance* to enable this task to be performed automatically.

3.1   Bendiness

An immediate observation from Figure 1 is that the word 'president' has a fairly consistent gradient whereas the gradients of the other three topic-specific words vary over time; they are more bendy. The gradient of the curve may be calculated as the derivative of the probability function $\frac{dP(w_i|t)}{dt}$ and we define a measure of *bendiness* as the variance of this:

$$\text{bendiness}(w_i) = \text{var}\left(\frac{dP(w_i|t)}{dt}\right) \tag{4}$$

where this is calculated numerically by sampling many points over the interval of the set of stories. Actually, the most bendy terms are singleton terms which cannot be used to characterise an event. Therefore we consider as topic specific terms the most bendy terms which have occurred in a minimum number of documents, here taken to be ≥ 5 although this would depend on the size of the corpus. Table 1 shows the bendiness measure for the four terms considered above.

Table 1. Bendiness of Specific Terms

Word	lewinsky	intern	court-martial	president
Bendiness	0.0118	0.00752	0.00249	0.000483

3.2   Shape Distance

Referring again to Figure 1 we can see that the two terms which relate to the same topic have broadly the same shape. This reflects the time and duration of the interval over which the event runs. We therefore define a measure of distance between each pair of terms which captures the extent that the two curves deviate from each other. If two curves have a short distance between them then they are more likely to relate to the same event. There may be cases where two words both relate to a common event but where one is used more frequently that the other. In this case, although the curves may have a similar shape, the magnitude of one may be greater than the other. Thus the curves are normalised to have a minimum value of 0 and a maximum value of 1. If the minimum and maximum values of a function f_i between t_1 and t_n (the times of the first and last story) are p and q respectively we define a normalised function as follows:

$$g_i(t) = \frac{f_i(t) - p}{q - p} \tag{5}$$

At any point t we can consider the difference between the functions as $g_1(t) - g_2(t)$ as a measure of the vertical distance between terms w_1 and w_2. To ensure that negative differences do not cancel out positive ones, we square the difference. We now define the total distance by integrating the square of differences and taking the square root. We also then normalise for the time interval so that the distance ranges between 0 and 1.

$$\text{shape\_distance}(w_1, w_2) = \sqrt{\frac{1}{t_N - t_1} \int_{t_1}^{t_N} (g_1(t) - g_2(t))^2 \, dt} \tag{6}$$

Analytical integration is intractable so a numerical approximation is used. The gives a pseudo-metric, which is sufficient for the clustering we employ below.

4   A New Clustering Model

Given the measures we have defined above we now seek to cluster the stories by event. This is essentially a three stage process:

1. Identify a minority of the terms which are topic specific;
2. Cluster these terms using QT (Quality Threshold) Clustering;
3. Use each word cluster to create a document cluster.

Essentially we are clustering the words and in particular the bendy words. We then use the word clusters, or a subset thereof, to identify document clusters. The document clustering is therefore indirect. We consider each of the three stages in turn.

4.1   Identifying the Event-specific Terms

We assume that stopwords and sparse terms have been removed during indexing. We then use the measure of bendiness to rank terms in the vocabulary according to how event specific they are deemed to be. It should be noted that a word that may be very event specific in one collection of stories may not be in another collection. To use the example data cited above, the word 'clinton' would be very frequent in the stories in 1998 from which the data was taken but this would not be the case in a different time window. The s most bendy terms are then selected where s is a parameter which is determined heuristically.

4.2   Clustering the Bendy Terms

We use QT clustering (full definition given in [6]) which has the advantage of being deterministic unlike k-means which depends on the initial allocation. It also does not require us to specify the number of clusters in advance. It requires a distance measure between each pair of objects and for this we use shape distance. For each term in the collection it creates candidate clusters based on all terms within a minimum distance of that term. The largest candidate is deemed a cluster and then the process is repeated for the remaining terms until none are left. The minimum distance (called somewhat misleadingly *diameter* in [6]) is set by the user and increasing it reduces the number of clusters. QT clustering is very computationally intensive when applied to large datasets since it is $\mathcal{O}(n^3)$ and thus reducing the number of words to cluster is clearly advantageous when dealing with large datasets hence the advantage of using just the bendy terms. Table 2 gives an example of bendy terms which have been assigned to the largest clusters; this is not the same set of stories as used above. We note that cluster

Table 2. Examples of Word Clusters from 100 Stories

Cluster	Bendy Words
1	nasa, give, john, return, astronaut, glenn, travel, space, orbit, human, etc.
2	site, iraqi, scott, monitor, inspector, bill, provide, spy, baghdad, house, etc.
3	call, home, press, put, send, announce, problem, hour, meet, morning, etc.
4	defense, face, theodore, sacramento, judge, thursday, unabomber, delay, etc.
5	president, test, continue, asia, crisis, clinton, situation, decision, week

3 seems to have picked up words which do not appear to relate to a particular topic but other clusters do.

4.3   Creating Document Clusters

The result of clustering is to produce a list of word clusters which are ranked in order of size. This list of words can be thought of as an index of an event. The stories can then be assigned to each event by techniques such as nearest neighbour, using a measure of document similarity such as the cosine measure.

5   Empirical Study

A pilot study was conducted to investigate the effectiveness of the approach proposed here. The news stories were taken from a standard TDT collection comprising CNN news broadcasts during the year of 1998. Each news broadcast was transcribed using automatic voice recognition and then has been separated into distinct news items. A subset of the stories (1432) were taken from a sample of news reports during each the day. Two results emerged:

- Using the continuous language models to define events and then using them to group stories performed significantly better than k-means clustering on the stories alone.
- The performance was unaffected by using only 150 of the most bendy terms compared to the whole vocabulary. Using the 150 least bendy terms showed marked degradation.

6   Conclusions

The use of continuous language models with a logistic polynomial function gives rise to two useful measures: bendiness and shape distance. Clustering the most bendy terms by their shape distance provides a promising technique for identifying events within a stream of news stories.

Acknowledgement. We would like to thank Leif Azzopardi and the Department of Computer Science at Glasgow University for supporting the completion of this work.

References

1. Allan, J., Carbonell, J., Doddington, G., Yamron, J., Yang, Y.: Topic Detection and Tracking Pilot Study Final Report. In: Proceedings of the DARPA Broadcast News Transcription and Understanding Workshop, pp. 194–218 (1998)
2. Bache, R., Crestani, F.: Estimating Real-valued Characteristics of Criminals from their Recorded Crimes. In: ACM 17th Conference on Information and Knowledge Management (CIKM 2008), Napa Valley, California (2008)
3. Bai, J., Nie, J., Paradis, F.: Text Classification Using Language Models. In: Asia Information Retrieval Symposium, Poster Session, Beijing (2004)
4. Dharanipragada, S., Franz, M., Ward, T., Zhu, W.: Segmentation and Detection at IBM - Hybrid Statisticsl Models and Two-tiered Clustering. In: Allan, J. (ed.) Topic Detection and Tracking. Kluwer Academic Publishers, Norwell (2002)
5. Clifton, C., Cooley, R., Rennie, J.: TopCat: Data Mining for Topic Identification in a Text Corpus. IEEE Transactions on Knowledge and Data Engineering 16(8) (2004)
6. Heyer, L., Kruglyak, S., Yooseph, S.: Exploring Expression Data: Identification and Analysis of Coexpressed genes. Genome Research 9, 1106–1115 (1999)
7. Kleinberg., J.: Bursty and Hierarchical Structure in Streams. In: Proc. 8th ACM SIGKDD Intl. Conf. on Knowledge Discovery and Data Mining, Edmonton, Alberta, Canada (2002)
8. Losada, D.: Language Modeling for Sentence Retrieval: A comparison between Multiple-Bernoulli Models and Multinomial Models. In: Information Retrieval Workshop, Glasgow (2005)
9. McCallum, A., Nigam, K.: A Comparison of Event Models for Naïve Bayes Text Classification. In: Proc. AAAI/ICML 1998 Workshop on Learning for Text Categorisation, pp. 41–48. AAAI Press, Menlo Park (1998)
10. Peng, F., Schuurmans, D.: Combining Naïve Bayes and n-gram Language Models for Text Classification. In: Sebastiani, F. (ed.) ECIR 2003. LNCS, vol. 2633, pp. 335–350. Springer, Heidelberg (2003)
11. Peng, F., Schuurmans, D., Wang, S.: Augmenting Naïve Bayes Classifiers with Statistical Language Models. Information Retrieval 7(3), 317–345 (2003)
12. Ponte, J.M., Croft, W.B.: A Language Modeling Approach to Information Retrieval. In: Proceedings of the Twenty First ACM-SIGIR, Melbourne, Australia, pp. 275–281. ACM Press, New York (1998)

Spoken Language Understanding via Supervised Learning and Linguistically Motivated Features

Maria Georgescul, Manny Rayner, and Pierrette Bouillon

ISSCO/TIM, ETI, University of Geneva
{Maria.Georgescul,Manny.Rayner,Pierrette.Bouillon}@unige.ch

Abstract. In this paper, we reduce the rescoring problem in a spoken dialogue understanding task to a classification problem, by using the semantic error rate as the reranking target value. The classifiers we consider here are trained with linguistically motivated features. We present comparative experimental evaluation results of four supervised machine learning methods: Support Vector Machines, Weighted K-Nearest Neighbors, Naïve Bayes and Conditional Inference Trees. We provide a quantitative evaluation of learning and generalization during the classification supervised training, using cross validation and ROC analysis procedures. The reranking is derived using the posterior knowledge given by the classification algorithms.

Keywords: N-best reranking, Comparative evaluation, Linguistically motivated features.

1 Introduction

Task failure due to automatic speech recognition (ASR) misrecognitions can lead to important dialogue misunderstandings. However, in many cases where the 1-best ASR hypothesis is incorrect, the ASR output list of n-best hypotheses contains the correct one.

In order to allow the dialogue system to effectively detect and handle errors that may have occurred, one of the solutions adopted is the use of confidence measures derived from n-best lists [1-3] or word lattices [4]. [5] proposed an approach based on dynamic Bayesian networks in order to incorporate auxiliary features such as pitch, energy, rate-of-speech, which are speaker-dependent or sensitive to environment prosodic conditions. In order to improve their spoken language understanding module, [6] used a generative approach to generate a list of n-best associations between words and concepts, which is then re-ranked using discriminative learning with tree kernels. Other successful n-best rescoring has been reported on in the literature used information derived from various stages of dialogue management [7-21].

In this paper, we make use of the ASR n-best list and linguistically motivated features in order to identify and reduce dialog understanding errors. We re-score the n-best recognition hypotheses using four machine learning techniques with features computed from acoustic confidence scores, surface semantics, and dialogue context.

While the evaluation of dialogue systems is in itself a challenging task and various approaches [22, 23] have been proposed in this active area of research, we simply

C.J. Hopfe et al. (Eds.): NLDB 2010, LNCS 6177, pp. 117–128, 2010.
© Springer-Verlag Berlin Heidelberg 2010

assess here the utility of n-best rescoring by a semi-automatically defined semantic error metric. That is, semantic correctness in a discourse context is considered as a binary measure, and several n-best hypotheses are typically correct for a given utterance, with none of them being "better" than the others. As exemplified in Table 1, an utterance U was tagged as "semantically correct" (SemC) if either a) the recognition result for U was the same as the transcription, or b) the natural language processing module could produce a non-trivial semantic representation for U, and this representation was the same as the one which would have been produced from the transcription.

Table 1. Example of 6-best list and the associated SemC score. The reference utterance transcription is "where is the next meeting".

Recognised words	SemC
Where is next meeting	Yes
Where is the next meeting	Yes
Where was the next meeting	No
Where was next meeting	No
Where is the sixth meeting	No
Where is that sixth meeting	No

Using the support vector machine (SVM) approach proposed by [24] for ranking, [19] performed n-best reranking using both word error rate (WER) and semantic error rate (SemER) as reranking target value. Their experimental results show that the performance of the reranking system, when using SemER as the reranking target value, is improved in terms of both WER and SemER. In this paper, we merely use SemER as reranking target during both phases of the experiments, the training phase and the evaluation phase. Since SemER is defined as having only 0 and 1 values, the ranking problem is thus reduced to a binary classification problem: semantically incorrect hypotheses constitute instances for the class labeled with "1", while semantically correct hypotheses belong to the class labeled with "0". During the testing phase of the classification process, we make use of the probabilities of correct class predictions in order to generate the n-best reranking from the classification result.

The transformation of the reranking problem into a classification problem could turn out to be suboptimal for two main reasons. First, categorization algorithms often optimize the model based on the observed class distribution and a classification loss function. Therefore, the classifier might generate results that are good with respect to the classification accuracy, but useless for ranking. Second, only the first in rank recognition hypothesis actually matters for our task, because the n-best reranking is used for 1-best hypothesis extraction. Thus it is essential whether according to the rescoring function the first-in-rank recognition hypothesis is correctly classified. In this study, we explore the impact of the previously mentioned generic problems via an empirical evaluation on n-best rescoring using four classification algorithms on a specialized corpus.

In Section 2, we describe the features and the data used in our experiments. The classification algorithms adopted in this study, the experimental settings and intermediate results obtained during parameter tuning are presented in Section 3. In Sections 4 and 5, we compare the experimental results of the n-best reranking obtained via various classification algorithms with the results obtained when using SVM learning designed for reranking. Concluding remarks are provided in Section 6.

2 Data Set and Linguistic Features Used for the Experiments

2.1 Data Set

For our experiments, we used data collected with an application which gives access via spoken queries to the content of a meeting database containing information regarding dates, times, locations, attendance, etc. The application used for data collection is described in [25]. The language model of the dialogue system was built using the Regulus toolkit [26] which constructs a PCFG grammar starting from a linguistically motivated feature grammar for English. The grammar contains a total of 1724 CFG productions. The interpretation of questions may depend on preceding context, in particular to determine referents for pronouns ("it", "he") and definite descriptions ("the meeting"), or to resolve ellipsis ("Is there a meeting on Monday?" ... "Tuesday?").

During data collection, five subjects were given simple information-retrieval problems to solve together with a few representative examples of system queries. The user utterances were recorded, manually transcribed, and analyzed by an offline version of the application to determine whether or not they were within the coverage of the grammar. This produced 527 transcribed utterances, of which 455 were in-coverage. Since grammar-based language models behave very differently on in-coverage and out-of-coverage data, we only used the in-coverage portion for our experiments.

Using a trigram model trained on 365 utterances randomly chosen from our corpus, the perplexity that has been obtained for a testing subset of 90 remaining utterances has a value of 7.13, corresponding to a cross-entropy of 2.83 bits per word. For purposes of comparison, the trigram perplexity of the Air Travel Information System (ATIS) corpus is around 12. The computation of the perplexity of the language model with respect to our test data was based on 474 words. The trigram language model was fitted using the toolkit proposed by [27].

Each in-coverage utterance was tagged as "semantically correct" or "semantically incorrect". For each utterance, we use the list of the 6-best recognition hypotheses provided by the speech recognition module. Therefore, each hypothesis was passed through various steps of the natural language processing module and it was also tagged as "semantically correct" or "semantically incorrect". The proportion of correct hypotheses over the total number of hypotheses, i.e. 45.49%, shows that the data is approximately balanced for the purposes of classification.

2.2 Feature Set

We use four types of information which are extracted from different levels of processing. We summarize them in the following.

First, we use the information derived directly from the recognizer. Specifically, we use two features: the recognition confidence score and the rank in the N-best list given by the recognition system. Although these two features are obviously correlated, the rank gives important extra information; empirically, the first hypothesis is much more likely to be correct than the second, even if the difference in confidence scores is minimal.

The second group of features encodes syntactic information, and is taken from the parsed representation of the hypothesis. For example, we defined features encoding whether an existential construction ("Is there...") is plausible given the syntactic evidence. Other features in this group classify the surface-semantic representation as one of a small number of types, for example "WH-question" or "Elliptical phrase". Here, the main intuition was that elliptical phrases are less frequent than full questions.

The third group is task specific being defined on the deep semantic representation of the query, and reflects the intuition that some queries are logically implausible for the application domain. In particular, one feature attempts to identify underconstrained queries; we defined a query to be underconstrained if it made no reference to dates, times, locations or people. Another feature attempted to identify queries where the tense is inconsistent with other temporal constraints. A typical example would be "What meetings were there next week?".

The fourth and final group consists of dialogue-level features, which are based on the system response that would be produced for each hypothesis. This time, the intuition is that some types of response are inherently more plausible than others. For example, other things being equal, an error or clarification response is less plausible than a simple answer; a large set of individuals is less plausible than a small set (usually this means the query is underconstrained); and a negative answer is less plausible than a positive one. We separated the space of possible responses into six types, "say yes", "say no", "say nothing found", "say list of n referents", "say clarification question" and "other", and defined one binary-valued feature for each type. As shown in [19], the dialogue-level features are among the most informative ones.

All the categorical features described above are transformed into separate equivalent numerical-valued features. That is, we convert any categorical feature having V different values into V separate binary-valued features. For instance, the six-category "response type" feature is represented as six binary-valued features. In the experiments, the classifiers are given as input the numerical-valued features.

3 Classification Learning Algorithms and Experimental Settings

The methods chosen for machine learning classification in this experimental comparison were: Support Vector Machines for binary Classification (SVC), Naïve Bayes, Conditional Inference Trees and Weighted k-Nearest Neighbours. The following should be noted about the implementations of each of the methods.

3.1 Support Vector Machine Classification

We performed parameter tuning for Support Vector Machines via grid search. We used the entire set of features (i.e. 34 features) and the Gaussian Radial Basis kernel function (RBF). For each possible value we perform five-fold cross validation and we measure both the train error and the cross validation error.

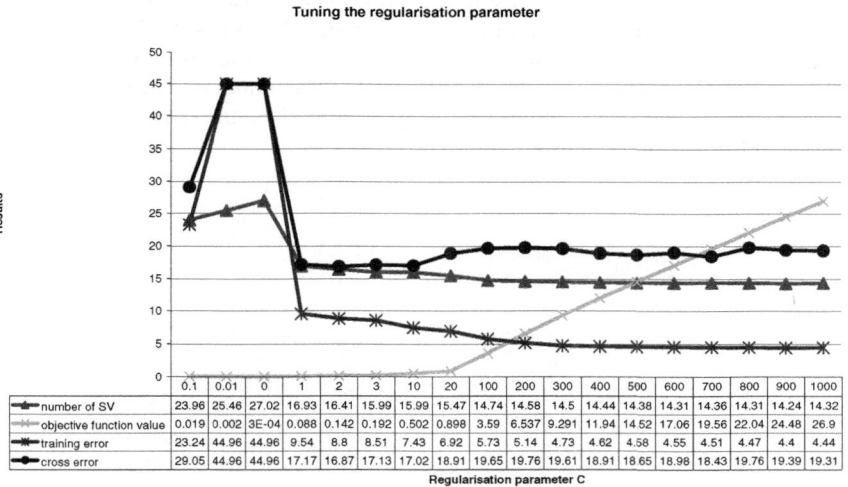

Tuning the regularisation parameter

	0.1	0.01	0	1	2	3	10	20	100	200	300	400	500	600	700	800	900	1000
number of SV	23.96	25.46	27.02	16.93	16.41	15.99	15.99	15.47	14.74	14.58	14.5	14.44	14.38	14.31	14.36	14.31	14.24	14.32
objective function value	0.019	0.002	3E-04	0.088	0.142	0.192	0.502	0.898	3.59	6.537	9.291	11.94	14.52	17.06	19.56	22.04	24.48	26.9
training error	23.24	44.96	44.96	9.54	8.8	8.51	7.43	6.92	5.73	5.14	4.73	4.62	4.58	4.55	4.51	4.47	4.4	4.44
cross error	29.05	44.96	44.96	17.17	16.87	17.13	17.02	18.91	19.65	19.76	19.61	18.91	18.65	18.98	18.43	19.76	19.39	19.31

Regularisation parameter C

Fig. 1. Plotting the SVC training error rate, the cross error rate on testing data, the normalized number of support vectors and the normalized values of the objective function when tuning the regularisation parameter and using the RBF kernel with a fixed value for his parameter σ.

Figure 1 exemplifies the SVC results obtained when tuning the regularisation parameter C via grid-search and the value of the RBF kernel parameter σ was fixed to 0.5. For plotting purposes the values of the objective function were divided by 10^4, while the number of support vectors was divided by 10^2.

As we observe from Figure 1, high values for C (i.e. when the C value is around 10^3), SVC yields excellent results as regarding the training accuracy (corresponding to an error rate of around 4%), but the cross validation error is around 19%, which suggests that values higher than 10^3 for C imply over-fitting. The optimal classification results were obtained for $\sigma = 0.5$ and C = 9, i.e. a cross validation error of 15.87% and a train error of 11.10%.

After SVC parameter tuning, we used the optimal parameter values and we studied the relative classification effectiveness of each feature via a backward sequential search method, i.e. by verifying whether cross-validation error could be decreased if one of the features was excluded. The experimental results shown that the classification cross error could not be significantly decreased in this manner, with the single exception of a feature encoding information regarding dialogue move state, which when excluded gave a slight decrease in the cross error, from 15.87% to 15.17%. When excluding any other feature, the cross error was not decreased. Therefore, we kept the entire set of features for the next phase of the experiments, i.e. for training and testing SVC and the other classifiers.

3.2 Weighted K-Nearest Neighbors

We train the Weighted k-Nearest Neighbors (WKNN) classifier proposed by [28] via leave-one-out cross-validation using the following kernels: the rectangular, the triangular, the Epanechnikov and the Gauss kernel. The WKNN with the rectangular kernel corresponds to standard unweighted kNN. In comparison with the standard kNN, the WKNN method was designed such that the number of nearest neighbors is implicitly taken into account in the weights, and thus too large values of k are adjusted automatically to a lower value.

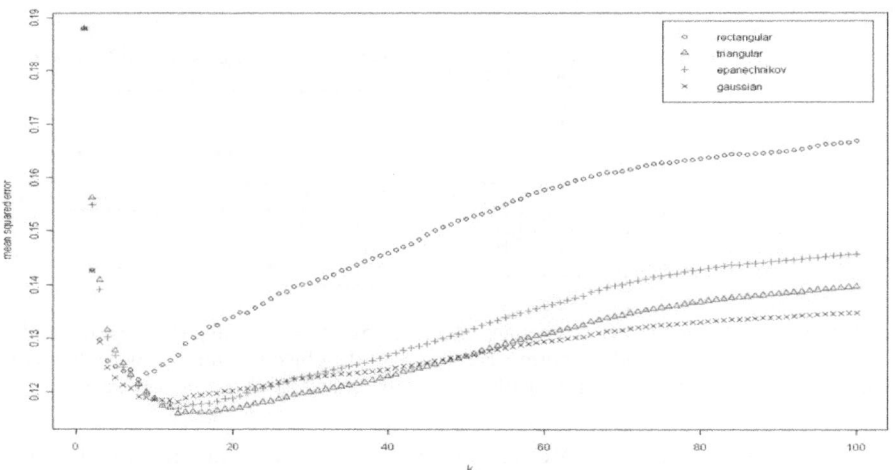

Fig. 2. Plotting the WKNN misclassification error for various kernels and varying the number of nearest neighbors from 0 to 100

As exemplified in Figure 2, our experiments show that for values of k larger than 20, the mean squared error obtained via leave-one-out cross-validation is increasing in a sharper way for the rectangular kernel (i.e. standard unweighted kNN) than for the other kernels. The experiments also showed that the choice between the triangular, Epanechnikov and the Gauss kernel is not crucial, since the resulting training accuracies are rather close for these kernels. Therefore, during the testing phase, we used only the Gaussian kernel and the kernel parameter value was selected automatically by WKNN.

In a first step of the testing phase, the $k = 15$ nearest neighbors in the training set vectors are calculated in terms of the Euclidean distance. In a second step, the classification of each test example is done by taking into account the maximum of summed kernel densities and the class of the nearest observations from the learning set.

3.3 Conditional Inference Trees

We investigate the classification and reranking prediction accuracy of Conditional Inference Trees (CIT) proposed in [29]. The conditional inference trees are constructed using c_{quad}-type test statistic with $\alpha = 0.05$ in order to test the null hypothesis of independence between all input variables and the response. For the computation of the distribution of the test statistic, we applied the Bonferroni correction.

3.4 Naive Bayes

Training data was used to estimate the conditional a-posterior probabilities of a categorical class variable given independent predictor variables using the Bayes rule. The classifier assumes Gaussian distribution of metric predictors given the target class. For these experiments, we used Laplace smoothing.

3.5 Generating the n-Best Rescored List from Classification

As mentioned in the introduction, transforming the result of binary classification into a ranking might be problematic. The classifiers used in our experiments yield an instance probability or score, i.e. a numeric value that represents the degree to which an instance is a member of a class. Therefore, we simply derived the n-best reranking from the posterior knowledge given by the classification algorithms.

During the experiments with SVC, class probabilities were produced using Platt's algorithm [30]. Platt's algorithm is based on fitting a sigmoid function to the classifier's decision values: $P(\ y = 1 | \ f\ ) = \dfrac{1}{1 + e^{Af + B}}$, where the parameters A and B are estimated by minimizing the negative log-likelihood function.

4 Empirical Classification Results

In this section, we compare the performance of each algorithm for classification on each test partition. From each classifier we used the output data frame of numeric predictors. These numeric predictions were used with a threshold of 0.5, in order to produce a discrete binary classification. That is, if the classifier numeric prediction for a hypothesis was above the threshold, then it produced a "1" label (i.e. the hypothesis was classified in the class labeled with "1"), else a "0" label.

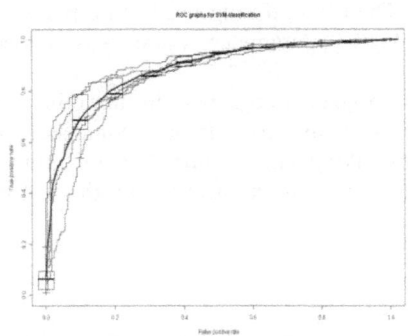 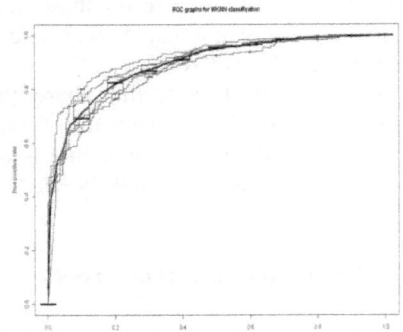

Fig. 3. ROC[1] graphs for SVC   **Fig. 4.** ROC graphs for CIT classification

Since the classification accuracy resulting for a threshold of 0.5 is not necessarily optimal, we also provide the ROC curves resulting for various thresholds in Figures 3, 4, 5 and 6. The y-axis represents the true positive rate (tpr), while the x-axis represents the false positive rate (fpr). In each figure, there are five curves resulting from 5-fold cross-validation runs. The mean performance (obtained via vertical averaging) is plotted using blue color and the variation around the average curve is visualized by box plots.

From the ROC graphs of each classifier we observe that the classification results of SVC and WKNN are the nearest to the y-axis indicating low false positive rates. Comparing Figure 6 with Figure 3, Figure 4 and Figure 5, we observe that only the error curves for Naïve Bayes are in the proximity of the diagonal line (tpr = fpr) showing that SVC, CIT and WKNN generally perform better and more consistently than Naïve Bayes. That is, the overall accuracy of SVC and WKNN is not dependent on a decision threshold chosen. Figure 6 also shows that for one of the test folds, using small threshold values involves that Naive Bayes classification performance gets very close to the performance of a baseline random classification. Thus Naïve Bayes classification proves to be sensitive to the threshold value and to the data set that has been used for training and testing.

Since we are primarily interested in finding out whether each classifier produces calibrated probability estimates for rescoring, we analyze in the next section whether the samples in the first three training and test sets are likewise less representative for the reranking task.

5 Empirical n-Best Rescoring Results

Figure 7 shows the empirical rescoring results obtained on each test partition when using classification likelihoods in order to generate a n-best reranking model. We observe that in average the best rescoring results are obtained when using SVC likelihoods of class membership for each hypothesis, the n-best re-scoring having an average SemER of 11.97%. Therefore, the SVC generates results that are optimal not only with respect to the classification accuracy, but also for reranking

---

[1] Receiver Operating Characteristics (ROC).

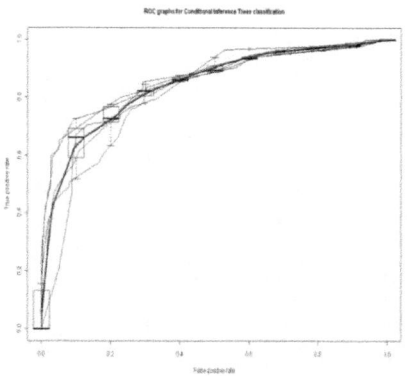

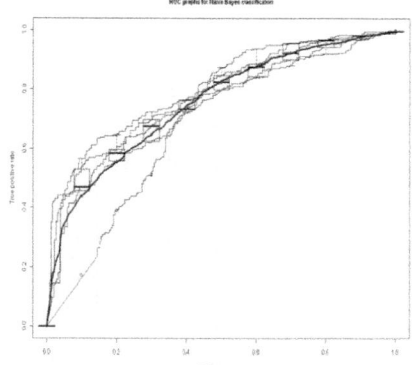

Fig. 5. ROC graphs for WKNN classification

Fig. 6. ROC graphs for Naïve Bayes classification

With regard to the error scores of n-best rescoring for each individual fold, we observe that the first partition round of cross-validation provides the worse model generalization. Therefore, samples in the first training set are less representative for the n-best rescoring task.

Summarizing the results obtained for rescoring, Figure 8 shows the overall reranking performance results obtained from the four different classifiers and the performance that was obtained in [19] with a support vector machine model tuned for reranking. The baseline represents the results obtained by choosing the first in rank hypothesis, as provided by ASR when no reranking was applied. The "oracle" represents the ideal reranking algorithm which would always select from the n-best list the hypothesis that has the smallest SemER.

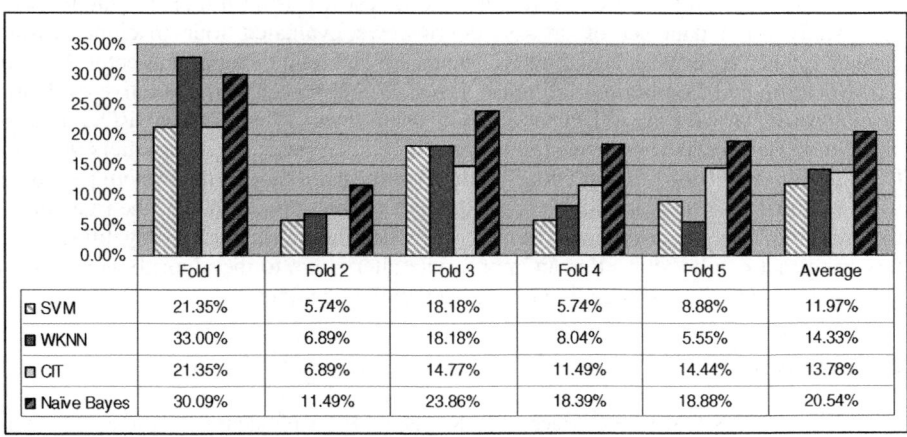

	Fold 1	Fold 2	Fold 3	Fold 4	Fold 5	Average
SVM	21.35%	5.74%	18.18%	5.74%	8.88%	11.97%
WKNN	33.00%	6.89%	18.18%	8.04%	5.55%	14.33%
CIT	21.35%	6.89%	14.77%	11.49%	14.44%	13.78%
Naïve Bayes	30.09%	11.49%	23.86%	18.39%	18.88%	20.54%

Fig. 7. Reranking error rates on each testing fold during 5-fold cross-validation

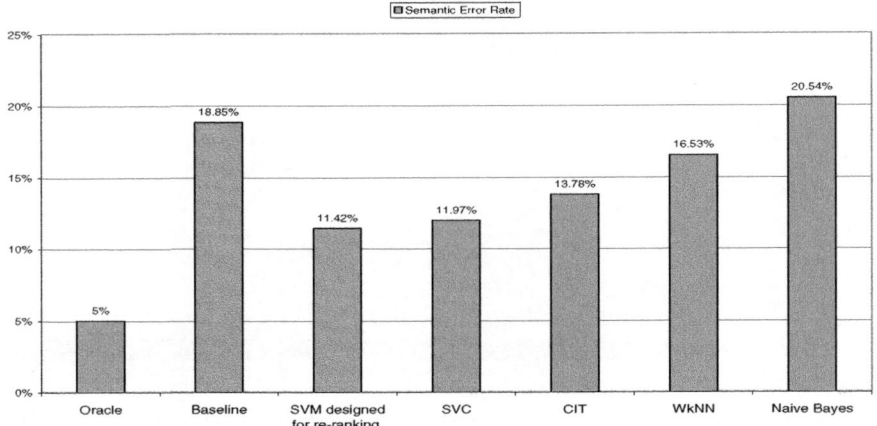

Fig. 8. The rescoring SemER mean obtained via 5-fold cross validation

As we can see from Figure 8, there is a significant difference between the accuracy obtained from reranking based on SVM and the baseline, in terms of average SemER. The Naïve Bayes classification method performed worse than the baseline, while the other three classifiers were able to reduce the number of dialog understanding errors by reranking the n-best list using linguistic features. Therefore, good reranking results are achieved via SVC, WKNN and CIT even though, the algorithms were optimized for classification. This shows that for our specialized corpus using classification likelihoods is effective in order to transform a classification result into a ranking.

6   Conclusions

We have considered the task of reranking the n-best list using linguistic features. As a case study on a data set of spoken queries, we evaluated four machine learning classifiers: Naïve Bayes, Weighted K-Nearest Neighbours, Support Vector Machine Classification and Conditional Inference Trees. We performed a quantitative evaluation of classification learning and generalization using cross validation and ROC analysis. The Naïve Bayes classifier proved unsuccessful, but the other three classifiers were able to reduce the number of dialog understanding errors, even though the system parameters were optimized for classification. Thus, the experiments revealed that the reduction of the ranking learning problem to binary classification leads to a competitive n-best rescoring that can be viewed as an appropriate alternative to the algorithms previously designed for speech understanding in a spoken dialogue system.

References

1. Wessel, F., Schlüter, R., Macherey, K., Ney, H.: Confidence Measures for Large Vocabulary Continuous Speech Recognition. IEEE Transactions on Speech and Audio Processing 9 (2001)
2. Gandrabur, S., Foster, G., Lapalme, G.: Confidence Estimation for NLP Applications. ACM Transactions on Speech and Language Processing 3, 1–29 (2006)

3. Torres, F., Hurtado, L., García, F., Sanchis, E., Segarra, E.: Error handling in a stochastic dialog system through confidence measures. Speech Communication 45, 211–229 (2005)
4. Mangu, L., Brill, E., Stolcke, A.: Finding consensus in speech recognition: word error minimization and other applications of confusion networks. Computer Speech and Language 14, 373–400 (2000)
5. Stephenson, T.A., Doss, M.M., Bourlard, H.: Speech Recognition With Auxiliary Information. IEEE Transactions on Speech and Audio Processing 12 (2004)
6. Dinarelli, M., Moschitti, A., Riccardi, G.: Re-Ranking Models For Spoken Language Understanding. In: Proceedings of the 12th Conference of the European Chapter of the ACL, Athens, Greece, pp. 202–210 (2009)
7. McNeilly, W.P., Kahn, J.G., Hillard, D.L., Ostendorf, M.: Parse Structure and Segmentation for Improving Speech Recognition. In: Proceedings of the IEEE/ACL Workshop on Spoken Language Technology, Aruba (2006)
8. Chotimongkol, A., Rudnicky, A.I.: N-best Speech Hypotheses Reordering Using Linear Regression. In: Proceedings of the Seventh European Conference on Speech Communication and Technology (EuroSpeech), Aalborg, Denmark, pp. 1829–1832 (2001)
9. Walker, M., Wright, J., Langkilde, I.: Using Natural Language Processing and Discourse Features to Identify Understanding Errors in a Spoken Dialogue System. In: Proceedings of the 17th International Conference on Machine Learning (2000)
10. Brill, E., Florian, R., Henderson, J.C., Mangu, L.: Beyond N-Grams: Can Linguistic Sophistication Improve Language Modeling? In: Proceedings of the International Conference On Computational Linguistics (COLING), Montreal, Canada, pp. 186–190 (1998)
11. Jonson, R.: Dialogue Context-Based Re-ranking of ASR Hypotheses. In: Proceedings of the Spoken Language Technology Workshop, Aruba, pp. 174–177 (2006)
12. Bohus, D., Rudnicky, A.: Constructing Accurate Beliefs in Spoken Dialog Systems. In: IEEE SPS 2005 Automatic Speech Recognition and Understanding Workshop, San Juan, Puerto Rico (2005a)
13. Bohus, D., Rudnicky, A.I.: Error Handling in the RavenClaw Dialog Management Framework. In: Proceedings of the Conference on Human Language Technology and Empirical Methods in Natural Language Processing (HLT/EMNLP), Association for Computational Linguistics, Morristown, NJ, USA, Vancouver, British Columbia, Canada, pp. 225–232 (2005)
14. Bohus, D., Rudnicky, A.: A "K Hypotheses + Other" Belief Updating Model. In: AAAI Workshop on Statistical and Empirical Approaches for Spoken Dialogue systems, Boston, USA (2006)
15. Williams, J.: Exploiting the ASR N-Best by tracking multiple dialog state hypotheses. In: Interspeech, Brisbane, Australia (2008)
16. Gabsdil, M.: Classifying Recognition Results for Spoken Dialogue Systems. In: Proceedings of the 41st Annual Meeting on Association for Computational Linguistics, Sapporo, Japan, pp. 23–30 (2003)
17. Gabsdil, M., Lemon, O.: Combining Acoustic and Pragmatic Features to Predict Recognition Performance in Spoken Dialogue Systems. In: Proceedings of the 42nd Meeting of the Association for Computational Linguistics (ACL), Barcelona, Spain, pp. 343–350 (2004)
18. Higashinaka, R., Nakano, M., Aikawa, K.: Corpus-based discourse understanding in spoken dialogue systems. In: Proceedings the 41st Annual Meeting on Association for Computational Linguistics (2006)

19. Georgescul, M., Rayner, M., Bouillon, P., Tsourakis, N.: Discriminative Learning Using Linguistic Features to Rescore N-best Speech Hypotheses. In: Proceedings of the IEEE Workshop on Spoken Language Technology, Goa, India (2008)
20. Raymond, C., Béchet, F., Camelin, N., De Mori, R., Damnati, G.: Sequential decision strategies for machine interpretation of speech. IEEE Transactions on Audio, Speech and Language Processing 15 (2007)
21. Dinarelli, M., Moschitti, A., Riccardi, G.: Re-ranking models based-on small training data for spoken language understanding. In: Proceedings of the Conference on Empirical Methods in Natural Language Processing, Singapore, pp. 1076–1085 (2009)
22. Tetreaulta, J.R., Litman, D.J.: A Reinforcement Learning approach to evaluating state representations in spoken dialogue systems. Speech Communication 50, 683–696 (2008)
23. Thomson, B., Yu, K., Gašić, M., Keizer, S., Mairesse, F., Schatzmann, J., Young, S.: Evaluating semantic-level confidence scores with multiple hypotheses. In: Proceedings of Interspeech, Brisbane, Australia (2008)
24. Joachims, T.: Optimizing Search Engines Using Clickthrough Data. In: Proceedings of the 8th ACM International Conference on Knowledge Discovery and Data Mining (KDD), Edmonton, Alberta, Canada (2002)
25. Tsourakis, N., Georgescul, M., Bouillon, P., Rayner, M.: Building Mobile Spoken Dialogue Applications Using Regulus. In: Proceedings of the 6th International Conference on Language Resources and Evaluation (LREC), Marrakech, Morocco (2008)
26. Rayner, M., Hockey, B.A., Bouillon, P.: Putting Linguistics into Speech Recognition. Center for the Study of Language and Information (2006)
27. Clarkson, P., Rosenfeld, R.: Statistical Language Modeling Using the CMU Cambridge Toolkit. In: Proceedings of the ESCA Eurospeech (1997)
28. Hechenbichler, K., Schliep, K.: Weighted k-Nearest-Neighbor Techniques and Ordinal Classification. Institut für Statistik, Ludwig-Maximilians University München, pp. 1–16 (2004)
29. Hothorn, T., Hornik, K., Zeileis, A.: Unbiased Recursive Partitioning: A Conditional Inference Framework. Journal of Computational and Graphical Statistics 15, 651–674 (2006)
30. Platt, J.C.: Probabilistic outputs for support vector machines and comparison to regularized likelihood methods. MIT Press, Cambridge (2000)

Topology Estimation of Hierarchical Hidden Markov Models for Language Models

Kei Wakabayashi and Takao Miura

Dept.of Elect. & Elect. Engr., HOSEI University
3-7-2 KajinoCho, Koganei, Tokyo, 184–8584 Japan

Abstract. Estimation of topology of probabilistic models provides us with an important technique for many statistical language processing tasks. In this investigation, we propose a new topology estimation method for *Hierarchical Hidden Markov Model* (HHMM) that generalizes Hidden Markov Model (HMM) in a hierarchical manner. HHMM is a stochastic model which has powerful description capability compared to HMM, but it is hard to estimate HHMM topology because we have to give an initial hierarchy structure in advance on which HHMM depends. In this paper we propose a recursive estimation method of HHMM submodels by using frequent similar subsequence sets. We show some experimental results to see the effectiveness of our method.

Keywords: Model Selection, Hierarchical Hidden Markov Model.

1 Introduction

Recently much attention have been paid on knowledge extraction techniques by analyzing huge documents statistically. In statistical natural language processing approach, a probabilistic model plays an important role in many domains[6]. When probabilistic model structure is given, parameters of model can be learned by maximizing likelihood of documents. By a word *Model selection*, we mean a task to estimate such a model structure from given documents.

However, one can't apply maximizing likelihood approach easily to model selection because of overfit problem. As an alternative to likelihood, *AIC* (Akaike Information Criteria) has been proposed to measure goodness for probabilistic model structure[1], which is defined as expected likelihood of unavailable future data that is derived approximately based on data distribution estimated from given samples. By maximizing AIC, one may estimate good model structure for the domain. However, it is said that maximizing AIC method can't be applied properly when the models include *unobserved variables*[4]. To obtain structures suitable for such model families, alternative methods to maximize AIC should be discussed.

One of the most useful models which involve unobserved variables is Hidden Markov Model (HMM). HMM is a stochastic model with unobserved states under simple Markov transition. It is well known that HMM can be employed for many natural language processing tasks such as speech recognition, handwritten

C.J. Hopfe et al. (Eds.): NLDB 2010, LNCS 6177, pp. 129–139, 2010.

character recognition, part of speech tagging, topic segmentation and document classification[8]. Although some generalization of HMM has been proposed like parallel HMM[7], coupled HMM[2] and incremental HMM[9], it is still hard to capture time (interval) dependent context explicitly because of simple Markov assumption.

Fine[3] has proposed *Hierarchical Hidden Markov Model* (HHMM) as hierarchical generalization of HMM. HHMM topology can be described by means of tree structures where each node represents sub-HHMM. In HHMM, states in upper level of hierarchy correspond to several observations while each HMM state corresponds to single observation. This means that HHMM can capture global context so that it models sequential data more precisely compared to HMM. There have been some studies to show effectiveness of HHMM description capability in several domains[5]. However, in these studies, an HHMM structure is given in advance. In fact, the more complicated structure HHMM has, the harder we estimate an HHMM topology.

In this investigation, we propose a method of estimating HHMM topology from given sample sequences. Conventional model selection approach has difficulties on 2 aspects. One aspect comes from the fact that HHMM contains unobserved variables and conventional information criteria such as AIC can't be applied correctly. Another is that it is expensive to enumerate candidates of structure. In conventional model selection approach, we enumerate candidates for models and select one which maximizes information criteria. However, an HHMM case becomes computationally intractable since HHMM topology is described as a tree structure. Therefore, we should apply another approach to HHMM topology estimation without enumeration of candidates of topology.

In this paper we propose a new approach of estimating HHMM topology by estimating HHMM submodels recursively. To estimate submodels, we examine frequent subsequences in a document. When a submodel structure exists, a document must contain output sequences of the submodel frequently. However, output of submodels depend probabilistically. In this paper we define similarity of subsequences by using edit distance, and define frequency of *subsequence set* on the basis of the similarity. We show a method to extract frequent subsequence sets from documents approximately and estimate HHMM topology by using extracted frequent subsequence sets.

In this work, we review Hierarchical Hidden Markov Model in section 2. In section 3 we define frequency of subsequence set and show our HHMM topology estimation method. Section 4 contains experimental results. We conclude our discussion in section 5.

2   Hierarchical Hidden Markov Model

Figure 1 shows an example of HHMM structure. HHMM states are organized hierarchically. Let q_i^d be a state of d hierarchical level. HHMM can have 2 types of state, *internal states* that contain sub-HHMM as output, and *production states* that output observation directly. In figure 1, internal states are q_1^1, q_1^2 and q_2^2, production

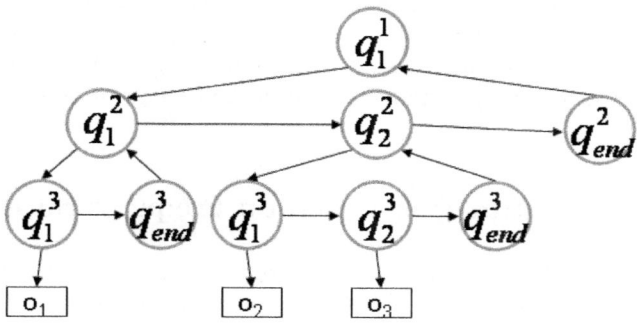

Fig. 1. Hierarchical Hidden Markov Model

states are q_1^3 and q_2^3. Parameters pertaining to internal state q^d are $\pi^{q^d}(q_i^{d+1})$ which is an initial state probability of submodels, and $a_{ij}^{q^d}$ which is a transition probability of submodel states from q_i^{d+1} to q_j^{d+1}. Parameters pertaining to production state q^d are $b^{q^d}(k)$ which is output probability of observation k.

In HHMM, each observation corresponds to a collection of states of each hierarchical level. For example, in figure 1, an observation o_2 corresponds to q_1^1, q_2^2 and q_1^3. Whenever next observation appears, transition on bottom state occurs with transition probability. HHMM submodels has only 1 *final state* q_{end}^d. Once transition to final state occurs, the submodel terminates and the control goes back to the parent. Then transition on the parent occurs. For instance, after o_2 appears, transition on bottom state from q_1^3 to q_2^3 occurs with probability $a_{12}^{q_2^2}$, then state q_2^3 generates new observation o_3 with output probability $b^{q_2^3}(o_3)$. In this example, transition from q_2^3 to final state q_{end}^3 occurs with probability $a_{2end}^{q_2^2}$, and the submodel of q_2^2 terminates. Then transition from q_2^2 to other state occurs. When transition to q_{end}^2 occurs, top level submodel terminates and whole output of the HHMM terminates.

To use HHMM, following 3 subtasks should be involved.

(1) For given HHMM λ and sequence O, we calculate probability $P(O|\lambda)$.
(2) For given HHMM λ and sequence O, we find the most likely state sequence corresponds to O.
(3) For given HHMM structure and sequence O, we estimate HHMM parameters which maximize likelihood of O.

Fine[3] gives the algorithms for obtaining solutions of these subtasks. For (1), they have proposed *generalized forward algorithm* which calculate $P(O|\lambda)$ with $O(NT^3)$ time complexity by using dynamic programming technique. For (2), *generalized Viterbi algorithm* has proposed which find the most likely state sequence also with $O(NT^3)$ time complexity by using dynamic programming. For (3), *generalized Baum-Welch algorithm* has proposed by using EM algorithm technique. This algorithm gives local optimum parameters depending on initial parameters.

Since HHMM has hierarchical structures, one can model sequential data more precisely compared to conventional HMM. However, clearly it is hard to decide HHMM topology because we need submodel relationships described as tree structure. In the following sections we discuss how to estimate HHMM topology initially.

3   Topology Estimation Based on Frequent Subsequences

The basic approach of our HHMM topology estimation method comes from recursive submodel estimation. When a submodel structure exists, sample data must contain output subsequences from the submodel frequently. Note that these subsequences are not necessarily same sequence because they are assumed as output of probabilistic model. In this section, we show a heuristic method of HHMM submodel estimation by finding *frequent similar subsequence sets*. Firstly we define frequency of similar subsequence set. And we discuss that finding frequent similar subsequence sets is difficult task since it requires large computational complexity. Then we show our proposal method of finding similar subsequence sets approximately by using dotplot approach.

3.1   Frequency of Subsequence Set

In this investigation we define similarity of subsequences on the basis of *edit distance*. Edit distance between sequence s_1 and s_2 is defined as sum of costs of edit operations which make s_1 into the identical sequence to s_2. In definition of edit distance, 3 kinds of edit operation is allowed: insertion, deletion and substitution. For example, to obtain edit distance between $s_1 = \{ABDAEF\}$ and $s_2 = \{BCDBE\}$, following edit operation sequence is assumed to make s_1 into s_2.

$$op = \{delete : A, substitute : BtoB, insert : C, substitute : DtoD, \qquad (1)$$
$$substitute : AtoB, substitute : EtoE, delete : F\}$$

Then we define edit distance as follows.

$$d(s_1, s_2) = \frac{\sum_i Cost(op_i)}{|op|} \qquad (2)$$

Let $Cost(x)$ be a cost function. If x is an operation which substitutes a character with identical one, $Cost(x)$ returns 0. Otherwise $Cost(x)$ returns 1. $|op|$ means length of an operation sequence op. Therefore, in this example $d(s_1, s_2) = \frac{4}{7}$. We define similarity between s_1 and s_2 as follows.

$$sim(s_1, s_2) = 1 - d(s_1, s_2) \qquad (3)$$

In this investigation we assume that pairs of subsequences, which have high similarities with each other, tend to be outputs from same HHMM submodels.

We examine frequent sets of subsequences in document as HHMM submodels. We define generalized frequency of subsequence set S as follows.

$$f(S) = \sum_{i=1}^{|S|} \prod_{j=1}^{|S|} sim(i,j) \tag{4}$$

When all subsequences in sets are identical, we can say that the frequency of sequence is same to the number of elements of sets naturally. In this paper, we say subsequence set S is frequent when $f(S) \geq \sigma$ for given threshold value σ.

It is hard computationally to find frequent subsequence sets because that calculation of $f(S)$ also means the calculation of edit distance which requires $O(mn)$ complexity generally. Therefore, one have to enumerate all possible subsequence pairs and calculate these similarities. It requires $O(n^4)$ complexity where the length of a given sequence is n. This means a naive approach should be avoided.

One idea to reduce computational complexity arises when we ignore edit distance calculation for subsequence pairs which are far from each other. Let M be $max_{i,j}sim(s_i, s_j)$ in a subsequence set S. When S is frequent, M must be greater than certain lower bound M_b depending on threshold σ. When subsequence s is added to S, $f(S)$ can't be reduced whenever similarities between s and all subsequences in S go beyond M_b. M_b can be obtained as M which give $f(S) = \sigma$ exactly when all similarities in S are M.

$$f(S) = \sum_{i=1}^{|S|} \prod_{j=1}^{|S|} M_b \tag{5}$$

$$= \sum_{i=1}^{|S|} M_b^{|S|-1}$$

$$= |S|M_b^{|S|-1} = \sigma \tag{6}$$

Taking logarithm on both sides we have:

$$log|S| + (|S| - 1)logM_b = log\sigma \tag{7}$$

Here we differentiate this equation with respect to $|S|$,

$$logM_b + \frac{1}{|S|} = 0 \tag{8}$$

$$|S| = \frac{1}{-logM_b} \tag{9}$$

this means $f(S) - \sigma$ becomes minimum when $|S| = \frac{1}{-logM_b}$. Therefore,

$$\frac{1}{-logM_b}M_b^{\frac{1}{-logM_b}-1} = \sigma \tag{10}$$

Although this equation is difficult to solve analytically, approximation solution can be obtained since it is monotone increasing with respect to M_b between 0.0 to 1.0. For instance, $M_b = 0.9032$ when $\sigma = 4$, $M_b = 0.8683$ when $\sigma = 3$.

3.2   Dotplot

Here we introduce a notion of *dotplot* approach to reduce computational complexity to find frequent subsequence sets. Initially dotplot is an approach to visualize similarities between sequences by plotting dots on two-dimensional space. Figure 2 shows a dotplot matrix over a sequence $s = \{ABCDECDEABCECD\}$. In dotplot approach, a sequence s is plotted on vertical and horizontal axis. Black dots in figure 2 show the situation that symbols on vertical and horizontal axis are identical. Black dots may be seen as slant lines in dotplot, for example coordinate (6,3) to (8,5) or (3,6) to (5,8), they mean identical subsequence $\{CDE\}$.

A slant line from (9,1) to (11,3) which corresponds to subsequence $\{ABC\}$ and a slant line from (12, 5) to (14,7) which corresponds to $\{ECD\}$ can be considered as a same line with 1 gap. This means that subsequence $s_{9,14} = \{ABCECD\}$ can be considered as edited sequence from subsequence $s_{1,7} = \{ABCDECD\}$ by applying one edit operation of deletion. In this paper we call the vertical sequence $s_{1,7}$ as *pattern*, and horizontal sequence $s_{9,14}$ as *cooccurrent subsequence* of pattern $s_{1,7}$. In our proposal method we find such gapped slant line.

Additionally, similarity between sequences $sim(s_{1,7}, s_{9,14})$ can be derived from following equation easily:

$$\frac{dotnum}{max(|s_{1,7}|, |s_{9,14}|)} \tag{11}$$

Let *dotnum* be number of dots in the slant line. This equation doesn't always give the same value to similarity given by definition, when optimal edit operation sequence includes both of insertion and deletion. However in this investigation we apply the value given by this equation as approximation similarity for purpose of decreasing complexity.

3.3   Extraction of Frequent Subsequence Set

Here we show how to find frequent subsequence sets approximately from given sequence D. In this approach we enumerate all possible patterns which can be element of frequent subsequence set with possible cooccurrent subsequences by examining dotplot space once. Then we extract frequent subsequence sets from enumerated patterns and cooccurrent subsequences.

First of all, let a pattern set P be empty, and a processing line l be 1. We examine the dotplot of line l and add pattern $D_{l,l}$ to P. We add all dots in line l to cooccurrent subsequence of pattern $D_{l,l}$. When $l \geq 2$, we add pattern $D_{i,l}$ with copy of cooccurrent subsequence of $D_{i,l-1}$ to P for all i. Then we evaluate whether each dot in line l constitutes a slant line or not for each cooccurrent subsequences. Let (a, b) be a coordinate pair of a last cooccurrent subsequence and (x, l) be coordinate of dot. When $x > a \wedge x \leq a + (l - b)$, we add the dot

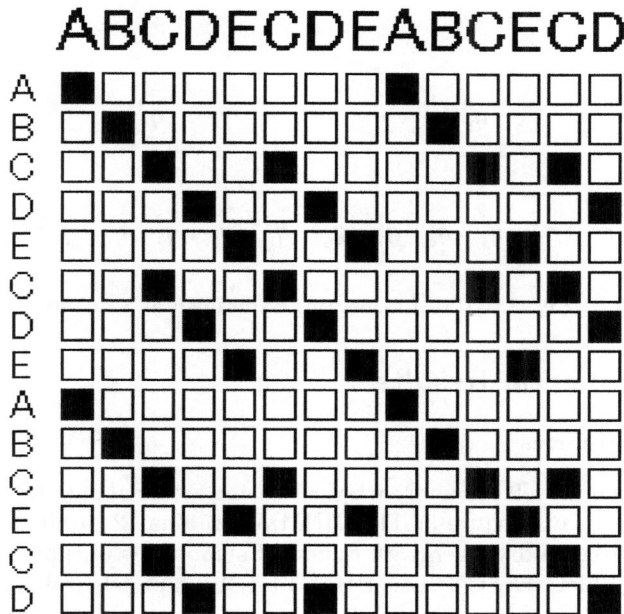

Fig. 2. Dotplot made with sequence $\{ABCDECDEABCECD\}$

to last of the cooccurrent subsequence. We repeat this process for line l until we examine all possible patterns and cooccurrent subsequences.

In this method we assume all similarities of subsequences in frequent subsequence sets are greater than M_b. By this assumption we can remove cooccurrent subsequences which has lower similarity than M_b at the end of line l process. By this assumption we can improve computational and space complexities.

By using obtained pattern set P and cooccurrent subsequences, we extract frequent subsequence sets. First we sort P by pattern length in descending order. Then we make a subsequence set S consisting of longest pattern sequences and its cooccurrent subsequences. Then we calculate the frequency $f(S)$. If $f(S) \geq \sigma$, we add S to frequent subsequence set S_{freq} and remove all pattern sequences which include subsequences of elements of S from pattern set P. We repeat this process for all P, then we obtain a frequent subsequence set S_{freq} approximately.

We see that the extracted submodel includes sub-submodels. Therefore we apply this process recursively to sequences which made by concatenating elements of extracted subsequence sets with $\sigma = f(S)$.

Finally, we construct HHMM topology by using recursive frequent subsequence set. Each frequent subsequence set corresponds to one HHMM submodel which has the number of states:

$$\frac{\sum_{i=1}^{|S|} |S_i|}{|S|} \tag{12}$$

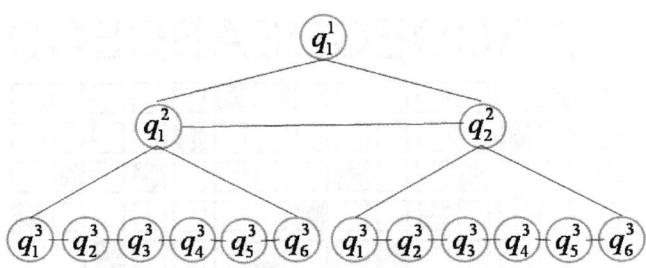

Fig. 3. HHMM Generates Experimental Sequence

4   Experimental Results

4.1   Preliminaries

Here we show how well our approach works using artificial data to obtain HHMM structure shown in a figure 3. This HHMM contains 12 production states and 3 internal states, and we generate 6 kind of symbols as output. The maximum state depth of hierarchy is 3. This HHMM has 2 submodels which generate likely output: $\{0, 1, 2, 3, 4, 5\}$ and $\{3, 1, 5, 0, 2, 4\}$ respectively. We generate 3 sequence data by using this HHMM.

We show part of the generated sequences in the top square in figure 5. Figure 4 shows dotplot of a part of generated sequence. We extract frequent subsequence set from this sequence with the threshold of frequency $\sigma = 3.0$.

4.2   Results

Here we show the hierarchy of extracted frequent subsequence sets in figure 5. In this figure, we illustrate a given sequence in a top square and extracted sequences in other squares. Underlines of sequence mean extracted subsequences. Extracted hierarchy appears deeply compared to real HHMM, but the subsequences in the extracted sets of bottom hierarchy look similar to likely sequences of true HHMM submodels.

Figure 6 shows HHMM topology estimated from extracted frequent subsequence sets. Estimated topology has 2 significant submodels. We can see these 2 submodels correspond to 2 real HHMM submodels respectively. However, estimated topology is more complex compared to true HHMM topology.

4.3   Discussions

Here we discuss what experimental results mean. Firstly let us discuss the extracted frequent subsequence sets. In every set, the elements are very similar subsequences with each other. On the other hand, many observations are not extracted by any sets. This is because we put our focus on extracting sets including

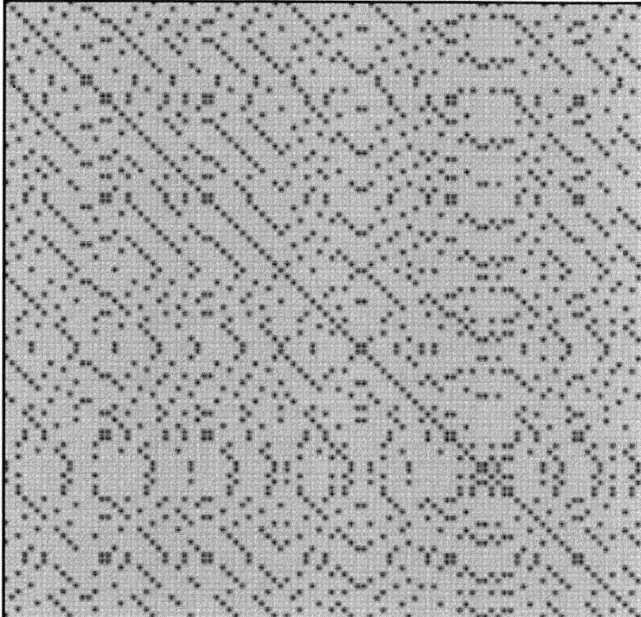

Fig. 4. Dotplot made by Experimental Sequence

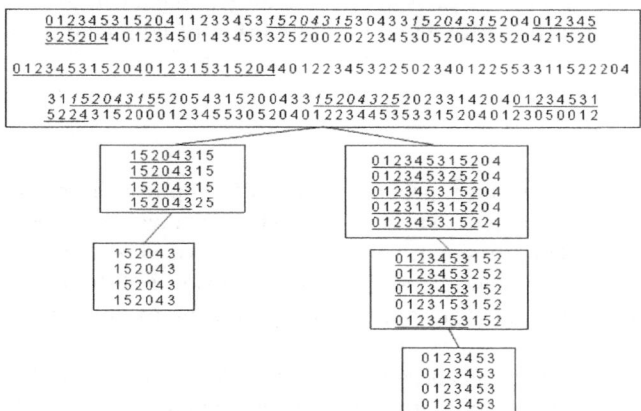

Fig. 5. Hierarchy of Extracted Frequent Subsequence Sets

similar subsequences when we define frequency, although the approach makes extraction coverage worse very often. Since we like to estimate HHMM topology, we need only frequent characteristic subsequences and we guess infrequent subsequences are useless. Therefore each extracted subsequence set is desirable to estimate submodels.

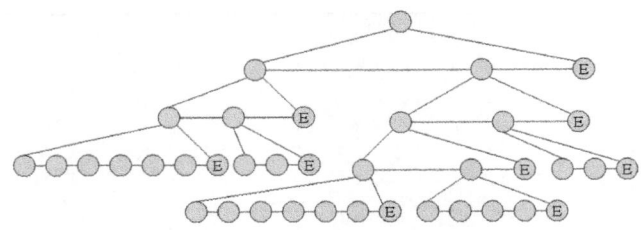

Fig. 6. Estimated Topology

However, the length of the extracted subsequences are identical in each set. This result also comes from frequency definition, since sequences of various length reduce frequency of the set significantly. Generally, length of sequences generated from HHMM submodel varies individually. In fact, many observations which have not extracted are part of characteristic subsequences have various length, for instance, $\{0, 1, 2, 2, 3, 4, 5\}$. This is why extraction coverage is significantly low. We can say that sequence similarity defined by using edit distance is not really good measure of likelihood of HHMM when the sequences have various length.

We have estimated topology which has 2 significant submodels. This result shows effectiveness of our approach for estimating hidden HHMM topology. To estimate language models, we should give only frequent structures because infrequent patterns can be learned automatically by using parameter estimation algorithm. Since we like to give submodels for significant frequent sequences, we believe our approach is promising for estimation of language model topology of HHMM.

5   Conclusion

In this investigation we have proposed estimation approach for HHMM topology based on extracting frequent subsequence sets. We have discussed an approximation method to extract frequent subsequence sets based on dotplot approach. We have examined experimental results and shown the usefulness of our approach. Experimental results show our approach is promising to estimate language model topology of HHMM.

In this work we have considered HHMM topology estimation. But it is possible to use our approach to generate several likely topologies for model selection by information criterion.

References

1. Akaike, H.: A new look at the statistical model identification. IEEE Trans. 19(6)
2. Brand, M., Oliver, N., Pentland, A.: Coupled hidden markov models for complex action recognition. In: CVPR 1997: Proceedings of the 1997 Conference on Computer Vision and Pattern Recognition (CVPR 1997), Washington, DC, USA, p. 994. IEEE Computer Society, Los Alamitos (1997)

3. Fine, S., Singer, Y.: The hierarchical hidden markov model: Analysis and applications. In: Machine Learning, pp. 41–62 (1998)
4. Katsuyuki, H.: On the problem in model selection of neural network regression in overrealizable scenario. Neural Computation 14(8), 1979–(2002)
5. Luhr, S., Bui, H.H., Venkatesh, S., West, G.A.: Recognition of human activity through hierarchical stochastic learning. In: IEEE International Conference on Pervasive Computing and Communications, vol. 0, p. 416 (2003)
6. Manning, C., Manning, C.D., Utze, H.S., The, M., Press, M., Lee, L.: Foundations of statistical natural language processing (1999)
7. Vogler, C., Metaxas, D.: Parallel hidden markov models for american sign language recognition. In: IEEE International Conference on Computer Vision, vol. 1, p. 116 (1999)
8. Wakabayashi, K., Miura, T.: Topics identification based on event sequence using co-occurrence words. In: Kapetanios, E., Sugumaran, V., Spiliopoulou, M. (eds.) NLDB 2008. LNCS, vol. 5039, pp. 219–225. Springer, Heidelberg (2008)
9. Wakabayashi, K., Miura, T.: Data stream prediction using incremental hidden markov models. In: Pedersen, T.B., Mohania, M.K., Tjoa, A.M. (eds.) Data Warehousing and Knowledge Discovery. LNCS, vol. 5691, pp. 63–74. Springer, Heidelberg (2009)

Speaker Independent Urdu Speech Recognition Using HMM

Javed Ashraf, Naveed Iqbal, Naveed Sarfraz Khattak, and Ather Mohsin Zaidi

College of Signals, National University of Sciences and Technology (NUST)
Rawalpindi 46000, Pakistan
{javed.ashraf,khattakn}@mcs.edu.pk, naveedrao@gmail.com,
athar-mcs@nust.edu.pk

Abstract. Automatic Speech Recognition (ASR) is one of the advanced fields of Natural Language Processing (NLP). Recent past has witnessed valuable research activities in ASR in English, European and East Asian languages. But unfortunately South Asian Languages in general and "Urdu" in particular have received very less attention. In this paper we present an approach to develop an ASR system for Urdu language. The proposed system is based on an open source speech recognition framework called Sphinx4 which uses statistical based approach (HMM: Hidden Markov Model) for developing ASR system. We present a Speaker Independent ASR system for small sized vocabulary, i.e. fifty two isolated most spoken Urdu words and suggest that this research work will form the basis to develop medium and large size vocabulary Urdu speech recognition system.

Keywords: Speech recognition, Urdu language, Hidden Markov Model, CMU Sphinx4.

1   Introduction

There are between 60 and 80 million native speakers of standard Urdu (Khari Boli). According to the SIL Ethnologue (1999 data), Hindi/Urdu is the fifth most spoken language in the world [13]. According to George Weber's article Top Languages: The World's 10 Most Influential Languages in Language Today, Hindi/Urdu is the fourth most spoken language in the world, with 4.7 percent of the world's population, after Mandarin, English, and Spanish [1].

Despite a large number of populations speaking Urdu, very little research work has been done related to ASR for Urdu language. Most of the previous work related to SR for Urdu language is restricted to digits only using ANN. However, modern general-purpose speech recognition systems are generally based on Hidden Markov Models (HMMs). This research work focuses on recognition of Urdu words using HMMs.

2   Related Work

Centre for Research in Urdu Language Processing (CRULP), Lahore, Pakistan [2] and National Language Authority (NLA), Islamabad, Pakistan [3] are the most prominent

C.J. Hopfe et al. (Eds.): NLDB 2010, LNCS 6177, pp. 140–148, 2010.
© Springer-Verlag Berlin Heidelberg 2010

names among those who are doing dedicated research in NLP for Urdu Language. CRULP and NLA have done research in many areas of Urdu Language Processing. However not much published research work is available related to ASR for Urdu language.

M.U Akram and M.Arif [4] try to analyse and implement Pattern-Matching and Acoustic Modelling approaches to SR for Urdu language. They propose an Urdu SR system using the later approach. They have used ANN to classify a set of frames into phonetic based categories at each frame. To search the best sequence path for the given word to be recognised they used Viterbi search algorithm.

S.M Azam et al. [5] use ANN to propose SR system for Urdu. They have developed SR system for isolated Urdu digits only. Moreover the proposed system works only for one speaker.

Abdul Ahad et al. [6] also use ANN using Multilayer Perceptron (MLP). Again they propose a system for Urdu digits for mono-speaker database only.

S K Hasnain et al. [7] also use ANN to implement their proposed system. They suggest feed-forward ANN model implemented using Matlab and Simulink. Again the model is developed only for isolated Urdu digits.

3   HMM Based Speech Recognition

Traditional SR software falls into one of three categories [8]. These categories are: Template-based approaches, Knowledge-based approaches and Statistical-based approaches. Template-based approaches compare speech against a set of pre-recorded words. Knowledge-based approaches involve the hard-coding of known variations of speech into a system. Both of these methods become impractical for a larger number of words. Finally in Statistical-based approaches, (e.g.HMM) variations in speech are modelled statistically using automatic learning procedures. This approach represents the current state of SR and is the most widely used technique today.

HMM are statistical models which output a sequence of symbols or quantities. Speech signal could be viewed as a piecewise stationary signal or a short-time stationary signal. That is, one could assume in a short-time in the range of 10 milliseconds, speech could be approximated as a stationary process. Speech could thus be thought of as a Markov model for many stochastic processes [9].

HMM can be trained automatically and are simple and computationally feasible to use. In speech recognition, the HMM would output a sequence of n-dimensional real-valued vectors (with n being a small integer, such as 10), outputting one of these every 10 milliseconds. The vectors would consist of cepstral coefficients, which are obtained by taking a Fourier transform of a short time window of speech and de-correlating the spectrum using a cosine transform, then taking the first (most significant) coefficients. The HMM will tend to have in each state a statistical distribution that is a mixture of diagonal covariance Gaussians which will give a likelihood for each observed vector. Each word, or (for more general SR systems), each phoneme, will have a different output distribution; a HMM for a sequence of words or phonemes is made by concatenating the individual trained HMMs for the separate words and phonemes [9].

We propose an Urdu ASR system using open source SR framework called Carnegie Mellon University (CMU) Sphinx4 which is also based on HMM. Sphinx has never been used to develop Urdu ASR system. This research work will open new doors for researchers of Urdu ASR systems.

3.1 CMU Sphinx4 Framework

Sphinx-4 is a state-of-the-art speech recognition system written entirely in the Java programming language. It was created via a joint collaboration between the Sphinx group at CMU, Sun Microsystems Laboratories, Mitsubishi Electric Research Labs (MERL), and Hewlett Packard (HP), with contributions from the University of California at Santa Cruz (UCSC) and the Massachusetts Institute of Technology (MIT) [10, 11].

3.2 Sphinx4 Architecture

The Sphinx-4 architecture has been designed with a high degree of modularity. There are three main blocks in the design: the frontend, the decoder, and the knowledge base (KB). These are controllable by an external application. The *Frontend* module takes in speech and parameterizes it into a sequence of output features.

The *KB* or Linguist provides the information the decoder needs to do its job. It is made up of three modules: Acoustic Model, Dictionary and Language Model. *Acoustic Model* contains a representation (often statistical) of a sound, created by training using acoustic data. *Dictionary* has the pronunciation of all the words to be recognised. *Language Model* contains a representation (often statistical) of the probability of occurrence of words. The decoder is the main block of Sphinx-4 and performs the actual recognition. It comprises a graph construction module, which translates any type of standard language model provided to the KB by the application into an internal format, and together with information from the dictionary, and structural information from one or more sets of acoustic models, constructs a Language HMM. The latter is then used by the search module to determine the most likely sequence of words that could be represented by a series of features [12, 13].

3.3 SphinxTrain

SphinxTrain is the acoustic training toolkit for CMU Sphinx (versions 1, 2, 3 and 4). It is a suite of programs, scripts and documentation for building acoustic models from speech data for the Sphinx recognisers [17].

4 Urdu Words Recognition Application

4.1 Brief Overview of Urdu Language

Urdu is written right-to left in an extension of the Persian alphabet, which is itself an extension of the Arabic alphabet. Urdu uses characters from the extended Arabic character set used for Persian. It further extends this set to represent sounds which are present in Urdu but not it Arabic or Persian, including aspirated stop and alveolar consonants, and long vowels [14]. Altogether there are 58 letters in Urdu, given in Figure 1 ([15]; other sources may give slightly different set).

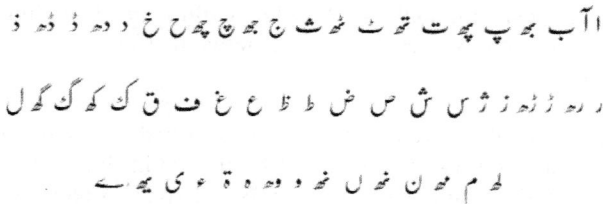

Fig. 1. Urdu Character Set

4.2 Roman Urdu

Urdu is occasionally also written in the Roman script. Roman Urdu is the name used for the Urdu language written with the Roman alphabet. Roman Urdu has been used since the days of the British Raj, partly as a result of the availability and low cost of Roman movable type for printing presses. The use of Roman Urdu was common in contexts such as product labels. Today it is regaining popularity among users of text-messaging and Internet services and is developing its own style and conventions. We have used Roman Urdu script for this research work. However, bare transliterations of Urdu into Roman letters omit many phonemic elements that have no equivalent in English or other languages commonly written in the Latin alphabet such as ژ خ غ ط ص ڈ or ق [16].

4.3 System Components

The system components are shown in figure 2 below:

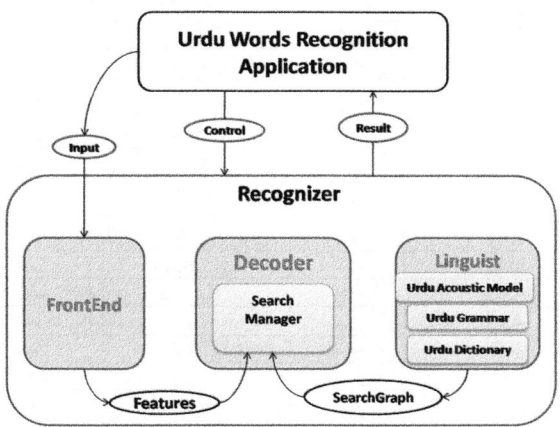

Fig. 2. System Components

4.3.1 Dictionary/Lexicon

The Lexicon development process consists of defining a phonetic set and generating the word pronunciation for training acoustic and language models. Our dictionary contains fifty two most commonly used spoken Urdu words including ten Urdu digits (0 to 9). The pronunciations of each word are shown in Table 1.

4.3.2 Acoustic Model

Acoustic Model contains a representation of a sound, created by training using acoustic data. Sphinx-4 uses models created with SphinxTrain. Two main acoustic models that are used by Sphinx-4, TIDIGITS and Wall Street Journal (WSJ), are already included in the "lib" directory of Sphinx-4 binary distribution. These acoustic models are also used for other Latin languages such as French or Italian due to similarities between these languages and English language from phoneme point of view. However, as mentioned before, there are great differences between English and Urdu language. We also tried to use English acoustic model provided with Sphinx4 libraries to recognise Urdu words but as expected the results were very poor (WER obtained was about 60%). Developing speech corpus or acoustic model is not a trivial task and requires resources to be allocated. Unfortunately such resources were not available for this research. For this research we managed to get recording of ten samples of each utterance from ten different persons. The speech recording and breaking each sample of fifty two words into one word file was very time consuming process.

Table 1. Urdu Words with their pronunciation

Arabic Urdu Script	Pronunciation (Roman Urdu)	Arabic Urdu Script	Pronunciation (Roman Urdu)	Arabic Urdu Script	Pronunciation (Roman Urdu)
ایک	Eyk	دانیں	Daen	نیچے	Neechey
دو	Dou	دروازه	Darwaza	نیلا	Neela
تین	Tiyn	ڈگری	Degree	آف	Off
چار	Chaar	دیکھو	Dekho	آن	On
پانچ	Paanch	دُشمن	Dushman	أوپر	Ooper
چھ	Cheh	فرش	Farsh	پانی	Pani
سات	Saat	فائر	Fire	پیچھے	Peechey
آٹھ	Aath	ہیٹر	Heater	پیلا	Peela
نو	Nau	جاؤ	Jaao	پین	Pen
صفر	Sifar	کمره	Kamrah	فون	Phone
آگے	Aage	کھانا	Khana	رک	Ruk
آؤ	Aao	کھول	Khol	سبز	Sabz
اےسی	AC	کتاب	Kitab	سیاه	Siah
الماری	Almari	کرسی	Kursi	سفید	Sufaid
بائیں	Baen	لال	Lal	ٹی وی	TV
بند	Band	لائٹ	Light	زمین	Zameen
بجلی	Bijli	لکھو	Likho		
بستر	Bister	مدد	Madad		

Feature Extraction

For every recording in the training corpus, a set of feature files was computed from the speech training data. Each utterance of a word was transformed into a sequence of feature vectors using the front-end executables provided with Sphinx-4 training package. The detailed process can be found from [10] and [12]. The training was performed using 5200 utterances of speech data from 10 different speakers (52 (words) x10(samples) =520 utterances by each speaker).

Transcription File

Each recording is accurately transcribed; any error in the transcription will mislead the training process later. This process is done manually, i.e. we listen to the recordings then we match exactly what we hear, into text. Even the silence and the noise should also be represented in the transcription file [18].

4.3.3 Language Model

Sphinx-4 supports three types of language models. The Context-Free-Grammar (CGF) is usually used in speech applications based on command and control. The n-gram grammars are used for free speech form. The simplest type of language model is called "wordlist" grammar in which instead of breaking a word into sub-units or phonemes, complete word is taken as one phoneme. For example "Chaar" (English: four) is taken as one phoneme. The wordlist grammar is recommended for isolated word recognition. Being an isolated Urdu words recognition application, we have used wordlist grammar in our proposed system.

5 Experimental Results

5.1 Experimental Setup

We performed two types of experiments for evaluating the performance of our system. First type of test involved evaluating the performance of the system using speakers involved in training. Second type of test was carried out using speakers who were not involved in the training phase. We observed the behaviour of the system for 1,2,4,8 and 16 Gaussian models.

5.2 Evaluation Criteria

The evaluation was made according to recognition accuracy and computed using Word Error Rate (WER) defined by the equation 1 below, which aligns a recognised word string against the correct word string and compute the number of substitutions (S), deletions (D), and insertions (I) from the number of correct words(N):

$$WER = (S + D + I)/N *100 \qquad (1)$$

WER of 10% means that there was an error (substitution, deletion or insertion) of one word out of ten words spoken by a speaker.

5.3 Results

We found that 8 Gaussian models produced the best results in both the tests. Recognition results for first test are shown in table 2. Five different speakers who were also part of training phase participated in this test. Each speaker was asked to utter, at random, twenty Urdu words thrice. The results are shown for each speaker.

Table 2. Results for the first test of Urdu Words Recognition Application

Speaker	WER (%)			
	Test1	Test2	Test3	Mean WER
Speaker1	10	0	10	6.66
Speaker2	10	10	10	10.00
Speaker3	0	0	10	3.33
Speaker4	0	0	0	0.00
Speaker5	10	10	0	6.66
			Mean	5.33

Recognition results for the second test are shown in table 3. Five different speakers who did not participate in the training phase were involved in this test. Again, each speaker was asked to utter, at random, any twenty Urdu words thrice. The results are shown for each speaker. Note that the WER increased in this case because of new speakers.

Table 3. Results for the second test of Urdu Words Recognition Application

Speaker	WER (%)			
	Test1	Test2	Test3	Mean WER
Speaker1	10	10	10	10.00
Speaker2	10	10	20	13.33
Speaker3	10	0	10	6.66
Speaker4	20	10	10	13.33
Speaker5	10	10	10	10.00
			Mean	10.66

5.4 Findings of the Experiments

There are very important findings that we would like to discuss here:

1. Words with more distinct phones produced less accuracy in recognition.
2. Words with similar type of phones were recognised interchangeably, for example system got confused with words saat, chaar and aath.
3. We first experimented with recognition of ten digits only and then increased the vocabulary to fifty two words. We found that Wordlist grammar is good for small size vocabulary, however as we increased number of words the accuracy of recognition decreased.

4. As mentioned earlier, transliterations of Urdu into Roman letters omit many phonemic elements that have no equivalent in English, for example letter 'ت' in word تین, ٹہ in word آٹه,etc. Transcription in Latin Urdu should increase the accuracy of recognition.

6 Conclusions and Future Work

In this research work we have developed first Urdu ASR system based on CMU Sphinx-4. The recognition results are much better as compared to Urdu ASR systems developed earlier. We demonstrated that how Sphinx-4 framework can be used to build small to medium sized vocabulary Urdu ASR system. The acoustic model for this ASR system is entirely developed by the authors of this work. Since this research was first of its kind we kept it to isolated words recognition. We identified the steps involved to develop such system. This system can be used in applications where small vocabulary Urdu speech recognition is required. Moreover, this research work could form the basis for further research in Urdu ASR systems.

An immediate future work is to increase the vocabulary size to medium (about 200 words). Later on research can focus on developing Large Vocabulary Continuous Speech Recogniser (LVCSR) system for Urdu language. We have already started our work on former. We are also planning to continue our research to build LVCSR for Urdu language using Arabic (also called Perso - Arabic script) phonemes instead of roman Urdu phonemes to cater for Urdu phonemes missing in the roman Urdu.

Acknowledgements. This research work was possible due to easily available open source CMU Sphinx Speech Recogniser.

References

1. Most Widely Spoken Languages, Saint Ignatius,
   http://www2.ignatius.edu/faculty/turner/languages.htm
2. Centre for Research in Urdu Language Processing (CRULP),
   http://www.crulp.org
3. National Language Authority (NLA), Islamabad, http://www.nlauit.gov.pk
4. Akram, M., Arif, M.: Design of an Urdu Speech Recognizer based upon acoustic phonetic modelling approach. In: IEEE INMIC 2004, pp. 91–96 (2004)
5. Azam, S., Mansoor, M., Shahzad, Z.A., Mughal, M., Mohsin, S.: Urdu Spoken Digits Recognition Using Classified MFCC and Backpropgation Neural Network. In: IEEE Computer Graphics, Imaging and Visualisation (CGIV (2007)
6. Ahad, A., Fayyaz, M., Tariq, A.: Speech Recognition using Multilayer Perceptron. In: Students Conference, ISCON apos 2002, August 16-17, IEEE vol. 1, pp. 103–109 (2002)
7. Hasnain, S., Samiullah, K., Muhammad: Recognizing Spoken Urdu Numbers Using Fourier Descriptor and Neural Networks with Matlab. In: Second International Conference on Electrical Engineering, ICEE 2008, March 25-26, pp. 1–6 (2008)
8. Tebelskis, J.M.: Speech Recognition Using Neural Networks. Doctoral Thesis. UMI Order Number: GAX96-22445., Carnegie Mellon University (1995)

9. Speech Recognition,
   http://en.wikipedia.org/wiki/Speech_recognition
10. Walker, W., et al. : Sphinx-4: A Flexible Open Source Framework for Speech Recognition,
    http://cmusphinx.sourceforge.net/sphinx4/doc/
    Sphinx4Whitepaper.pdf
11. Lamere1, P., et al.: Design of the CMU Sphinx-4 Decoder
12. CMU Sphinx-4, http://cmusphinx.sourceforge.net/sphinx4
13. Satori, H., Harti, M., Chenfour, N.: Arabic Speech Recognition System Based on CMU
    Sphinx. In: 3rd International Symposium on Computational Intelligence and Intelligent
    Informatics – ISCIII 2007, Agadir, Morocco (2007)
14. Hussain, S.: Letter to Sound Rules for Urdu Text to Speech System. In: The Proceedings
    of Workshop on Computational Approaches to Arabic Script-based Languages, COLING
    2004, Geneva, Switzerland (2004)
15. National Language Authority, http://www.nla.gov.pk
16. Urdu, http://en.wikipedia.org/wiki/Urdu
17. ShinxTrain,
    http://cmusphinx.sourceforge.net/html/
    download.php/#SphinxTrain
18. Hussein, H., Raed, A.: Arabic Speech recognition using Sphinx Engine. Int. J. Speech
    Technol. 9, 130–150 (2006)

Second-Order HMM for Event Extraction from Short Message

Huixing Jiang[1], Xiaojie Wang[1], and Jilei Tian[2]

[1] Center of Intelligence Science and Technology, Beijing University of Posts and
Telecommunications, Beijing, China
jhx0129@163.com, xjwang@bupt.edu.cn
[2] Nokia Research Center, Beijing, China
jilei.tian@nokia.com

Abstract. This paper presents a novel integrated second-order Hidden
Markov Model (HMM) to extract event related named entities (NEs)
and activities from short messages simultaneously. It uses second-order
Markov chain to better model the context dependency in the string sequence. For decoding second-order HMM, a two-order Viterbi algorithm
is used. The experiments demonstrate that combing NE and activities as
an integrated model achieves better results than process them separately
by NER for NEs and POS decoding for activities. The experimental results also showed that second-order HMM outperforms than first-order
HMM. Furthermore, the proposed algorithm significantly reduces the
complexity that can run in the handheld device in the real time.

1 Introduction

Short Message Service (SMS) is a widely-used communication service in mobile
communication system, allowing the interchange of short text messages among
mobile devices. It has been the most increasing mobile service in recent years.
The amounts of SMS in Asia/Pacific reach 1.9 trillion in 2009, a 15.5 percent
increase from 2008, according to Gartner, Inc. In 2010, SMS volumes are fore-
casted to surpass 2.1 Trillion, a 12.7 percent increase from 2009[1]. More users
more often use SMS make SMS based information extraction and user behavior
recognition play a key role to enhance the user experience in the mobile communication. Naturally it is the key enabling technique for creating key business
value due to its huge amount of user base and high frequency of the usage.

Information abstraction has paid more attentions on recognition of named
entities recognition (NER) including person names, location names, organization
names, as well as date, time, and so forth. NEs, consisting of a large of useful
information in SMSs, is helpful on answering wh-questions such as who, when and
where. Besides of this kind of information, the activity that normally expresses
the users intention is also useful information in short message. Activities are
normally described by verbs in sentences.

Some applications need both of NE and activity information in SMS simultaneously. Event abstraction is a typical example. The task wants to find what is

C.J. Hopfe et al. (Eds.): NLDB 2010, LNCS 6177, pp. 149–156, 2010.

the activity, who is the participants of the activity, when and where the activity happens. For abstraction of both NE and activity information in SMS, the cascaded employment of NER and POS tagger is a natural choice. There are lots of choices. Models like maximum entropy (ME), conditional random fields (CRF), support vector machine (SVM) and hidden Markov model (HMM) have been widely used for NER [2–5, 9]. A Part-Of-Speech (POS) tagger can help to identify verbs as labels of activities. First-order HMM is commonly applied for POS tagging and performs very well[6, 7]. Therefore, a solution to event abstraction is to use a NER model and a POS model in a cascaded way. Some works have also been done to integrate NER and POS based on a first-order cascaded HMM[13]. There are several problems in above cascaded way of extracting NEs and verbs separately. First, the computational complexity of the approach is very high, particularly for implementation in mobile devices. Its employment of at least two statistical models like HMMs takes much memory footprint and computational power. Secondly, a POS tagger will label all POSs (at least 26 POSs in Chinese POS tagging task), but only the verb-related tags are needed in our task. Thirdly, except for cascaded HMM, since the two types of information are sequentially identified one-by-one, normally, POS taggers will use information from labeled by NER, therefore, the accumulation of errors is unavoidable.

In this paper, we propose an integrated second-order HMM which can label NEs and verbs simultaneously. It can also dramatically reduce the computational complexity compared to the cascaded approach. It is shown the promising performance in the following experiments. The complexity of our second-order HMM is low. It can be easily implemented even in mobile devices.

The second-order HMM can be backed off to first-order HMM, and then use Viterbi algorithm to decode[8]. But decoding in this way cannot fully utilize second-order HMMs capability of making use of more context, which brings higher performance. In this paper, to decode second-order HMM, we use a second-order Viterbi algorithm similar to [9]. It first uses forward dynamic programming (DP) to create word lattice[10], which is powerful and convenient to store all paths' information. And then it searches the best path backward.

The rest sections of the paper are organized as follows. A second-order integrated HMM is described in Sect. 2. Section 3 presents the decoding algorithm of the model. The experimental results and discussion are described in Sect. 4. The conclusions are drawn in Sect. 5.

2   Second-Order Integrated HMM

Identification of NEs and verbs simultaneously in a sentence can be also viewed as a problem of sequence labeling. We first define a tag set as $TS = \{$B_Nh, I_Nh, B_Ns, I_Ns, B_Nt, I_Nt, B_Vb, I_Vb, O$\}$, shown in Table 1. It include 9 tags which can help to abstract person name (Nh), location name (Ns), organization name (Nt) and verb (Vb). It is worth to point out that these BIO tags in our model are

Table 1. The tag set for second-order HMM

Label	Explanation
B_Nh	beginning word of a person name
I_Nh	inner word of a person name
B_Ns	beginning word of a location name
I_Ns	inner word of a location name
B_Nt	beginning word of a organization name
I_Nt	inner word of a organization name
B_Vb	beginning word of a verb
I_Vb	inner word of a verb
O	other word

applied on Chinese characters, for the experiments in this paper are conducted on Chinese which is a character-based language.

For a given sentence, $X = x_1 x_2 \ldots x_T$, where $x_i, i = 1, \ldots, T$ is a Chinese character. Suppose the label sequence of X is $S = s_1 s_2 \ldots s_T$, where $s_i \in TS$ is the tag of x_i. Then what we want to find is an optimal tag sequence S^* which is defined in (1).

$$S^* = \arg\max_S P(S|X) \propto \arg\max_S P(X|S)P(S) \tag{1}$$

This is the standard HMM. Where, first-order Markov assumption is generally used in both emission probability and transition probability. In this paper, we keep the first-order Markov assumption in emission probability as shown in (2).

$$P(X|S) = \prod_T^{t=1} p(x_t|s_t) \tag{2}$$

For transition probability, since first-order model cannot give satisfied results, second-order Markov model is used to estimate probability of the sentence as described in (3).

$$P(S) = \prod_{t=1}^{T} p(s_t|s_{t-2}s_{t_1}) \tag{3}$$

Combing with (2) and (3), (1) can be rewritten as (4) where negative logrithm is also applied.

$$S^* = \arg\min_S (\sum_{t=1}^{T} -\log p(s_t|s_{t-2}s_{t-1}) + \sum_{t=1}^{T} -\log p(x_t|s_t)) \tag{4}$$

For a normally used first-order HMM, Viterbi algorithm is well ready for its decoding. But for above second-order HMM, the standard Viterbi algorithm cannot be used directly. Next section presents an efficient decoding algorithm.

3   Decoding of Second-Order Integrated HMM

We use a second-order Viterbi algorithm to decode the second-order HMM. Firstly, a forward DP is used to create a lattice of tags for each input sentence. Then, a backward search process is carried to find the best tag sequence according to the labels marked in the forward DP.

We add s_0 at the beginning ($t = 0$) of S and s_e at the end ($t = T + 1$) of S. Without loss of generality, we assume each time we have m nodes.

For descriing the forward DP, we define following notations:

We denote the distance of a path which begins from s_0, passes node j at $t - 1$ and ends at node k at t by $dis_t^{j,k}$. For calculating it in second-order HMM, we need $dis_t^{i,j,k}$ which denotes the distance of a path which begins from s_0, passes node i and j at $t - 2$ and $t - 1$ separately, and ends at node k at t. Then we can calculate $dis_t^{j,k}$ as shown in (5) and (6).

$$dis_t^{i,j,k} = dis_{t-1}^{i,j} - \log p(tag_t^k | tag_{t-1}^j, tag_{t-2}^i) - \log p(x_t | tag_t^k) \qquad (5)$$

$$dis_t^{j,k} = \min_i dis_t^{i,j,k} \qquad (6)$$

$$L_t^{j,k} = \arg\min_i dis_t^{i,j,k} \qquad (7)$$

At the same time, we use $L_t^{j,k}$ (given in (7)) to mark the position of node in time $t - 2$ of the path which is the minimum. It is shown in Fig.1.

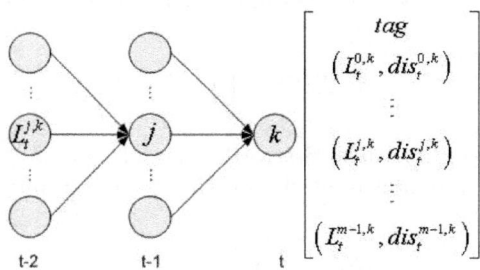

Fig. 1. The left part shows the path reached the k node in current time t through the $L_t^{j,k}$ node in time $t - 2$ and j node in time $t - 1$; the right represents the content stored in the k node in time t and $dis_t^{j,k}$ is the distance of the path described in the left

When $t = 1$, we initialize the nodes by:

$$dis_{t=1}^{0,k} = -\log p(tag_1^k | s_0) - \log p(x_1 | tag_1^k \qquad (8)$$

$$L_{t=1}^{0,k} = 0 \qquad (9)$$

When $t = T + 1$, the forward process ends at node s_e. Equations (6) and (7) can be rewritten as:

$$dis_{T+1}^{j,e} = \min_i dis_T^{i,j} \tag{10}$$

$$L_{T+1}^{j,e} = \arg\min_i dis_T^{i,j} \tag{11}$$

The process of forward DP can be divided to the following 3 steps:

1. Initialization at $t = 14$. For node $k(k = 1, \ldots, m - 1)$, we initial the node by (8) and (9).
2. Recursion at $t = 1, \ldots, T$. For node $k(k = 1, \ldots, m - 1)$, we calculate the distances reached to the current node by (5), (6) and (7).
3. Termination at $t = T + 1$. For node s_e, we get the minimum distance in all paths calculated by (10) and (11).

After building lattice in forward process, we search backward the best path in the lattice. we define a triples $[p, p\_1, p\_2]$ as middle variables to record. Beginning at node s_e, we get the position j at time $t - 1 = T$ and the position $L_{T+1}^{j,e}$ at time $t - 2 = T - 1$. We assume $p\_1 = j$, $p\_2 = L_{T+1}^{j,e}$. Then we set $p = p\_1$, $p\_1 = p\_2$. And searching the node p of time $t = t - 1 = T$, we can get the position $p\_2 = L_t^{p-1,p}$, iteratively until $p = 1$. In this searching process, we record the tag of each node we passed. By reversing the tags sequence, it is the optimal tag sequence S^*.

4   Experiments

In our experiments, we first build a cascaded baseline model (denoted by CM) where NEs are extracted by a NER and only verbs are extracted by a POS tagger sequently. Then the performance of the two-order HMM is compared with the CM. We also compare second-order HMM with first-order HMM (denoted by FHMM).

The experiments were carried out on 1000 (27064 Chinese characters) pieces of manually labeled short messages, five-fold cross validation is used for each experiment. The results are presented in Table 2.

Comparing the results of different models shown in Table 2, it can easily be seen that the second-order HMM outperforms CM. Actually it improves correct rate, recall rate and f-value at 8.58% 0.91% and 3.31%, respectively. And the computation cost in integrated model is nearly half of the time cost and lower a quarter memory cost than using NER and POS tagger respectively. We can also see that the time and memory cost of SHMM model is almost the same low with the FHMM model.

Comparing the results of FHMM and SHMM in Table 2, it can be seen that first-order model outperforms second-order model. Considering the corpus of short messages we used is too small, we think the poor result of second-order model may be caused partially by data sparseness. Two factors are checked. One is the smoothing, and other is the scale of corpus.

Table 2. F-values of the different models where add-one smooth used in HMM

Model	F-value	Time(s)	Memory(kB)
CM	0.5194	10	1852
SHMM	0.5525	5	459.7
FHMM	0.6341	4	456.5

We first repeat the same experiments for FHMM and SHMM using KN smoothing[14] which is more elaborate than add-one (the above experiments are all on it).

The results are shown in Table 3. Comparing Table 2 and 3, we get f-value 6.70% improvement when we use KN smoothing on second-order HMM. We can

Table 3. F-values of KN smoothing integrated models

Model	F-value
SHMM	0.6195
FHMM	0.6192

also clearly find that high order model is affected by data sparseness problem dramatically. By identifying the data sparseness problem, we consider to use big corpus in experimental design. We then repeat the same experiments for FHMM and SHMM on a big corpus. Since we have not big annotated corpus on short message, we use one part of a Chinese Peoples Daily news corpus includes two millions Chinese characters as training data, and test on another part of the corpus includes 122141 characters. The result shown in Table 4 indicates that high order model outperforms the low order model when the corpus is big.

Table 4. Second-order integrated model using add-one smoothing

Model	Second-order	First-order
F-value	0.6966	0.6811

From above experiments, we find second-order HMM for SMS is benefit from deliberately chosen of smoothing and big corpus. Since we have no big SMS corpus, to make use of the benefit from big Chinese Peoples Daily news corpus, we therefore consider the linear integration estimation of parameters by (12):

$$P(S|X) = \lambda P_1(S|X) + (1 - \lambda)P_2(S|X) \tag{12}$$

Where $P_1(S|X)$ is calculated on short messages corpus, $P_2(S|X)$ is on Chinese People's Daily corpus. We test the parameter on SMS set.

We can find the best f-value is at neither $\lambda = 0$ nor $\lambda = 1$. $\lambda = 0$ means that only People's Daily corpus is used. Although it achieve 0.6966 (f-value) when it is tested on data from same corpus, but it works not so good on SMS corpus

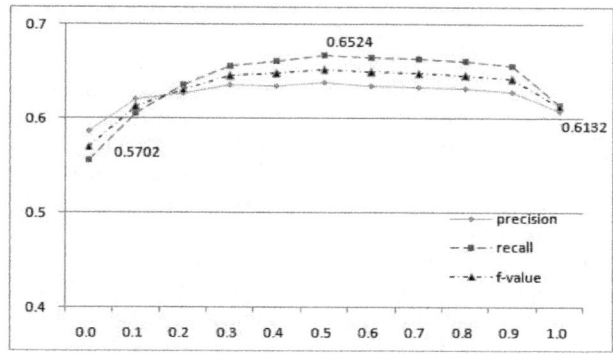

Fig. 2. x-axis is the values axis calculated in equation (12)

independently. While $\lambda = 1$ means that only SMS corpus is used, it backs to SHMM model in Fig.2. The best value 0.6524 is achieved at $\lambda = 0.5$. Comparing with results in Table 3, we can find an improvement of 3.29 points for second-order model by combining large scale of People's Daily corpus.

5 Conclusions and Future Works

This paper has presented a second-order integrated HMM and applied in event extraction from short message. It considers NEs and user activities as an unified identification task and the result shows that the integrated model outperforms than when we extract NE by NER and Verb by POS tagger separately. The new word lattice makes calculating the second-order transition probabilities in dynamic propagation easily and the second-order HMM outperforms than the first-order HMM. For the data sparseness problem, we consider using better smoothing technology and simultaneously using linear integration technology to expand the scale of corpus. As expected we get good performance.

Furthermore, the proposed integrated model significantly reduces the complexity that is able to run in the handheld device in the real time.

The word lattice build in section 3.1 has more information than we used to find the best result. We can use backward A^* algorithm[11, 12] to find top N-sequences. In the future, we will consider using more information in forward dynamic programming in building word lattice, for example, we can expand the emission probability in each node to a features function.

References

1. Ingelbrecht, N., Gupta, S.B.M., Hart, T.J., Shen, S., Sato, A.: Forecast: Mobile Messaging, Major Markets Worldwide, 2004–2013 (2009),
   http://www.gartner.com
2. Yuanyong Feng, L.S., Zhang, J.: Early Results for Chinese Named Entity Recognition Using Conditional Random Fields Model, HMM and Maximum Entropy. In: Proceeding of NLP-KE 2005, pp. 549–552 (2005)

3. Yimo Guo, H.G.: A Chinese Person Name Recognization System Based on Agent-based HMM Position Tagging Model. In: Proceedings of the 6th World Congress on Intelligent Control and Automation, pp. 4069–4072 (2006)
4. Alireza Mansouri, L.S.A., Mamat, A.: A New Fuzzy Support Vector Machine Method for Named Entity Recognition. In: Computer Science and Information Technology (ICCSIT 2008), pp. 24–28 (2008)
5. Hongping Hu, H.Z.: Chinese Named Entity Recognition with CRFs: Two Levels. In: International Conference on Computational Intelligence and Security 2, CIS 2008, vol. 2, pp. 1–6 (2008)
6. Helmut Schmid, F.L.: Estimation of conditional probabilities with decision trees and an application to fine-grained POS tagging. In: Proceedings of the 22nd International Conference on Computational Linguistics (2008)
7. Manju, K.S.S., Idicula, S.M.: Development of A Pos Tagger for Malayalam-An Experience. In: 2009 International Conference on Advances in Recent Technologies in Communication and Computing, pp. 709–713 (2009)
8. Juan, W.: Research and Application of Statistical Language Model. In: Beijing University of Posts and Telecommunications 2009, pp. 81–82 (2009)
9. Thede, S.M., Harper, M.P.: A Second-Order Hidden Markov Model for Part-of-Speech Tagging. In: Proceedings of the 37th Annual Meeting of the Association for Computational Linguistics, pp. 175–182 (1999)
10. Richard Zens, H.N.: Word Graphs for statistical Machine Translation. In: Proceedings of the ACL Workshop on Building and Using Parallel Texts, pp. 191–198 (2005)
11. Franz Josef Och, N.U., Ney, H.: An Efficient A* Search Algorithm for Statistical Machine Translation. In: Proceedings of the ACL Workshop on Data- Driven methods in Machine Translation, Toulouse, France, vol. 14, pp. 1–8 (2001)
12. Ye-Yi Wang, A.W.: Decoding algorithm in statistical machine translation. In: Proceedings of the 35th Annual Meeting of the Association for Computational Linguistics and Eighth Conference of the European Chapter of the Association for Computational Linguistics, pp. 366–372 (1997)
13. Yu, H.: Chinese Lexical Analisis and Named Entity Identification Using Hierachical Hidden Markov Model, Beijing University of Chemical Technology (2004)
14. Stanley, F., Chen, J.G.: An Empirical Study of Smoothing Techniques for Language Modeling. In: Technical Report TR-10-98, Computer Science Group, Harvard University (1998)

Goal Detection from Natural Language Queries

Yulan He

Knowledge Media Institute, The Open University
Walton Hall, Milton Keynes MK7 6AA, UK
y.he@open.ac.uk

Abstract. This paper aims to identify the communication goal(s) of a user's information-seeking query out of a finite set of within-domain goals in natural language queries. It proposes using Tree-Augmented Naive Bayes networks (TANs) for goal detection. The problem is formulated as N binary decisions, and each is performed by a TAN. Comparative study has been carried out to compare the performance with Naive Bayes, fully-connected TANs, and multi-layer neural networks. Experimental results show that TANs consistently give better results when tested on the ATIS and DARPA Communicator corpora.

Keywords: Goal detection, Tree-Augmented Naive Bayes networks (TANs), natural language query.

1 Introduction

Understanding natural language queries normally requires a two-step pipeline process, mapping a input word string onto a sequence of predefined semantic labels or concepts, and predicting users' goals based on the current semantic concepts and the current system's belief. This paper focuses on the second step, goal detection from natural language queries. Existing approaches to this problem include Bayesian Networks (BNs) [1–3] which take the mapped semantic concepts as inputs. Other methods operate directly on words instead of semantic concepts though the extension to semantic concepts is straightforward, such as vector-based methods [4, 5].

Bayesian Networks (BNs) have been used to infer information goals from users' free-text queries in information retrieval [6]. Meng *et al.* [2, 7] proposed a modified Naive Bayes networks to infer users' information goals in spoken dialogue applications and Keizer [1, 3] used more general BNs to perform the same tasks for Dutch dialogues. In these approaches, a user's input query is first transformed into a sequence of semantic concepts by a semantic parser. These concepts serve as the input to the BNs for inferring the query's communication goal.

Goal detection can also be considered as similar to topic identification in information retrieval. Users' utterances and corresponding goals are represented by vectors in a vector space. The detected goal is the one whose corresponding vector is the closest to the vector representing the incoming query. Traditional methods which are widely used in information retrieval can be used to detect goals such as latent semantic analysis [4, 8] and Kohonen's Self-Organizing Maps (SOM) [5].

When comparing the existing approaches to goal detection, BNs appears to be the most promising one due to the following reasons. First it provides a soft decision and

C.J. Hopfe et al. (Eds.): NLDB 2010, LNCS 6177, pp. 157–168, 2010.

thus multiple information goals can be detected. Second, various knowledge sources such as the semantic concepts extracted, the system's beliefs, users' beliefs etc can be integrated into one probability model and the most probable information goal can be found through belief updating.

Existing BN-based approaches to goal detection either use Naive Bayes [2] or a more general BN with network topology defined manually [1, 3]. The basic Naive Bayes classifier learns from training data the conditional probability of each semantic concept C_i given the goal G_u[1], $P(C_i|G_u)$. Classification is done by picking the goal with the highest posterior probability of G_u given the particular instance of concepts $C_1 \cdots C_n$, $P(G_u|C_1 \cdots C_n)$. The strong independence assumption made in Naive Bayes networks is that all the concepts C_i are conditionally independent given the value of the goal G_u. A natural extension of Naive Bayes Networks, Tree-Augmented Naive Bayes networks (TAN) [9], relax this conditional independence assumption by allowing dependency links between between concepts. They are however still a restricted family of Bayesian networks since the goal variable has no parents and each concept has as parent the goal variable and at most one other concept. An example of such a network is given in Figure 1 where each concept may have one augmenting edge pointing to it.

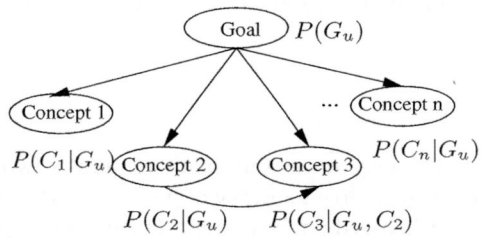

Fig. 1. Example of a Tree-Augmented Naive Bayes Network

It was reported that TAN maintains the robustness and computational complexity of Naive Bayes, and at the same time displays better accuracy [9]. Thus, this paper explores learning TAN topology automatically from data and uses the learned TAN for goal detection. The rest of the paper is organized as follows. The approach of using TAN for goal detection from natural language queries is presented in section 2 where the procedures for network topology learning, parameter learning and inference are described. The experimental setup and evaluation results using TAN networks for goal detection are presented in sections 3 and 4 respectively. Finally, section 5 concludes the paper.

2   Tree-Augmented Naive Bayes Networks for Goal Detection

A TAN network can be fully specified by its topology and conditional probability distribution at each node. Training therefore encompasses these two aspects, topology

---

[1] Capital letters such as C, G are used for variable names, and lowercase letters such as c, g are used to denote specific value taken by those variables. Sets of variables are denoted by boldface capital letters such as **C, G**, and assignments of values to the variables in these sets are denoted by boldface lowercase letters **c, g**.

learning and parameter learning, which are described in section 2.1 and 2.2 respectively. Section 2.3 presents the inference procedure in TAN.

2.1 Topology Learning

In goal detection discussed here, one TAN was used for each goal and the semantic concepts which serve as input to each TAN were selected based on the mutual information (MI) between the goal and the concept. Dependencies between concepts were then added based on the conditional mutual information (CMI) between concepts given the goal.

The procedure for building a TAN network for a goal G_u is based on the well-known Chow-Liu algorithm [10] and is summarized in Algorithm 1.

Algorithm 1. Procedure for building a TAN network.

1: Select the top n semantic concepts based on the MI value between the concept C_x and the goal G_u, $MI_{(C_x;G_u)}$, where

$$MI_{(C_x;G_u)} = \sum_{C_x,G_u} P(C_x, G_u) \cdot \log \frac{P(C_x|G_u)}{P(G_u)} \qquad (1)$$

2: Compute $CMI_{(C_x;C_y|G_u)}$ between each pair of concepts, $x \neq y$.

$$CMI_{(C_x;C_y|G_u)} = \sum_{C_x,C_y,G_u} P(C_x, C_y, G_u) \cdot \log \frac{P(C_x, C_y|G_u)}{P(C_x|G_u)P(C_y|G_u)} \qquad (2)$$

3: Build a fully connected undirected graph in which the vertices are the concepts $C_1, \cdots, C_n$. Annotate the weight of an edge connecting C_x to C_y by $CMI_{(C_x;C_y|G_u)}$.
4: Build a maximum weighted spanning tree which is a spanning tree of a weighted connected graph such that the sum of the weights of the edges in the tree is maximum.
5: Transform the resulting undirected tree to a directed one by choosing a root variable and setting the direction of all edges to be outward from it.
6: Construct a TAN model by adding a vertex labeled by G_u and adding an arc from G_u to each C_x.

In the TAN approach proposed in [9], it is assumed that all the attribute nodes (or concept nodes here) are fully connected. This may force dependencies between random variables in the network which are in fact statistically independent and thus degrades TAN network performance. Therefore, such a constraint is relaxed here. There are many ways to relax this constraint. One simple approach is only keeping the dependency links between concept nodes if the CMI between them is greater than certain threshold. A more formal approach is to add dependency links if they result in an increase of the quality measure of the network topology [11], which is described in Algorithm 2.

Let $\mathcal{D}$ be training data and S be a Bayesian network structure, a quality value that is function of the posteriori probability distribution $P(S|\mathcal{D})$ can be assigned to each network. Such values can then be used to rank Bayesian network structures, this is Bayesian quality measure. Quite a number of methods for Bayesian quality measure

---

Algorithm 2. Procedure of adding dependency links.

1: Compute $CMI_{(C_x;C_y|G_u)}$ between each pair of concepts $\{C_x, C_y\}$, $x \neq y$, add undirected edges between C_x and C_y to a graph, and annotate the weight of the edge by $CMI_{(C_x;C_y|G_u)}$.
2: Build a maximum weighted spanning tree A.
3: Sort the edges in the tree A in descending order according to the weights.
4: Construct a network B by adding the goal vertex G_u and its corresponding concepts vertices $\{C_1, C_2, \cdots, C_n\}$. Also add an arc from G_u to each C_x in the network B.
5: For each sorted edge in A, add it to the network B. Compute the network quality measure score for B given a training data set $\mathcal{D}$, $Q(B, \mathcal{D})$. If this score increases compared to the one computed for the previous network topology, then keep the edge; else, remove this edge.

---

have been proposed [12, 13]. The Bayesian Dirichlet (BD) metric [12] has been used here because it has the lowest numerical complexity [11].

$$Q(S^h, \mathcal{D}) = \log P(S^h) + \sum_{i=1}^{n} [\sum_{j=1}^{q_i} [\log \frac{\Gamma(\alpha_{ij})}{\Gamma(\alpha_{ij} + N_{ij})} + \sum_{k=1}^{r_i} \log \frac{\Gamma(\alpha_{ijk} + N_{ijk})}{\Gamma(\alpha_{ijk})}]] \quad (3)$$

where

- S^h denotes the hypothesis that the training data $\mathcal{D}$ is generated by network structure S;
- n is the total number of training cases, q_i is the number of parent configurations of variable C_i and r_i is the number of configurations of variable C_i, i.e. $q_i = \prod_{C_j \in \Pi_{C_j}} r_j$ where Π_{C_i} denotes the parents of C_j;
- N_{ijk} is the number of cases in $\mathcal{D}$ where the random variable C_i is in configuration k and its parents, Π_{C_i}, are in configuration j, N_{ij} is the number of cases in $\mathcal{D}$ where parents of the random variable C_i are in configuration j, i.e. $N_{ij} = \sum_{k=1}^{r_i} N_{ijk}$;
- α_{ijk} is the parameter of the associated Dirichlet prior of θ_{ijk}, the distribution of the variable C_i in configuration k and its parents, Π_{C_i} in configuration j, $\alpha_{ij} = \sum_{k=1}^{r_i} \alpha_{ijk}$ and $\alpha_{ijk} > 0$;
- Γ is the gamma function

$$\Gamma(z) = \int_0^{\infty} t^{z-1} e^{-t} \mathrm{d}t \quad (4)$$

Lonczos' approximation of function $\log \Gamma(z)$ [14] is used in the implementation here.

In practice, the prior network structure probability $P(S^h)$ is unknown and assumed to be constant over all possible network structures S^h.

In previous work [2, 7], the minimum description length (MDL) metric is used to learn the network topology. The MDL scoring function is in fact an approximation of the BD metric. As the number of training cases tends to infinity, MDL would give the same score as the BD metric, assuming Dirichlet distribution with uniform priors on network structures.

2.2   Parameter Learning

Consider a finite set $\mathbf{U} = C_1, \cdots, C_n, G$, where the variables $C_1, \cdots, C_n$ are the semantic concepts and G is the goal variable. Formally, a Bayesian network for $\mathbf{U}$ is a pair $B = \langle S, \Theta \rangle$, where S is a *directed acyclic graph (DAG)* (such as Figure 1) whose vertices correspond to the variables, $C_1, \cdots, C_n, G$, and whose edges represent direct dependencies between the variables, Θ represents the set of parameters that quantify the network. It contains a parameter $\theta_{c_i | \Pi_{c_i}} = P_B(c_i | \Pi_{c_i})$ for each possible value c_i of C_i and Π_{c_i} of Π_{C_i}, where Π_{C_i} denotes the set of parents of C_i in $\mathbf{U}$. Since the goal variable G is the root, $\Pi_G = \oslash$. A Bayesian network B defines a unique joint probability distribution over $\mathbf{U}$ given by

$$P_B(C_1, \cdots, C_n, G) = P(G) \prod_{i=1}^{n} P_B(C_i | \Pi_{C_i}) = P(G) \prod_{i=1}^{n} \theta_{C_i | \Pi_{C_i}} \qquad (5)$$

The TAN parameters can be estimated as $\theta(c | \Pi_c) = \hat{P}_\mathcal{D}(c | \Pi_c)$ where $\hat{P}_\mathcal{D}(\cdot)$ denotes the *empirical distribution* defined by frequencies of events in a training data set $\mathcal{D}$, namely, $\hat{P}_\mathcal{D}(A) = \frac{1}{N} \sum_j \delta_A(\mathbf{u}_j)$ for each event $A \subseteq \mathbf{u}$, where $\delta_A(\mathbf{u}) = 1$ if $\mathbf{u} \in A$ and $\delta_A(\mathbf{u}) = 0$ otherwise. Here, N is the total number of training instances. It has been shown in [9] that the above estimation leads to "overfitting" the model and is unreliable especially in small data sets due to many unseen events and the Dirichlet probability distribution is normally incorporated to estimate parameters which in fact performs a smooth operation [15]. In the context of learning Bayesian networks, a different Dirichlet prior may be used for each distribution of a random variable C_i given a particular value of its parents. Thus,

$$\theta^s(c | \Pi_c) = \frac{N \cdot \hat{P}_\mathcal{D}(\Pi_c)}{N \cdot \hat{P}_\mathcal{D}(\Pi_c) + N^0_{c | \Pi_c}} \cdot \hat{P}_\mathcal{D}(c | \Pi_c) + \frac{N^0_{c | \Pi_c}}{N \cdot \hat{P}_\mathcal{D}(\Pi_c) + N^0_{c | \Pi_c}} \cdot \hat{P}_\mathcal{D}(c) \qquad (6)$$

where N is the total number of training instances, $N^0_{c | \Pi_c}$ is the confidence associated with the prior estimate of $P(c | \Pi_c)$. This application of Dirichlet priors mainly affects the estimation in those parts of the conditional probability table that are rarely seen in the training data. In the application here, $N^0_{c | \Pi_c}$ is set to be the same for all probability distributions and its optimal value is determined experimentally.

2.3   Inference in TAN Networks

A method called *Probability Propagation in Trees of Clusters (PPTC)* [16] is used to compute the probability $P(G_u = g | \mathbf{C} = \mathbf{c})$ where g is a value of a variable G_u and $\mathbf{c}$ is an assignment of values to a set of variable $\mathbf{C}$. PPTC works in two steps, first converting a TAN network into a secondary structure called a junction tree, then computing the probabilities of interest by operating on the junction tree. The overall procedure of the PPTC algorithm is shown below:

Algorithm 3. Probability Propagation in Trees of Clusters (PPTC) algorithm.

1: *Graphical Transformation.* Transform the *directed acyclic graph (DAG)* of a Baysian network into a junction tree.

2: *Initialization.* Quantify the junction tree with belief potentials, where the potential $\Phi_\mathbf{C}$ of a set of variables $\mathbf{C}$ is defined as a function that maps each instantiation $\mathbf{c}$ into a nonnegative real number $\Phi_\mathbf{C}(\mathbf{c}) \rightarrow \mathcal{R}$. The result is an inconsistent junction tree.

3: *Global Propagation.* Perform message passing on the junction tree potentials such that all the potentials are locally consistent, which results in a consistent junction tree.

4: *Marginalization.* From the consistent junction tree, compute $P(G_u)$ for each goal G_u.

---

3   Experimental Setup

Experiments have been conducted on both the ATIS corpus and the DARPA Communicator Travel Data. Two sets of TAN networks were therefore trained on them separately.

For the ATIS corpus, altogether 4978 training utterances were selected from the Class A (context-independent) training data in the ATIS-2 and ATIS-3 corpora and both the ATIS-3 NOV93 and DEC94 datasets were used as the test sets. Each training utterance contains a goal and a set of semantic concepts. For example,

```
show flights between toronto and san francisco
Goal    : FLIGHT
Concepts: FLIGHT FROMLOC.CITY_NAME TOLOC.CITY_NAME
```

Performance was measured in terms of goal detection accuracy, which is calculated as the number of correctly detected goals divided by the total number of utterance goals. The semantic concepts of each utterance were derived semi-automatically from the SQL queries provided in ATIS. There are in total 68 distinct concepts. 16 goals were defined by enumerating the key attribute to be fetched from each SQL query.

The DARPA Communicator training set contains 12702 utterances while the test set consists of 1178 utterances. The Phoenix parse result for each utterance was transformed into a frame structure which consists of a goal and a number of semantic concepts and these frames were then manually checked and corrected. There are in total 105 distinct concepts. The Phoenix parser defines only six topics or goals for the DARPA Communicator data: *Air, Car, Hotel, Query, Respond, Return.* The goal *Air* is rather too broad so this was further divided into sub-classes such as *Date_Time, Location, Origin, Destination, Depart_Time, Return_Time* etc. After this refinement, 16 goals in total were defined which are listed in Table 1 together with the ATIS goals.

4   Experimental Results

This section presents experimental results for both the ATIS and DARPA Communicator corpora using TAN networks for goal detection. Naive Bayes networks and neural networks have also been trained and evaluated on the same data to provide baselines for performance comparison.

Table 1. Goals defined for ATIS and DARPA Communicator Data

Data Set	Goals
ATIS	abbreviation, aircraft, airfare, airline, airport, airport_distance, city, class_service, flight, flight_distance, flight_no, ground_service, location, meal, quantity, restriction
DARPA	air, airline, airport, arrive_time, car, city, date_time, depart, depart_time, destination, fare, flight_time, hotel, origin, return_time, stop

4.1 Optimal Value of Dirichlet Prior

A parameter setting to be determined is the value of the Dirichlet prior, i.e. the term $N^0_{c|\Pi_c}$ in Equation 6. As the same Dirichlet prior will be used through all probabilistic distributions, the subscript of $N^0_{c|\Pi_c}$ may be dropped and the Dirichlet prior can be simply denoted as N^0. A number of different values have been tried and the goal detection accuracy vs different settings of the Dirichlet prior is plotted in Figure 2. It can be observed that the best performance is obtained when N^0 is set to 5 for both the ATIS NOV93 and DEC94 test sets. For the DARPA Communicator data, the choice of N^0 did not affect the goal detection accuracy much. One main reason is that the total number of training instances in the Communicator data is 12702, which is much larger than the number of training cases used in ATIS, 4978. When the number of training instances is large, the bias caused by the Dirichlet prior would essentially disappear. Recall from section 2.2 that the application of Dirichlet prior mainly affects the estimation of the events that are rarely seen in the training data. Nevertheless, the detection accuracy at N^0 set to 10 is slightly better than that obtained with the smaller values of N^0, therefore, this value is chosen for all the subsequent experiments.

4.2 Comparison with Naive Bayes

TAN networks without dependencies between concepts are just Naive Bayes networks. Experiments were conducted to observe the effect of adding dependency links between

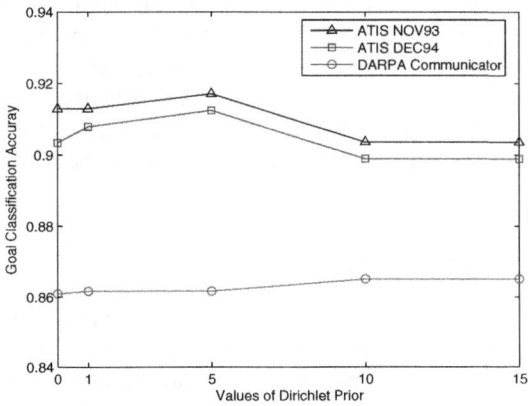

Fig. 2. Goal detection accuracy vs Dirichlet prior value

semantic concepts. Figure 3 shows the TAN errors versus the Naive Bayes errors by varying the input dimensionality of the networks. Each point represents a network input dimension setting, with its x coordinate being the goal detection error using the TAN networks and y coordinate being the goal detection error according to the Naive Bayes networks. The diagonal line in the figure denotes the locations where TAN and Naive Bayes produce the same errors. Thus points above the diagonal line mean TAN outperforms Naive Bayes. It can be observed that most points are above the diagonal line in Figure 3 (a) and (b). That is, under most circumstances, TAN outperforms Naive Bayes for both the NOV93 and DEC94 test sets.

Similar experiments were also conducted on the DARPA Communicator data. Figure 3 (c) shows that all the dots are above the diagonal line which means that TAN always outperforms Naive Bayes for the DARPA Communicator test set.

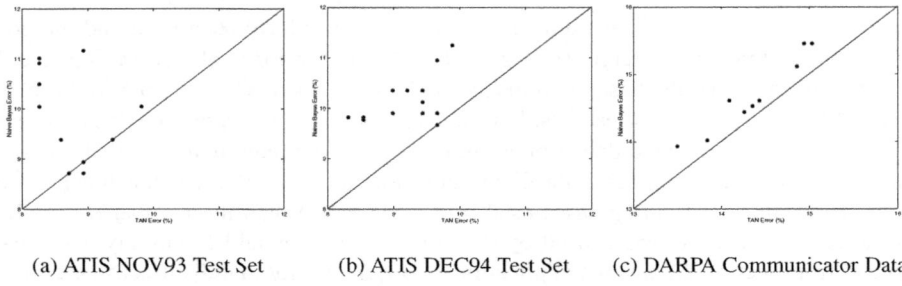

(a) ATIS NOV93 Test Set (b) ATIS DEC94 Test Set (c) DARPA Communicator Data

Fig. 3. Comparison of TAN vs Naive Bayes on goal detection error by varying network input dimensionality

4.3   Comparison with Fully-Connected TAN Networks

As mentioned earlier, all the attribute nodes in the TAN networks proposed in [9] are fully connected, such a constraint is relaxed in the work here where not all concept pairs have dependency links between them. Experiments have been conducted to compare the performance of the TAN networks built according to the procedures proposed in section 2.1 with the fully-connected TAN networks as in [9]. Figure 4 shows that generally the TAN networks without full-connected dependency links between concept nodes perform better than the fully-connected TAN networks for both the ATIS and DARPA Communicator test sets. This confirms that some concept nodes are indeed only weakly coupled and the dependency links between them should be removed to achieve better goal detection accuracy.

4.4   Comparison with Neural Networks

From the perspective of pattern recognition, neural networks can be regarded as an extension of Bayesian inference. Indeed, Bayesian methods have been applied to neural networks learning [17]. Thus, neural networks and Bayesian networks share certain similarities. It is therefore interesting to compare their performance when applied in goal detection.

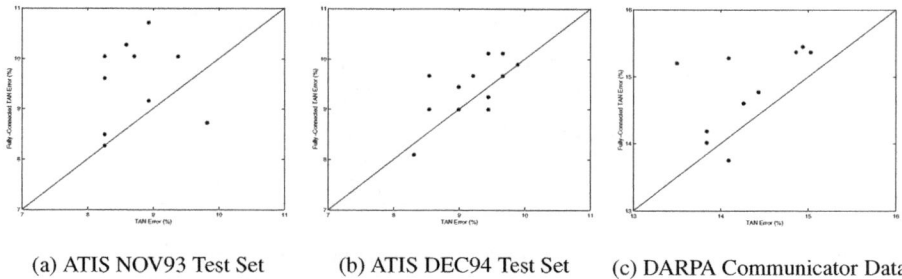

(a) ATIS NOV93 Test Set   (b) ATIS DEC94 Test Set   (c) DARPA Communicator Data

Fig. 4. Comparison of TAN vs Fully-Connected TAN on goal detection error by varying network input dimensionality

In the experimental work reported here, goal detection was performed by a single multilayer perceptron (MLP) neural network. This is in contrast to the setup used with Bayesian networks where an individual network was used for each possible goal. The number of input units to the MLP was determined by the number of distinct semantic concepts whilst the number of output units was set by the number of goals defined.

Training of the MLP network was conducted using the neural network software package, Stuttgart Neural Network Simulator (SNNS) [18] which supports a variety of learning algorithms. The original training data was divided into two parts, 90% of the data forms the training set while the remaining 10% forms the validation set. Various network topologies were tried such as one hidden layer with different numbers of units (e.g. 2, 4, 8, 16, 32, 64), or two hidden layers with different combinations of units in each layer (e.g. 2-2, 4-2, 4-4, 8-2, 8-4 etc). Learning algorithms tested include the backpropagation algorithm with or without momentum term and the quickprop algorithm. Learning parameters such as the learning rate and the momentum term were tuned to give the best configuration. For each training scenario, the validation data was used to stop the learning process if its minimum square error (MSE) saturated.

It was found that for ATIS the best results on the validation data were obtained using an MLP with one hidden layer of 16 units, trained using the standard backpropagation algorithm with the learning rate 0.2. For the DARPA Communicator data, an MLP with one hidden layer of 8 units, trained using backpropagation with momentum term, with learning rate 0.3 and momentum term 0.3 gave the best results on the validation data.

The goal detection errors of the MLP networks when tested on the ATIS NOV93, DEC94 test sets and the DARPA Communicator data test set are shown in Table 2. For comparison purposes, the test results on the same test sets using Naive Bayes, fully-connected TAN and TAN networks are also given. It can be observed that the MLP performs worse than the TAN for both the ATIS test sets. However, it gives the same result as the TAN for the DARPA Communicator test set.

4.5   Additional Context for Goal Detection

All the above experiments on goal detection were based solely on the extracted semantic concepts of the current utterance without considering the historical context of the dialogues. The DARPA Communicator data consists of a set of dialogues within which

Table 2. Goal detection accuracy using different algorithms

Algorithm	ATIS NOV93	ATIS DEC94	DARPA Test Data
Naive Bayes	91.1%	90.2%	86.1%
Fully-Connected TAN	90.9%	91.4%	83.9%
MLP	88.1%	86.9%	86.5%
TAN	91.7%	91.2%	86.5%

each utterance is a part of one conversation. Experiments were therefore conducted on the DARPA Communicator test set to measure the impact of including the historical context. Table 3 gives the results of including the semantic concepts of the immediately preceding utterance, the previously detected goal or both, based on the reference parse results and the parse results generated by the hidden vector state (HVS) semantic parser [19]. The semantic concepts extracted from the reference parse results are assumed to be all correct. The HVS parsing error for the DARPA Communicator data is 11.9%.

It can be observed that by including the semantic concepts or both goal and concepts from the previous utterance, more noise seems to have been introduced and thus goal detection accuracy degrades. This is partly due to the propagation of the previous semantic tagging errors. The analysis done here was limited to only the user side of the dialogue. It is likely that larger improvements in goal detection accuracy could be gained if the Bayesian networks were conditioned on the previous system output but this information was unfortunately not accessible for the data sets used here.

Table 3. Goal detection accuracy based on different contexts

Context	HVS Parse Results	Reference Parse Results
Current concepts	86.5%	90.2%
+ prev. concepts	83.5%	86.8%
+ prev. goal	86.3%	90.0%
+ both	83.4%	86.6%

5   Conclusion

This paper has discussed goal detection from natural language queries using Tree-Augmented Naive Bayes networks (TANs). Experiments have been conducted on the ATIS and DARPA Communicator corpora. Goal detection accuracy rates of 91.7% and 91.2% were obtained for the ATIS-3 NOV93 and DEC94 test sets respectively, which are better than the earlier results obtained using Naive Bayes for goal detection where 87.3% accuracy was obtained for the ATIS NOV93 test set [2].

Performance has been compared on the effectiveness of concept dependencies in TAN and Naive Bayes networks. It has been found that simple models such as Naive Bayes classifiers perform surprisingly well on both the ATIS and the Communicator data. Capturing dependencies between semantic concepts improves goal detection accuracy slightly. However, fully-connected TAN networks gave worse classification accuracy as more noise was introduced to the networks. This is more significant for the Communicator data.

A neural network approach, multilayer perceptron (MLP) neural network, has also been investigated. Though both MLP and TAN aim to create models which are well-matched to the data, they seem to occupy opposite extremes of data modeling spectrum. The idea behind MLP training is to find a single set of weights for the network that maximize the fit to the training data, perhaps modified by some sort of weight penalty to prevent overfitting. It can be interpreted as variations to maximum likelihood estimation. The individual relations between the input variables and the output variables are not developed by engineering judgment so that the model tends to be a black box or input/output table without analytical basis. Bayesian network learning is based on a different view in which probability is used to represent uncertainty about the relationship being learned. Prior knowledge about what the true relationship might be can be expressed in a probability distribution over the network weights that define this relationship. After presenting training data to the Bayesian networks, probabilities are redistributed in the form of posterior distributions over network weights. Therefore, unlike neural networks which are prone to overfitting, Bayesian networks automatically suppress the tendency to discover spurious structure in data [17]. Experimental results show that if there exist a large amount of training data, the neural network approach may give the same results as the Bayesian network approaches.

Finally, it would be interesting to speculate on performance with more complex tasks. The experimental results reported in this paper only focussed on individual utterance goal detection. It might be possible to incorporate top level query information such as context of the previous utterance, discourse information, intonation etc, and predict topics of the whole dialogue session. But that would require the availability of history context.

References

1. Keizer, S., Akker, R., Nijholt, A.: Dialogue act recognition with bayesian networks for Dutch dialogues. In: 3rd SIGdial Workshop on Discourse and Dialogue (2002)
2. Meng, H., Wai, C., Pieraccini, R.: The use of belief networks for mixed-initiative dialog modeling. IEEE Transactions on Speech and Audio Processing 11(6), 757–773 (2003)
3. Keizer, S., Akker, R.: Dialogue act recognition under uncertainty using Bayesian networks. Natural Language Engineering 13(04), 287–316 (2006)
4. Bellegarda, J., Silverman, K.: Natural language spoken interface control using data-driven semantic inference. IEEE Trans. on Speech and Audio Processing 11(3), 267–277 (2003)
5. Lagus, K., Kuusisto, J.: Topic identification in natural language dialogues using neural networks. In: 3rd SIGdial Workshop on Discourse and Dialogue (2002)
6. Heckerman, D., Horvitz, E.: Inferring information goals from free-text queries: a Bayesian approach. In: Proceedings of the 14th Conference on Uncertainty in Artificial Intelligence, pp. 230–237 (1998)
7. Meng, H., Siu, K.: Semi-automatic acquisition of domain-specific semantic structures. IEEE Trans. on Knowledge and Data Engineering (2001)
8. Hui, P., Lo, W., Meng, H.: Usage patterns and latent semantic analyses for task goal inference of multimodal user interactions. In: Proceedings of International Conference on Intelligent User Interfaces, IUI (2010)

9. Friedman, N., Geiger, D., Goldszmidt, M.: Bayesian network classifiers. Machine Learning 29(2), 131–163 (1997)
10. Chow, C., Liu, C.: Approximating discrete probability distributions with dependence tree. IEEE Trans. on Information Theory 14, 462–467 (1968)
11. Sacha, J.: New synthesis of Bayesian network classifiers and interpretation of cardiac SPECT images. PhD thesis, University of Toledo (1999)
12. Heckerman, D., Geiger, D., Chickering, D.: Learning Bayesian networks: the combination of knowledge and statistical data. Machine Learning 20, 197–243 (1995)
13. Castillo, E., Gutierrez, J., Hadi, A.: Expert Systems and Probabiistic Network Models. Springer, New York (1996)
14. Press, W., Flannery, B., Teukolsky, S., Vetterling, W.: Numerical Recipes in C: The Art of Scientific Computing. Cambridge University Press, Cambridge (1992)
15. Heckerman, D.: A tutorial on learning Bayesian networks. Technical Report MSR-TR-95-06 (1995)
16. Huang, C., Darwiche, A.: Inference in belief networks: A procedural guide. International Journal of Approximate Reasoning 15(3), 225–263 (1996)
17. MacKay, D.: Probable networks and plausible predictions - a review of practical bayesian methods for supervised neural networks. Network: Computation in Neural Systems 6, 469–505 (1995)
18. Zell, A., Mache, N., Huebner, R., Schmalzl, M., Sommer, T., Korb, T.: SNNS: Stuttgart Neural Network Simulator, Institute for Parallel and Distributed High Performance Systems. University of Stuttgart, Germany (1998), http://www-ra.informatik.uni-tuebingen.de/SNNS/
19. He, Y., Young, S.: Semantic processing using the hidden vector state model. Computer Speech and Language 19(1), 85–106 (2005)

Parsing Natural Language into Content for Storage and Retrieval in a Content-Addressable Memory

Roland Hausser

Universität Erlangen-Nürnberg
Abteilung Computerlinguistik (CLUE)
rrh@linguistik.uni-erlangen.de

Abstract. This paper explores the possibility of applying Database Semantic (DBS) to textual databases and the WWW. The DBS model of natural language communication is designed as an artificial cognitive agent with a hearer mode, a think mode, and a speaker mode. For the application at hand, the hearer mode is used for (i) parsing language data into sets of proplets, defined as non-recursive feature structures, which are stored in a content-addressable memory called Word Bank, and (ii) for parsing the user query into a DBS schema employed for retrieval. The think mode is used to expand the primary data activated by the query schema to a wider range of relevant secondary and tertiary data. The speaker mode is used to realize the data retrieved in the natural language of the query.

Keywords: NLP, DBS, Word Bank, Content-Addressable Memory.

1 Four Levels of Abstraction for Storing Language Data

A written text may be electronically stored at different levels of abstraction. At the lowest level, the pages may be scanned into the computer as bitmaps. This preserves the appearance of the page (which may be important, as in a medieval manuscript), but does not allow any text processing.

At the second level, the bitmap representation of the letters is transferred automatically into a digital representation (e.g., ASCII or Unicode) by means of an OCR software. The result allows text processing, such as automatic search based on letter sequences, simultaneous substitution of letter sequences, the movement of text parts, etc.

At the third level, the digital letter sequences are enriched with a markup, for example in XML (preferably in stand-off), which characterizes chapter and/or section headings, the paragraph structure, name and address of the author, bibliography, etc., depending on the kind of text, e.g., newspaper article, novel, play, dictionary, etc. As a result, the text may be printed in different styles while maintaining the encoded text structure. Furthermore, the markup may be extended to a semantic characterization of content, for example the text's domain, thus supporting retrieval.

At the fourth level of abstraction, the text is represented as content. The content is automatically derived from the letter sequence by means of a syntactic-semantic parser. The resulting output depends on the underlying linguistic theory.

C.J. Hopfe et al. (Eds.): NLDB 2010, LNCS 6177, pp. 169–176, 2010.

2   The Retrieval Problem

Today's search engines build their indices on the basis of significant letter sequences (words) occurring in the document texts. Though automatic and highly effective, such a second level approach has the drawback that the indexing based on significant word distributions is not as precise as required by some applications.

Consider a car dealer interested in the question of whether a university professor would be likely to drive a BMW. A search with Google (2009-12-06CET12:45:23) using the query *professor drives BMW* returns 454 000 sites, beginning as follows:

> Green Car Congress: *BMW* Developing Steam Assist *Drive* Based on ...
> —*Professor* Burkhard Göschel, BMW Board of Management. BMW designed the components of this drive system to fit in existing model series. ...

This and the next few hundred documents retrieved are not what the user had in mind. Leaving aside how to formulate a better query, one approach to improve recall and precision is a third level XML markup conforming to RDF [2], intended for the Semantic Web. This requires skilled work from the persons posting their documents. The alternative proposed here is an automatic fourth level derivation of content based on syntactic-semantic parsing. This method improves recall and precision by coding and utilizing the grammatical relations in the sentence sequence.

3   Data Structure of Proplets

In DBS, propositional content is coded as an order-free set of flat (non-recursive) feature structures[1] called *proplets*, serving as the abstract data type. As a simple example, consider the following representation of *Julia knows John.* as a content:

3.1   Set of Proplets Coding Content

$$
\begin{bmatrix} \text{noun: Julia} \\ \text{cat: nm} \\ \text{fnc: know} \\ \text{prn: 625} \end{bmatrix}
\begin{bmatrix} \text{verb: know} \\ \text{cat: decl} \\ \text{arg: Julia John} \\ \text{prn: 625} \end{bmatrix}
\begin{bmatrix} \text{noun: John} \\ \text{cat: nm} \\ \text{fnc: know} \\ \text{prn: 625} \end{bmatrix}
$$

In a proplet, the lexical and the compositional aspects of meaning are systematically distinguished. The lexical aspect is represented by the core value, e.g., *know*, of the core attribute specifying the part of speech, e.g., *verb*. The compositional aspect is represented by the continuation attribute(s), e.g., *arg*, and its continuation value(s), e.g., *Julia John*, which code the compositional semantic relations between proplets, namely functor-argument and coordination structure, intra- as well as extrapropositionally. The order-free proplets of a proposition are held together by a common *prn* (for proposition number) value, here *625*.

---

[1] This is in contrast to the feature structures of HPSG or LFG, which are recursive in order to model phrase structure trees, based on the principle of substitution. Cf. [6], 3.4.5.

4 Coordinate-Addressable vs. Content-Addressable Memory

The representation of content as a set of proplets raises the question of how they should be stored. The most basic choice is between a *coordinate*-addressable and a *content*-addressable memory (cf. [3] for an overview). Though peppered with patents, the content approach is much less widely used than the coordinate approach, and employed in applications for the super-fast retrieval of fixed content.

A coordinate-addressable memory resembles a modern public library in which books can be stored wherever there is space (random access) and retrieved using a separate index (inverted file) relating a primary key (e.g., author, title, year) to its location of storage (e.g., 1365). A content-addressable memory, in contrast, is like a private library in which books with certain properties are grouped together on certain shelves, ready to be browsed without the help of a separate index. For example, at Oxford University the 2 500 volumes of Sir Thomas Bodley's library from the year 1598 are still organized according to the century and the country of their origin.

As an initial reaction to a content-addressable approach, main stream database scientists usually point out that it can be simulated by the coordinate-addressable approach, using well-established relational databases. The issue here, however, is whether the formal intuitions of the content-addressable approach can be refined naturally into an efficient retrieval method with good recall and precision.

5 Structure of a Word Bank

In DBS, the storage and retrieval of a content like 3.1 uses the letter sequence of the core values for completely determining the proplets' location:

5.1 Word Bank Storing Content 3.1

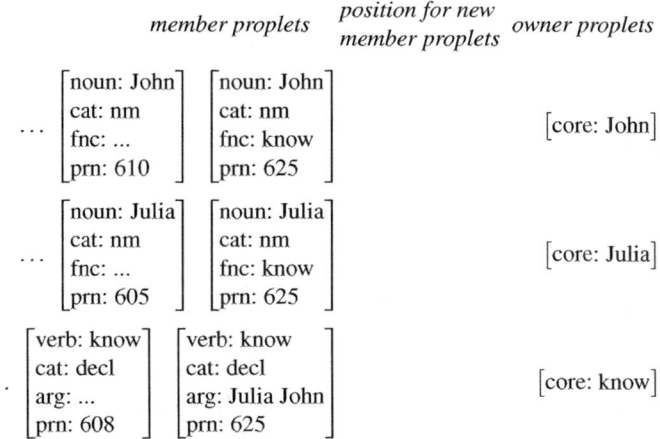

This conceptual schema resembles a classic network database with owner records and member records. It is just that the records are represented equivalently as proplets.

A sequence of member proplets followed by an owner proplet is called a *token line*. The proplets in a token line must all have the same core value and are in the temporal order of their arrival (reflected by the value of a proplet's *prn* attribute).

The sorting of proplets into a Word Bank is simple and mechanical. It is content-addressable in that no separate index (inverted file) is required. Instead, any incoming proplet is always stored in the penultimate position of the corresponding token line. Like the XML markup of the Semantic Web, such a DBS content representation may be added to the documents as standoff markup in accordance with the TEI guidelines.

A Word Bank is scalable (a property absent or problematic in some other content-addressable systems): the cost of insertion is constant, independent of the size of the stored data, and the cost of retrieving a specified proplet grows only logarithmically with the data size (external access) or is constant (internal access). External access to a proplet requires (i) its core and (ii) its *prn* value, e.g., *know 625*.[2] Most retrieval operations, however, require internal access, as in the navigation from one proplet to the next (e.g., 6.2). Because content is fixed, internal access may be based on pointers, resulting in a major speed advantage over the coordinate-addressable approach.

6   Cycle of Natural Language Communication

Having outlined the specifics of the content-addressable DBS memory, let us review the cognitive operations which are based on it. We begin with the cycle of natural language communication [1], consisting of the hearer mode, the think mode, and the speaker mode. Consider the following hearer mode derivation:

6.1   DBS Hearer-Mode Derivation of "Julia Knows John"

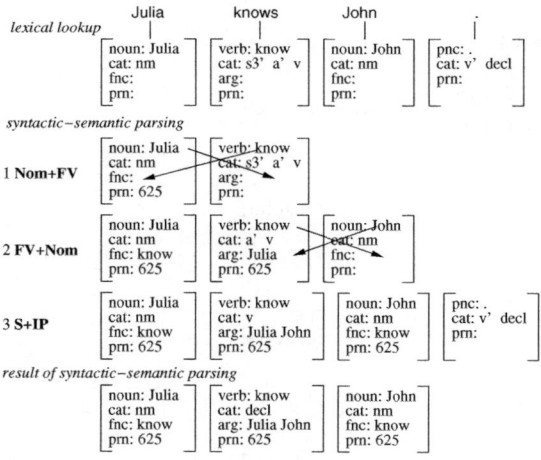

---

[2] The token line for any core value is found by using a trie structure [4]. Finding a proplet within a token line may be based on binary search ($O(log(n))$) or interpolation ($O(log(log(n)))$), where n is the length of the token line. The search uses the *prn* value of the address in relation to the strictly linear increasing *prn* values of the token line. Thus, there is no need for a hash function (which is unusual compared to most other content-addressable approaches).

The grammatical analysis is surface compositional in that each word form is analyzed as a lexical proplet (lexical lookup). The derivation is time-linear, as shown by the stair-like addition of one lexical proplet in each new line. Each line represents a derivation step, based on the application of the specified LA-hear grammar rule, e.g., *Nom+FV*. The rules establish semantic relations by copying values (as indicated by the diagonal arrows).

The result of the derivation is an order-free set of proplets, ready to be stored in the agent's content addressable memory (as shown in 5.1). Based on the semantic relations between the stored proplets, the second step in the cycle of natural language communication is a navigation which activates content selectively in the think mode:

6.2   DBS Think Mode Navigation

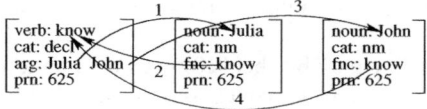

Using the *arg, fnc*, and *prn* values, the navigation proceeds from the verb to the subject noun (1), back to the verb (2), to the object noun (3), and back to the verb (4).

Such a think mode navigation provides the *what to say* for language production from stored content, while the third step in the cycle of communication, i.e., the speaker mode, provides the *how to say it* in the natural language of choice:

6.3   DBS Speaker Mode Realization

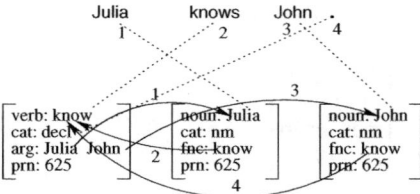

The surfaces are realized from the *goal proplet* of each navigation step, using mainly the core value. In [6], the DBS cycle of communication has been worked out in detail for more than 100 English constructions of intra- and extrapropositional functor-argument and coordination structures as well as coreference.[3]

7   Retrieving Answers to Questions

So far, the database schema of a Word Bank, i.e., ordered token lines listing connected proplets (cf. 5.1), has been shown to be suitable for (i) storage in the hearer mode (cf. 6.1) and (ii) visiting successor proplets (cf. 6.2) in the most basic kind of the think mode, with the speaker mode riding piggyback (cf. 6.3). Next we turn to another operation enabled by this database schema, namely (iii) retrieving answers to questions. This

---

[3] For a concise summary see [5].

operation is based on moving a query pattern along a token line until matching between the query pattern and a member proplet is successful.

Consider an agent thinking about girls. This means activating the corresponding token line, as in the following example:

7.1 Example of a Token Line

$$
\begin{array}{cccc}
& & \textit{member proplets} & \textit{owner proplet} \\
\begin{bmatrix} \text{noun: girl} \\ \text{fnc: walk} \\ \text{mdr: young} \\ \text{prn: 10} \end{bmatrix} &
\begin{bmatrix} \text{noun: girl} \\ \text{fnc: sleep} \\ \text{mdr: blond} \\ \text{prn: 12} \end{bmatrix} &
\begin{bmatrix} \text{noun: girl} \\ \text{fnc: eat} \\ \text{mdr: small} \\ \text{prn: 15} \end{bmatrix} &
\begin{bmatrix} \text{noun: girl} \\ \text{fnc: read} \\ \text{mdr: smart} \\ \text{prn: 19} \end{bmatrix} &
\begin{bmatrix} \text{core: } \textit{girl} \end{bmatrix}
\end{array}
$$

As indicated by the *fnc* and *mdr* values of the connected proplets (member proplets), the agent happened to observe or hear about a young girl walking, a blonde girl sleeping, a small girl eating, and a smart girl reading.

For retrieval, the content proplets of a token line may be checked by using a pattern proplet as the query; a pattern proplet has one or more variables as values. Consider the following example, in which a pattern proplet representing the query *Which girl walked?* is applied systematically to the content proplets in the token line 7.1:

7.2 Applying a Query Pattern

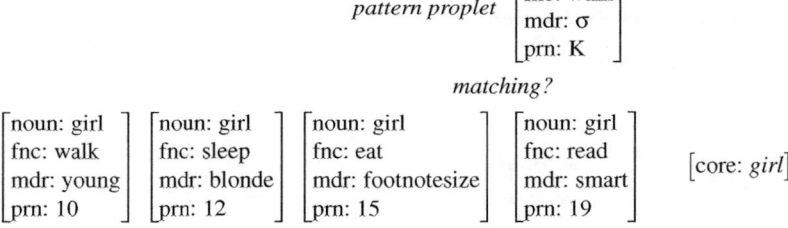

The indicated attempt at matching fails because the *fnc* values of the pattern proplet (i.e., *walk*) and of the content proplet (i.e., *read*) are incompatible. The same holds after moving the pattern proplet one content proplet to the left. Only after reaching the leftmost content proplet is the matching successful. Now the variable σ is bound to the value *young* and the variable *K* to the value *10*. Accordingly, the answer provided to the question *Which girl walked?* is *The young girl (walked)* (cf. [6], Sect. 5.1).

8 Pattern/Proplet Matching

A set of connected pattern proplets is called a DBS schema. Matching between a schema and a content is based on the matching between the schema's order-free set of pattern proplets and the content's order-free set of content proplets. Matching between an individual pattern proplet and a corresponding content proplet is based in turn on their non-recursive (flat) feature structures. Consider the following example:

8.1   Applying a Schema to a Content

$$
\begin{array}{l}
schema \\
level
\end{array}
\begin{bmatrix}
\text{noun: } \alpha \\
\text{cat: nm} \\
\text{fnc: know} \\
\text{prn: K}
\end{bmatrix}
\begin{bmatrix}
\text{verb: know} \\
\text{cat: decl} \\
\text{arg: } \alpha\ \beta \\
\text{prn: K}
\end{bmatrix}
\begin{bmatrix}
\text{noun: } \beta \\
\text{cat: nm} \\
\text{fnc: know} \\
\text{prn: K}
\end{bmatrix}
\quad
\begin{array}{l}
\textit{where } \alpha\ \varepsilon\ \{Julia,\ Suzy,\ ...\} \\
\textit{and } \beta\ \varepsilon\ \{John,\ Mary,\ Bill,\ ...\}
\end{array}
$$

matching and binding

$$
\begin{array}{l}
content \\
level
\end{array}
\begin{bmatrix}
\text{noun: Julia} \\
\text{cat: nm} \\
\text{fnc: know} \\
\text{prn: 625}
\end{bmatrix}
\begin{bmatrix}
\text{verb: know} \\
\text{cat: decl} \\
\text{arg: Julia John} \\
\text{prn: 625}
\end{bmatrix}
\begin{bmatrix}
\text{noun: John} \\
\text{cat: nm} \\
\text{fnc: know} \\
\text{prn: 625}
\end{bmatrix}
$$

For example, in the first pair of a pattern proplet and a content proplet, matching is successful (i) because they share the same set of attributes, and (ii) because the value *Julia* satisfies the restriction on the variable α. The simplicity of pattern/proplet matching is supplemented by the efficiency of finding a proplet or a set of proplets in the content-addressable memory of a Word Bank (cf. Sect. 5). For example, the yield of the schema in 8.1 is determined exactly by (i) accessing the token line of *know* (cf. 5.1), and (ii) by using the *arg* and the *prn* values of each proplet found there to access all and only propositions in which someone knows someone – resulting in practically[4] perfect recall and precision at very high speed.

9   Discussion

This brief outline of storing and retrieving level four language data in DBS raises the following questions. First, is it practically feasible to automatically parse large amounts of natural language into representations of content like 3.1? In this respect, DBS is in the same boat as competing approaches to representing content, such as truth-conditional semantics, phrase-structure analysis, and their combination. In response we point to the continuous expansion of functional completeness and data coverage in DBS, applied to a wide range of grammatical constructions and to some very different natural languages, such as Chinese, Russian, and Tagalog in addition to English, German, Bulgarian, etc.

Second, how does DBS compare with the other systems in terms of storage and retrieval? In truth-conditional semantics, content 3.1 would be represented as follows:

9.1   Content 3.1 as Logical Formula

$$
know'_{(e\backslash(e/t))}(Julia'_{(e)}, John'_{(e)}) \tag{1}
$$

This Montague-style formula characterizes functor-argument structure by means of complex categories subscripted to the items *know'*,*Julia'* and *John'*. In contrast to 3.1, which codes content as an (order-free) set of proplets at the word level, formula 9.1 applies to the sentence (proposition) level. This is because order within the formula cannot be changed without destroying either wellformedness or the original meaning. Consequently, 9.1 must be stored as a whole, which raises the question of what the primary key should be. For this, the sentence level has no obvious answer.

Similarly for content represented as a phrase or a dependency structure:

---

[4] Recall and precision are defined in terms of subjective user satisfaction. Cf. [8].

9.2 Content 3.1 as Phrase Structure

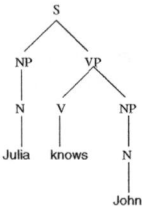

This two-dimensional representation defined in terms of the dominance and precedence of nodes represents the whole sentence structure as one unit – again raising the paradoxical question of what the primary key should be, for example for storage and retrieval in a tree-bank.

10 Conclusion

The DBS approach to practical applications of natural language processing is based on solving the most important theoretical question first. This is the question of how the mechanism of natural language communication works. It is answered in DBS by modeling the cycle of natural language communication in the form of an artificial agent with interfaces for recognition and action, and a hearer, a think, and speaker mode.

The application discussed in this paper is the storage and retrieval of language data in a textual database. For this the three most relevant properties of the overall system are (i) the efficiency of retrieval based on the pointers in a content-addressable memory, (ii) the semantic relations between proplets, implemented as the pointers on which most of the retrieval is based, and (iii) the ease of turning content into DBS schemata which provide for effective database querying based on pattern matching.

References

1. Hausser, R.: Database Semantics for natural language. Artificial Intelligence 130(1), 27–74 (2001)
2. Berners-Lee, T., Hendler, J., Lassila, O.: The Semantic Web. Scientific American 284(5), 34–43 (2001)
3. Chisvin, L., Duckworth, R.J.: Content-Addressable and Associative Memory. In: Yovits, M.C. (ed.) Advances in Computer Science, pp. 159–235. Academic Press, London (1992)
4. Fredkin, E.: Trie Memory. Communication of the ACM 3(9), 490–499 (1960)
5. Hausser, R.: Modeling Natural Language Communication in Database Semantics. In: Proceedings of the APCCM, ACS, CIPRIT, Wellington, New Zealand, vol. 96, pp. 17–26 (2009)
6. Hausser, R.: A Computational Model of Natural Language Communication. Springer, Heidelberg (2006)
7. Quillian, M.: Semantic memory. In: Minsky, M. (ed.) Semantic Information Processing, pp. 227–270. MIT Press, Cambridge (1968)
8. Salton, G.: Automatic Text Processing: The Transformation, Analysis, and Retrieval of Information by Computer. Addison-Wesley Longman Publishing Co., Inc., Boston (1989)
9. Hausser, R.: Complexity in Left-Associative Grammar. Theoretical Computer Science 106(2), 283–308 (1992)

Vague Relations in Spatial Databases

Michael J. Minock

Department of Computing Science, Umeå University
SE-901 87 Umeå, Sweden

Abstract. While qualitative relations (e.g. RCC8 relations) can readily be derived from spatial databases, a more difficult matter is the representation of vague
spatial relations such as 'near-to', 'next-to', 'between', etc. After surveying earlier
approaches, this paper proposes a method that is tractable, learnable and directly
suitable for use in natural language interfaces to spatial databases. The approach
is based on definite logic programs with contexts represented as first class objects
and supervaluation over a set of threshold parameters. Given an initial hand-built
program with open threshold parameters, a polynomial-time algorithm finds a
setting of threshold parameters that are consistent with a training corpus of vague
descriptions of scenes. The results of this algorithm may then be compiled into
view definitions which are accessed in real-time by natural language interfaces
employing normal, non-exotic query answering mechanisms.

1 Introduction

Typically spatial databases (see [1]) represent regions by adding geometric types
(POLYGON, CIRCLE, etc.) to the basic attribute types (e.g. INT, VARCHAR, DATE,
etc.). Such geometric types (e.g. as proposed in SQL/MM) have a host of well-defined
operators (e.g. *overlaps, contains, disjoint, strictly-above, does-not-extend-to-the-right-
of*, etc.) and functions (*distance, area*,etc.) that can be used as conditions or terms in
queries. The canonical types of queries are *point queries* (e.g. "what district am I currently in"), *range queries* (e.g. "what are the Chinese restaurants between 5th and 7th
avenues, 23rd and 40th streets?"), *nearest neighbors* (e.g. "Where is the nearest park?")
and *spatial joins* (e.g. "give Indian restaurants within 100 meters of a metro stop."). The
use of R-tree indexes [2] and their generalization ([3]) give log-based access to spatially
arranged objects, by indexing on the bounding rectangle of polygon regions.

While the basic spatial operators and functions have exact, highly optimized implementations, there has been less consideration given to supporting spatial predicates of
a more vague, linguistic nature (e.g. 'near-to', 'next-to', 'between' etc.) This is unfor-
tunate because for many applications, users are interested in interacting via such terms.
For example one may ask for pubs 'near to' a metro station, and descriptions of where
such pubs are could be couched in phrases like 'next to the church', 'in the park across
from the Church', etc. Of course it would be naive to reduce such vague relations to
simple qualitative predicates such as touches – there may be a small alley between the
church and the pub even though they are 'next-to' one another.

Obviously vague relations are *context dependent*. Two parks may be 'near to' one
another in the context of driving, while in the context of walking they are not. Given a

C.J. Hopfe et al. (Eds.): NLDB 2010, LNCS 6177, pp. 177–187, 2010.
© Springer-Verlag Berlin Heidelberg 2010

specific context, there might also be type dependencies. For example given the context of walking, when I say a person is 'near to' a metro station you have a vague idea of what I mean, and if I say a park is 'near to' another park, you also have a vague idea. And even in the same context the exact conditions to meet 'near to' might be different for parks, pubs, metro stations, churches, roads, etc.

These are difficult issues that have been addressed many times and in many guises, although never in a comprehensive way. To a degree this paper continues in this fragmentary tradition, but it does so from a practical stance. In short this paper shows how work in knowledge representation and logic programming can be the applied in implementable systems to support natural language interfaces to spatial databases. The approach is based on definite logic programs with contexts represented as first-class objects and a type of supervaluation over a set of threshold parameters. Given an initial hand-built program with open threshold parameters, a polynomial time algorithm finds a setting of the parameters that are consistent with a training corpus of vague spatial statements. The results of this algorithm may then be compiled into view definitions that may be integrated into normal SQL-based databases. And in turn such view definitions may be exploited by natural language interfaces employing, normal, non-exotic relational query answering mechanisms.

2   Prior Work

There has been a wealth of prior work addressing the representation of vagueness. One popular approach is the use of fuzzy logic, where objects have graded membership in predicates with a value of 0.0 standing for not being in a predicate and 1.0 standing for full membership. Thus if pedestrian A is within 1 meter of pedestrian B, then A may be said to be 'near to' B with membership 1.0, while if pedestrian A is within 100 meters of B, they may be said to be 'near to' each other with a membership 0.5. Pedestrians 1 km apart may be said to be 'near to' each other with weight 0.0. While the setting of such characteristic functions seems rather *ad hoc*, it is feasible that such functions could be properly balanced to give at least some support for the properties outlined in the introduction.

While many researchers have pursued such approaches [4–6], it should be remarked that most fuzzy approaches have focussed on what could be called *indeterminate* spatial relations [7] of fuzzy objects, rather than on vague relations that obtain between crisp objects. An indeterminate relation considers an object that has unknown shape, such as a river that may widen during rainy season or a forest whose boundary may taper off into a clearing. While such objects have unknown exact shape, there does exist some shape and thus, in principle, there are answers to basic spatial predicates (e.g. RCC8 relations[1]) and functions (e.g. distance) evaluated over them. A common way to represent such indeterminate regions is through the so-called *egg yoke* method [9].

---

[1] The well known region connection calculus (RCC8)[8] has eight exhaustive and mutually exclusive relations that may hold between regions: disconnected (DC), externally connected (EC), equal (EQ), partially overlapping (PO), tangential proper part (TPP), tangential proper part inverse (TPPi), non-tangential proper part (NTPP), and non-tangential proper part inverse (NTPPi).

The greatest possible extent of a region is represented as well as the most compact possible region and the calculus of the exact spatial predicates and functions gives rise to three-valued logics and intervals of possible values.

An alternative to fuzzy logic, and the tradition which guides the present work, is *supervaluation semantics* (see [10–14]). In its direct form, this semantics is higher-order, where lower-level predicates are parameterized by higher order variables. So for example we might state that B is nearer to A than C is to A via the statement $(\forall d)near_d(A, C) \Rightarrow near_d(A, B))$. Thus we are ranging over all different precisifications of nearness (e.g. 1-km-near, 1-meter-near, etc.) and in all cases if C is within the bound, then so is B. Hence B is nearer to A than C is to A. Such supervaluation semantics can be difficult to formalize, but through reification one may often express supervaluation in first-order formulas [15]. Related to a form of supervaluation, the work in [11] was one of the first attempts to fix parameters in rule based systems by letting domain knowledge interact with rules to give tighter bounds on intervals. Other important work is Bennett's work on VAL (Vague Adjective Logic) [15] and Pulman's work on vague predicates [13].

Of particular relevance to this work are systems that have attempted to build natural language interfaces to GIS systems. Although there have been numerous interfaces to geography databases and even to spatial databases that directly express qualitative (e.g. RCC8) relations, the interest here is on attempts to support vague natural language relations such as 'near to', 'next to' and 'between'. Surprisingly there has been very little, if any work toward this. Some recent work that flirted with this possibility is [16]. In that work the authors explicitly side step "vague, ambiguous and context dependent expressions" via a controlled language approach. Still they provide a very interesting definition of *between* that can be thought of as being on the cusp of becoming a vague relation.

3 A Concrete Example

Consider the scene depicted in figure 1. We are given complete information on the regions and types of five objects represented as:

Obj$(o_1, ((.15, .43), (.12, .5), (.17, .51)))$ $\wedge$
Obj$(o_2, ((.18, .4), (.18, .43), (.23, .43), (.18, .43)))$ $\wedge$
Obj$(o_3, ((.17, .35), (.17, .38), (.27, .37), (.27, .34)))$ $\wedge$
Obj$(o_4, ((.02, .04), (.17, .25), (.60, .21), (.4, .06)))$ $\wedge$
Obj$(o_5, ((.41, .11), (.41, .13), (.45, .13), (.45, .11)))$ $\wedge$
Metro(o_1) $\wedge$ Pub(o_2) $\wedge$ Church(o_3) $\wedge$
Park(o_4) $\wedge$ Pub(o_5).

Furthermore we have basic type hierarchy information also encoded[2]:

Building(X) $\leftarrow$ Metro(X)
Building(X) $\leftarrow$ Pub(X)
Building(X) $\leftarrow$ Church(X)

---

[2] Recall that the Horn clause $\leftarrow$ A(X) $\wedge$ B(X) means $(\forall X)(\neg A(X) \vee (\neg B(X))$.

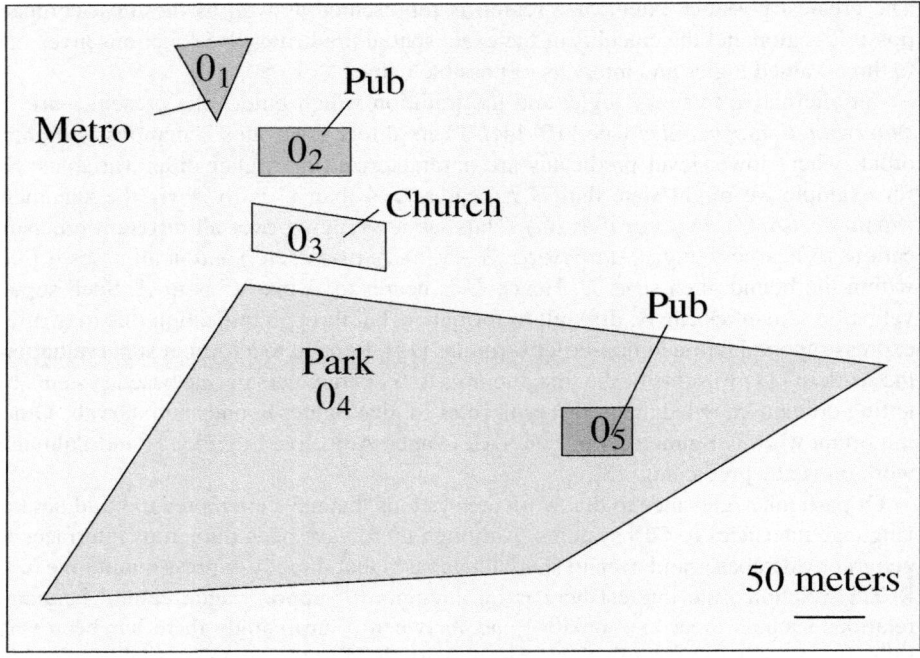

Fig. 1. A simple city scene

$\leftarrow$ Metro(X) $\wedge$ Pub(X)
$\leftarrow$ Metro(X) $\wedge$ Church(X)
$\leftarrow$ Metro(X) $\wedge$ Park(X)
$\leftarrow$ Pub(X) $\wedge$ Church(X)
$\leftarrow$ Pub(X) $\wedge$ Park(X)
$\leftarrow$ Church(X) $\wedge$ Park(X)

Based on geometric reasoning over polygon regions of objects, additional predicates and functions are defined. For example the function dist calculates the pairwise minimum distance between polygonal regions. Qualitative relations such as the RCC8 predicate *non-tangential proper part* (e.g. $\mathsf{NTPP}(reg(o_5), reg(o4))$) may also be derived. Now let us consider the parameterized definition that defines a vague predicate such as near_to.

near_to(X, Y, c_1) $\leftarrow$
      Building(X) $\wedge$ Metro(Y) $\wedge$
      dist(reg(X), reg(Y))$< n_1$

near_to(X, Y, c_1) $\leftarrow$ Park(X) $\wedge$ Metro(Y) $\wedge$
            dist(reg(X), reg(Y))$< n_2$

These definitions, using a PROLOG like syntax in which variables are capitalized, provide sufficient conditions for membership in the vague predicate in context c_1. (We will refrain from naming contexts, but you may assume c_1 is something like the context of 'walking'). Now if the context is c_1, and $n_1 = .06$ and $n_2 = .25$ then we get a reasonable definition of 'near to' that lets the first pub be 'near to' the metro station, the church not provably 'near to' the metro station, while the park is 'near to' the metro station. As a technical matter, note that the vague predicate is over objects, not regions. The function reg return an object's region.

Along with the vague predicate near_to we are also interested in its opposite, far_from. This will involve a similar type-based definition, and will involve additional parameters (n_3 and n_4). We also include the common sense predicate that objects can not be both 'near to' and 'far from' one another *within a given context*.

far_from(X,Y,c_1) ←
    Building(X) ∧ Metro(Y) ∧
    dist$($reg$(X),$reg$(Y))>n_3$

far_from(X,Y,c_1) ←
    Park(X) ∧ Metro(Y) ∧
    dist$($reg$(X),$reg$(Y))>n_4$

← near_to(X,Y,C) ∧ far_from(X,Y,C)

Continuing, let us define what two objects being 'next to' each other means in context c_1.

next_to(X,Y,c_1) ←
    near_to(X,Y,c_1) ∧
    near_to(Y,X,c_1) ∧
    DC$($reg$(X),$reg$(Y))$ ∧
    percent_visible$($reg$(X),$reg$(Y))>n_5$

Here we are using vague predicates (near_to) in the definition of a more abstract vague predicate (next_to). The function percentage_visible serves as a call to a lower level geometric routine that calculates the amount of visibility there is between two objects. Although not specified here, this may in fact implicitly take the database state as an argument to the calculation. Note also the use of DC (disconnected) one of the base RCC8 predicates over regions.

Now we extend the example even farther with between. One may argue with the particular formulation, but keep in mind that this definition applies only within the context c_1.

between(X,Y,Z,c_1) ←
    next_to(X,Y,c_1) ∧
    next_to(Y,Z,c_1) ∧
    percent_hull_overlap$($
    $($reg$(X),$reg$(Y),$reg$(Z)))> n_6$

In this definition we see the use of a function that takes three regions and calculates as an integer the percent that the convex hull determined by the first two regions overlaps with the third region. It may seem strange to include such specialized functions in a logic program. Instead one might propose to build up complex terms from more elementary functions, such as:

```
toInt(
    divide
        area(
            intersect(
                hull(reg(X),reg(Y))),
                reg(Z)),
        area(Z)) > n₆
```

As we shall see however, to get a decidable logic, we will need to impose rather serious type restrictions on functions.

3.1 Inducing Threshold Parameters

While a knowledge engineer might be able to build a knowledge base Σ of rules of the form above, they will have difficulty deciding the exact threshold parameters $n_1, n_2, \ldots$. Let us show instead how these parameters may be derived from a corpus of vague statements over scenes. We assume that such vague statements are made in some controlled or simple natural language (e.g. "the church is near to the park", etc.). These are then translated into formal expressions over the vague relations. For example the following vague statements might be associated with the scene in figure 1.

```
near_to(o₂,o₁,c₁) ∧
near_to(o₄,o₁,c₁) ∧
¬ near_to(o₂,o₅,c₁) ∧ ¬ far_from(o₂,o₅,c₁)
far_from(o₅,o₁,c₁)
```

Each scene asserts a large, though finite set of true atoms that may be derived from the spatial database. Thus in addition to the facts representing the regions and types of the five objects (see above) we have many facts such as:

```
5 < 12 ∧
DC(((.15,.43),(.12,.5),(.17,.51)),
    ((.02,.04),(.17,.25),(.60,.21),(.4,.06)))) ∧
. . .
```

Given that the i-th scene is represented with the conjunction $scene_i$ and the j-th vague statement over i-th scene is represented by the literal $vague_{i,j}$, we can represent a corpus Ω of statements over spatial databases as conjunction of rules $vague_{i,j} \leftarrow scene_i$.

Assuming that $SAT(\Sigma \wedge \Omega)$, we consider if $SAT(\Sigma \wedge \Omega \wedge n_1 = d_1)$ where d_1 is a data value constant. For example we may ask $SAT(\Sigma \wedge \Omega \wedge n_1 = 1.2)$. It isn't. Because $SAT(\Sigma \wedge \Omega)$, we will eventually find an equality constraint that is consistent (e.g. $n_1 = .06$). In this case we will add it as a fact in the knowledge base yielding Σ'. Then we may proceed onto the next constant n_2. In such a way we may find a setting for all our parameters $n_1, n_2, \ldots$. In the next section we will show that this algorithm is fixed parameter tractable.

4   Foundation

Definition 1. The vocabulary σ_D of a spatial database is $(\{o_1, ...\}, \{r_1, ...\},$ $\{.5, 1, 2, 3, ..., n_1, ...\}, \{B_1, ...\}, \{f_1, ...\}, \{g_1, ...\})$ where o_i and n_i are constant symbols, r_i is a specification of a (polygonal) region, B_i is a predicate symbol of arbitrary arity and f_i and g_i are function symbols of arbitrary arity.

As a point of nomenclature, the constant symbols may either be *data value constants* (e.g. .5,1,2,...) or they may be *parameter constants* (e.g. n_1, n_2,...).

Definition 2. A structure d over σ_D is $\langle \mathcal{O}, \mathcal{R}, \mathcal{Q}, \{o_i^d\}, \{r_i^d\}, \{n_i^d\}, \{B_i^d\}, \{f_i^d\}, \{g_i^d\}\rangle$ where $\mathcal{O}$ are objects, $\mathcal{R}$ are (polygonal) regions, $\mathcal{Q}$ are the rational numbers and:
$$o_i^d \in \mathcal{O}, r_i^d \in \mathcal{R}, n_i^d \in \mathcal{Q}$$
$$B_i^d \subseteq (\mathcal{O} \cup \mathcal{R} \cup \mathcal{Q})^k \text{ where } B_i \text{ has arity } k$$
$$f_i^d \text{ is } (\mathcal{O})^k \to \mathcal{R} \text{ where } f_i \text{ has arity } k$$
$$g_i^d \text{ is } (\mathcal{R})^k \to \mathcal{Q} \text{ where } g_i \text{ has arity } k$$

Definition 3. The vocabulary σ_V of vague relations is $(\{o_1, ...\}, \{c_1, ...\}, \{P_1, ...\})$, where o_i, and c_i are constant symbols and P_i is a predicate symbol of arbitrary arity.

Definition 4. A structure v over σ_V is $\langle \mathcal{O}, \mathcal{C}, \{o_i^v\}, \{c_i^v\}, \{P_i^v\}\rangle$ where $\mathcal{O}$ are objects, $\mathcal{C}$ are contexts and $P_i^v \subseteq (\mathcal{O} \cup \mathcal{C})^k$ where P_i has arity k

We denote $\sigma = \sigma_d \cup \sigma_v$ as the combined vocabulary of the spatial database and vague relations only when $\{o_1, ...\}$ in σ_d is exactly $\{o_1, ...\}$ in σ_v. Likewise a combined structure $u = d \cup v$ over the combined vocabulary σ is meaningful only when $o_i^d = o_i^v$ for all i.

The syntax and the semantics of the logic over σ are standard, defined in the normal way using connectors, variables, and quantifiers and the constants, functions, predicates of σ. The only special syntax is the specification of region constants which are denoted above as sequences of rational valued coordinates. In the discussion below we also assume the standard definitions of atoms (i.e. expressions $P_i(term_1, ..., term_n)$ or $B_i(term_1, ..., term_n)$), literals (i.e. atoms or their negation) and clauses (i.e. disjuncts of literals). Note that an $atom^V$ denotes an atom over a predicate in σ_V.

Definition 5. An admissible program $\mathcal{P}$ is a finite conjunction of universally quantified sentences of rules, integrity constraints and facts where:

A *rule* is: $atom^V \leftarrow atom_{i_1} \wedge ... \wedge atom_{i_n}$
An *integrity constraint* is $\leftarrow atom_{i_j}^V \wedge atom_{i_k}^V$
A *fact* is $literal_{i_j}$

Lemma 1. The Herbrand base of an admissible program $\mathcal{P}$ is finite

Proof
There are a finite number of constant symbols used in $\mathcal{P}$ and the type signatures on functions f and g restrict terms to nesting depth of two. Thus for a program of n constants, f f-functions and g g-functions where k is the maximum function arity, we have a Herbrand universe of terms of the size $O(f \cdot (g \cdot n^k)^k)$. Assuming the k is also the maximum predicate arity and we have p predicates, then we have a Herbrand universe (i.e. all the possible atoms) the size $O(p \cdot (f \cdot (g \cdot n^k)^k)^k)$.   □

Lemma 2. If $\mathcal{P}$ is an admissible program with n constants, m literals f f-functions and g g-functions where k is the maximum function or predicate arity, then $\mathcal{P}$ may be grounded to an equivalent propositional Horn clause program using $O(m \cdot (f \cdot (g \cdot n^k)^k)^k)$ propositional literals

Proof
Rule, integrity constraint and fact clauses can be written out as a series of grounded clauses that preserve the Horn property (i.e. they are a clause with at most one positive literal). □

Lemma 3. Given t objects and b base relations, f f-functions and g g-functions, a spatial database may be described with $O(t \cdot b(f \cdot (g \cdot n^k)^k)^k)$ facts.

Proof
A simple observation. □

Theorem 1. Checking the consistency of an admissible program is fixed parameter tractable given a maximum arity k for function or predicate symbols in σ.

Proof
A simple consequence of Lemma 1, Lemma 2 and Lemma 3. □

Finally we turn to the problem of finding a consistent set of equality statements between parameter constants and data value constants in a admissible program $\mathcal{P}$. Assume that $\mathcal{P}$ has m parameter constants $n_1,..,n_m$. This means that there exists an assignment $n_1 = d_i$ where d_i is a data value constant. If there are d distinct data value constants, then we can determine a constraint $n_1 = d_i$ through a $O(d)$ calls to a SAT oracle over consistency checking of admissible programs. Given this, once we find a consistent constraint, we can add it to $\mathcal{P}$ and continue on to the next parameter n_2. And so forth. Through this procedure we can obtain a consistent setting of $n_1,..,n_m$ in $O(d \cdot m)$ calls to an SAT oracle over consistency checking of admissible programs.

5   A Proof of Concept Prototype

A prototype for computing parameter constant assignments in admissible programs has been implemented. While the proof of decidability and fixed-parameter tractability is based on a propositional grounding, the prototype, written in LISP, uses the first order resolution theorem prover SPASS [17].

Given the knowledge base and a corpus of scene, vague statement pairs, the first operation is to extract all the numerical data constants and to build a formula expressing the unique names assumption (UNA) between all pairs as well as to assert all the initial $<$ and $>$ facts and the formulas:

$$X{<}Z \leftarrow X{<}Y \wedge Y{<}Z$$
$$X{>}Z \leftarrow X{>}Y \wedge Y{>}Z$$
$$\leftarrow X{<}Y \wedge Y{<}X$$
$$\leftarrow X{>}Y \wedge Y{>}X$$

Then we build a formula expressing the rule base and rules for each scene, vague predicate pair. All of these formulas are conjoined together and given to SPASS to see if a contradiction can be found. If none can be found then the knowledge-base and corpus are consistent, and a series of repeated SAT problems are presented to SPASS to derive a consistent setting of the threshold parameters. Once these parameters are found, the parameter settings are added as facts to the original rule base and a routine translates the rules into SQL based view definitions. For the simple city example developed above this results in an expression such as:

```
SELECT INTO NEAR_TO(X,Y,C)
  (SELECT XID,YID,c1
   FROM METRO,BUILDING,
      OBJECT AS X,OBJECT AS Y
   WHERE DISTANCE(X.Region,Y.Region)<.06
   BUILDING.ID = X.ID AND
   METRO.ID = Y.ID)
  UNION
    . . .
```

This view based definition of NEAR_TO(X,Y,C) that can be accessed via natural language through the natural language interface system C-Phrase [18].

Remarkably, the derivation of parameters for the example presented in this paper is solved in sub-second time on a PC with rather humble computational resources. Larger problems have also been solved quickly. The prospect that the approach will scale up to very large problems seems quite likely. Again one must remember that even if the compiling phase takes hours on powerful computers, the resulting view expressions can be integrated into very large spatial databases for rapid real-time query processing.

6   Discussion

Much of the linguistic literature explains vague spatial reference based on economy of expression; we must be vague, because a spatially precise language would be to verbose. The decision of how we cut up space into prepositions is based on affordances [19]. A well known example concerns when an object is said to be 'in' a person's hand. This is true when the object would not, on its own, roll out of their hand. Another example is that a person is said to be 'in' a car even if their head is sticking out the window; the car represents a shell that essentially contains them. While it is the province of knowledge representation to go about defining such conditions, the position here (inspired by [11]) is that such definitions will always be subject to numerical thresholds that will be very difficult to decide *a priori*. Indeed such numerical parameters should be derived from observation. The approach developed in this paper is consistent with this outlook.

An interesting question surrounds how a knowledge engineer is to construct consistent programs. In the extension of the program presented above, the engineer might include the following rule:

$$near\_to(X, Y, c_1) \leftarrow$$
$$Pub(X) \wedge Metro(Y) \wedge$$
$$dist(reg(X), reg(Y)) < n_8$$

This rule has the potential to cause problems in the knowledge base because a given pub and metro station may find their way to membership in `near_to` via this rule or its generalization presented earlier. This constrains consistent corpora considerably. One way to avoid this type of problem is to stipulate that the bodies of two rules must be pairwise disjoint under the substitution of the unifications of their heads. Interestingly, the fact that the base qualitative relations (e.g. RCC8) are pairwise disjoint is especially virtuous in this regard. It is immediately the case that bodies using different RCC relations to constraint the variables X and Y are disjoint. In short, the use of RCC8 relations, leads to more consistent programs.

Even if our programs are properly designed, let us consider how we will deal with the likely case of large corpora being inconsistent with a program. This is a tough problem, but at an algorithmic level it is not intractable to find some locally maximal subset of the corpus which is consistent with the program. Of course much work remains to be done to guide which type of parameter settings are preferable and in such a case this may determine which consistent subset of the corpus to focus on in the case of inconsistency.

Another issue that has been largely passed over in the body of this paper is that of context. Certainly including first class representations of context in the logic opens up many interesting possibilities. If context is taken as being very specific (i.e. every situation is it's own context) then after setting parameters, we may consider in polynomial time whether we can merge contexts. Also the representation of context in the vague conditions can lead to interesting dialog processes in the case of unknown context. For example, to the question "where are the parks near to me?" the system might request a clarification to determine context (e.g. "Are you driving, walking, or biking?").

7   Conclusions

The need to map vague spatial descriptions to precise logical formulas over spatial databases is a problem that is quite relevant as we develop natural language interfaces for communicating with anything, anywhere. The paper has presented a scalable approach to support vague spatial relations for querying spatial databases using natural language phrases such as 'near to', 'next to' and 'between'. In doing so, it has brought to bear work in supervaluation based approaches to vagueness and treatments of context. Initial experiments have been very encouraging and efforts are underway to turn our prototype into a system.

References

1. Shekhar, S., Chawla, S.: Spatial Databases: A Tour. Prentice-Hall, Englewood Cliffs (2003)
2. Guttman, A.: R-trees: A dynamic index structure for spatial searching. In: SIGMOD 1984, June 18-21, pp. 47–57. ACM Press, New York (1984)

3. Hellerstein, J., Naughton, J., Pfeffer, A.: Generalized search trees for database systems. In: VLDB, pp. 562–573 (1995)
4. Dilo, A., de By, R.A., Stein, A.: A system of types and operators for handling vague spatial objects. International Journal of Geographical Information Science 21(4), 397–426 (2007)
5. Altman, D.: Fuzzy set theoretic approaches for handling imprecision in spatial analysis. Journal of Geographical Information System IBM 8(3), 271–289 (1994)
6. Wang, F.: Handling grammatical errors, ambiguity and impreciseness in GIS natural language queries. Transactions in GIS 7(1), 103–121 (2003)
7. Roy, A.J., Stell, J.G.: Spatial relations between indeterminate regions. Int. J. Approx. Reasoning 27(3), 205–234 (2001)
8. Randell, D., Cui, Z., Cohn, A.: A spatial logic based on regions and connection. In: KR, pp. 165–176 (1992)
9. Cohn, A.G., Gotts, N.M.: Representing spatial vagueness: A mereological approach. In: KR, pp. 230–241 (1996)
10. Fine, K.: Vagueness, truth and logic. Synthèse 30, 263–300 (1975)
11. Goyal, N., Shoham, Y.: Reasoning precisely with vague concepts. In: AAAI, pp. 426–431 (1993)
12. Santos, P., Bennett, B., Sakellariou, G.: Supervaluation semantics for an inland water feature ontology. In: IJCAI, pp. 564–569 (2005)
13. Pulman, S.: Formal and computational semantics: a case study. In: Proceedings of the Seventh International Workshop on Computational Semantics: IWCS-7, Tilburg, The Netherlands, pp. 181–196 (2007)
14. Bennett, B., Mallenby, D., Third, A.: An ontology for grounding vague geographic terms. In: FOIS, pp. 280–293 (2008)
15. Bennett, B.: A theory of vague adjectives grounded in relevant observables. In: KR, pp. 36–45 (2006)
16. Mador-Haim, S., Winter, Y., Braun, A.: Controlled language for geographical information system queries. In: Proceedings of Inference in Computational Semantics (2006)
17. Weidenbach, C., Brahm, U., Hillenbrand, T., Keen, E., Theobalt, C., Topić, D.: SPASS version 2.0. In: 18th International Conference on Automated Deduction, Kopenhagen, Denmark, pp. 275–279 (2002)
18. Minock, M.: C-phrase: A system for building robust natural language interfaces to databases. Data and Knowledge Engineering 69(3), 290–302 (2010)
19. Pinker, S.: The Stuff of Thought. Penguin (2007)

Conceptual Modeling of Online Entertainment Programming Guide for Natural Language Interface

Harry Chang

AT&T Labs – Research, Austin, TX USA
harry_chang@labs.att.com

Abstract. This paper describes a new novel approach to the conceptual modeling of text-based electronic programming guide (EPG) for broadcast TV programs by using a large text corpus constructed from the EPG metadata source. Two empirical experiments are carried out to evaluate the EPG-specific language models created using the new algorithm in context of natural language (NL) based information retrieval systems. The experimental results show the effectiveness of the algorithm for developing low-complexity concept models with high coverage for the user's language models associated with both typed and spoken queries when interacting with a NL based EPG search interface.

Keywords: Natural language modeling, linguistic properties of entertainment language, electronic programming guide, metadata harvesting.

1 Introduction

To average TV and web users, the most common form of visual representation of a TV Electronic Programming Guide (EPG) is a two-dimensional grid as illustrated in Table 1. Besides *program title*, there is *other info* visible in the EPG grid such as *subtitle, description, genres, cast members, movie rating (PG-13, R)*, and etc.

Table 1. A common visual representation of digital TV guide and an associated taxonomy

Channel # and name	5:30pm	6:00pm	...	11:30pm
1: ABC	*program title [other info]*	...		
2: BBC	...	...		
3: CBS	...	...		
4: ...	...	...		
N: ...	...	...		

(Wednesday, 12/30/2009)

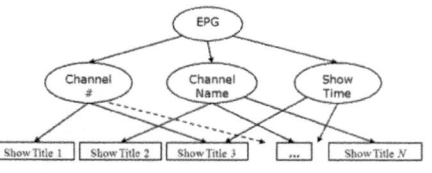

A simplified taxonomy with four concept nodes to characterize how average TV viewers to formulate their search requests

To search and browse through hundreds of channels and vast depositories of on-demand content over a rigid two-dimension grid of EPG can be tedious and frustrating using the limited capabilities of a traditional TV remote control. To overcome this challenge, researchers have been exploring the use of natural modalities such as speech

C.J. Hopfe et al. (Eds.): NLDB 2010, LNCS 6177, pp. 188–195, 2010.
© Springer-Verlag Berlin Heidelberg 2010

and pen input for media search [1], [2], and [3]. For speech input, a long term goal is to create a user-friendly Natural Language (NL) based search interface for TV-centric EPG. Our study approach focuses on constructing a set of EPG concept models with a low complexity that could improve NL-based search interface while providing a sufficient coverage for the user language model.

The conceptual modeling for information systems to support a NL-based user interface can be framed as two mutually-related NL problems. The first concerns learning and acquiring members of a given a taxonomic class constructed from a text corpus [4] and [5]. The second addresses the problems for mapping n-gram text entities to taxonomic classes using different class labeling algorithms based on domain-independent corpus such as WordNet [6] and [7] and/or domain-specific semantic parsers [8] and [9]. The paper describes a novel approach that addresses the both problems by analyzing the semantic information inherited in a highly-structured EPG text corpus incrementally built over a long period of time. By combining linguistic and statistical information extractable from the EPG metadata files, we hope to find a systemic process of generating the EPG-specific concept models to support a NL-based search interface. Section 2 discusses the method for creating a user-centered EPG taxonomy based on the EPG metadata structure and user mental models. Section 3 describes the EPG text corpus constructed for our studies and a simple algorithm to map n-gram text strings in the corpus to an ordered list of EPG concept models using the Zipf-Mandelbrot equation. Section 4 describes two experiments involving an NL-based EPG search interface, evaluating the concept recognition accuracies based on the semantic models derived from the EPG corpus. To our knowledge, there is only one prior study [10] applying the Zipf-Mandelbrot law to statistical-based modeling of usage patterns in a video search application.

2   Building Conceptual Model of EPG for NL-Based Search

In this section we describe a simple process for incrementally building a conceptual model of EPG using XML-based metadata objects and their text attributes. We also study common graphic user interface (GUI) used to present EPG information to TV consumers and gather related user data to seek a better match between our target taxonomy and the user's mental model in context of NL-based search interface.

Creating EPG Taxonomy. This process enables us to analyze and understand the semantic relationship among the metadata objects in an XML-based EPG data source. Intuitively, we expect a simple mental model to represent what average TV users may utilize in constructing their search expression at a given moment in television watching. To verify our thinking, we undertook a few informal experiments of observing how average TV users interact with a GUI-based EPG on TV and how they would interact with the same content but using a NL-based search interface. Altogether, 40 subjects with matching profiles of the average TV viewers in the U.S. were recruited to participate in the experiments. Based on our observations from the experiments, we create a simple EPG taxonomy with four high-frequency EPG concepts: *Program Titles*, *Channel Names*, *Genres* (including its descendent nodes), and various subsets related to *Person's Names*.

Semantic Analysis of EPG Concepts. After a simple lexicon analysis of the EPG metadata files for a few large TV markets in the U.S., we found that the Program Titles concept has a vocabulary of approximately 25,000 words, about 20% of the vocabulary for an EPG corpus. Yet, this vocabulary set has a highest hit rate on all other EPG concepts. Given its semantic diversity and the disproportionally-large visual space it occupies on the EPG grid, we believe that the language modeling of EPG at minimum should include Program Titles and Person's Names. The simple hypothesis is this: if these two EPG concepts can be effectively modeled with a data-driven taxonomy generated from the actual EPG metadata source, it could lead to a high-precision language model required by a NL-based search interface.

3   Language Model for TV Program Titles

3.1   EPG Corpus: EPG10

The EPG corpus used to study the language model for Program Titles is constructed from an EPG data source covering 10 U.S. cities over a 12-month period in 2008. For this paper, the corpus is referred to as the corpus EPG10. Only the *program titles* in the original EPG data source are included in the corpus. EPG10 has a vocabulary of 26,386 unique words and 47,057 unique sentences. The entire corpus contains over 86 millions of sentences instances with average sentence length as 3.1 words. Less than 20% of EPG10 sentences have more than 3 words. To minimize false positive errors with concept classifications, only true *substrings* of any EPG10 sentence are used to build n-gram data sets to be discussed in the next sections. As such, the program title "I Love Lucy" only has two bigram instances: *I Love*, and *Love Lucy*.

3.2   Linguistic Properties of TV Program Titles

One of the fundamental methods of linguistic analysis for textual content is to examine the usage of words/phrases such as their frequency f and relative rank r within a corpus-based domain. One such relationship between f and r is described by the equation (1), also known as the Zipf-Mandelbrot law [11], where c is a constant for the corpus, v is an integer between 1 to 50, and β takes a value of greater than 1.

$$f(r) = \frac{c}{(r+v)^{\beta}} \tag{1}$$

In our study the Zipf-Mandelbrot law is primarily used to segment the n-gram text strings in the EPG10 corpus by analyzing their corresponding frequency rank orders within the corpus. A simple classification algorithm is used to divide an n-gram data set into three subsets: *top-tier*, *middle-tier*, and *bottom-tier*. For the unigram data set, its three subsets are marked on the Zipf-Mandelbrot curve plotted in Figure 1. Its top-tier subset consists of those words with their ranking at $r <= 135$. As expected, many members in this top-tier of the unigram set are common prepositions words in English such as the, of, with, in, on, to, for, from, up, by, and etc.

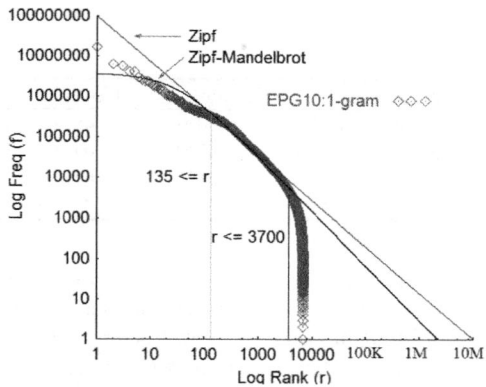

Fig. 1. Zipf-Mandelbrot curve for the unigrams of EPG10 with all unigram members divided into three groups: top-tier ($r <= 135$), middle-tier ($135 < r <= 3700$), and bottom-tier ($r > 3700$)

3.3   Finding Concept Identities for N-Gram Words

As common to many TV programs in the U.S., an n-gram entity associated with a Program Title concept may also contain words describing semantically other EPG concepts such as Actor Names. For example, *Seinfeld* is the title name of a popular TV program and the word is also the last name of the main cast. To address the ambiguity—the multiple concepts to which the same n-gram text may refer, each n-gram is assigned to one of three fussy sets depending upon their rank order within the corpus using a simple segmentation algorithm applied to the Zipf-Mandelbrot curve.

Definition. Let $F(w)$ be the corpus frequency for the n-gram words w and $f(w)$ be the equation as defined by the Zipf-Mandelbrot curve. The distance measure between the two frequency distribution curves is d which is computed as a relative deviation of predicted frequency by Zipf-Mandelbrot equation over the actual corpus frequency in percentage:

$$d = \frac{|F(w) - f(w)|}{F(w)} \qquad (2)$$

A threshold empirically derived from the distance curve is then used to determine the two boundaries between the three segments on the Zipf-Mandelbrot curve as illustrated in Figure 1. In our study, the threshold of $<= 20\%$ in the distance measure is selected. The n-gram words with their rank order in a middle-tier can be considered in a linguistic sense as the center of gravity for the Program Titles concept.

Coverage of Middle-tier. For the unigram data set in EPG10, its mid-tier region lies at $135 <= r <= 3700$, as marked on the Zipf-Mandelbrot curve in Figure 1. Similarly, we compute the boundaries for other *n*-grams ($n<=4$). Table 2 lists the key characteristics of the mid-tier regions for the first four n-grams in EPG10.

Concept Models. Our ultimate goal is to map any unknown n-gram words to the EPG taxonomy that is automatically generated from daily EPG data source. As a first step,

Table 2. The mid-tier regions of f curves for the n-gram measured based on the relative deviation from the Zipf-Mandelbrot curve

n-gram	Mid-tier	Coverage
1-gram	$135 <= r <= 3700$	49.9%
2-gram	$120 <= r <= 5000$	61.4%
3-gram	$50 <= r <= 3000$	65.3%
4-gram	$80 <= r <= 1800$	47.5%

our attempt is to determine if any incoming n-gram words from a NL-based search interface belong to its corresponding middle-tier data set. If yes, its primary semantic identity is classified as a Program Titles concept. For those n-gram words in the other two tiers, additional procedures may be required to determine which EPG taxonomy class will be the best to represent them. In addition to the Program Title concept model, we also create a simple People's Name model from the EPG metadata where it contains names associated with movie Actors/Directors and/or Cast members credited to a TV show listed in the original EPG data source for EPG10

4   Experiments

Two empirical studies are conducted to study the user language models specific to an EPG-based information system that supports a NL-based search interface. The first study uses a laboratory prototype accepting spoken expressions. The second study analyzes *typed* search queries at a public web site containing EPG information.

4.1   User Language Modeling of Spoken Expressions for EPG Search

Similar to the setup reported by Wittenburg et al. [1], the users of this experiment interact with the prototype using a multimodal TV remote with speech input capability. Six users from the same city participated in this experiment using a laptop-based prototype setup in their home for a period of 3 to 5 days. The prototype system supports a NL-based speech interface that allows the participants to search a TV-like EPG displayed on the screen via their speech input.

Data Summary. A total number of 588 voice input sessions are recorded from the users. The spoken utterances are transcribed and labeled manually according to the EPG taxonomy as defined in Section 2. There are 1,677 words transcribed with an average utterance length at 2.8 words. When excluding those expressions with a carrier phrase (e.g., "*show with ...*"), the spoken queries have 2.5 words on average. Table 3 lists the primary EPG concepts that make up the users' language model.

Evaluation. Each sentence in the data set is treated as unknown n-gram text strings. Let w_i be a text string with i words, let $PT_j = \{w_j \mid f(w_j) \subset middle\text{-}tier(w_j)\}$ be the j-gram text strings with their *frequency ranking* in their corresponding middle-tier regions as defined in Section 3.3. The concept identity of w_i is classified as Program Title concept, C(Program Title), if w_i can be consumed by at least one PT_k concept model as defined in Equation (3).

Table 3. The key statistics of the users' language model for spoken queries

EPG Concepts (N=588)	#	Percent	Accumulative Coverage
Program titles	125	21.3%	21.3%
Actors/Directors/Cast	336	57.1%	78.4%
Channel Names/Numbers	57	9.7%	88.1%
Genre{Movie, Sports, etc}	34	5.8%	93.9%
other	36	6.1%	100.0%

$$c(w_i) = C(\text{Program Title}) \quad \{ \begin{array}{ll} w_i \in PT_{j=1,\dots,i} & if\ i \leq 4 \\ w_i \in PT_4 & if\ i > 4 \end{array} \tag{3}$$

If w_i is not consumed by C(Program Title), its concept identity is determined by a simplest table lookup based on the four other EPG concept models: C(Actors), C(Directors), C(Cast), and C(Channel Name/Number). Each model is simply constructed using their corresponding data fields in the original EPG metadata file. Together, the three tables for People's Name contain approximately 150,000 entries. The table for Channel Names/Numbers concept contains close to 600 entries. In addition to the n-gram middle-tier data sets, the entire EPG10 corpus is used to determine if w_i belongs to Program Titles concept. The results measured in terms of standard *precision* (the percentages of classified concept identities that were correct) and *recall* (the percentages of all sentences mapped to the EPG taxonomy tree) are shown in Table 4.

Table 4. The performance metric for the spoken data set on the three concept models

EPG Concepts (N=588)	#	Model: PT			Model: EPG10		
		Mapped	Precision	Recall	Mapped	Precision	Recall
Program titles	125	59	100%	47.2%	100	100%	80%

PN: People's Names, **CN**: Channel Names/Numbers	Models: C(PN), C(CN)			
	#	Mapped	Precision	Recall
Actors/Directors/Cast	336	320	100%	95.2%
Channel Names/Numbers	57	51	100%	89.5%

4.2 User Language Model from Text Input

This experiment studies the *typed* queries received at a website exclusively for searching the EPG for a TV service provider. The data set was collected during a 2-week period in December 2008, from an estimated user population over 10,000. This data set is referred to as the **EPG-T** corpus in this paper.

Data Summary. The EPG-T corpus contains 493,342 *typed* queries. The corpus has 37,084 unique sentences from a vocabulary of 21,111 words. The users choose short sentences (1.7 words on average) versus with 2.5 words in spoken expressions in the first experiment. A simple lexicon analysis shows that over 99% of all words in the EPG-T corpus are covered by EPG10. Table 5 contains the coverage statistics of the middle-tier for the first three n-gram data sets using the 20% threshold. The work reported by Beck et al. [10] suggests that the Zipf-Mandelbrot law can be used effectively to model the usage patterns in a very similar domain (movie titles) when r is as low as 5000. To increase recall, we decided to *expand* middle-tier regions to include those n-gram texts with r as low as 5000.

Table 5. The coverage statistics for the mid-tier regions of the first three n-gram data

n-gram	Mid-tier (20% threshold)	Coverage	*Expanded* Middle-tier	Coverage
1-gram	$23 <= r <= 190$	28.7%	$23 <= r <= 5000$	66.5%
2-gram	$11 <= r <= 320$	42.1%	$11 <= r <= 5000$	70.1%
3-gram	$25 <= r <= 60$	17.6%	$25 <= r <= 5000$	55.7%

Test Set and Evaluation. Each word string w_i in the EPG-T corpus represents a baseline test sentence for the classification algorithm. The probability for w_i to have a perfect match with an EPG10 sentence is less than 20% if $i > 4$ (Section 3.1), and it is also an outlier relative to the TSEE corpus because less than 7% of all typed queries have more than 4 words. To increase the recall, a few substrings, $w_i(j)$, of each $w_{i>3}$ is selected for the classification where $w_i(j)$ contains the first j words in w_i and $j < i$. Each w_i including its expanded subset is submitted to the same classification algorithm as defined in Equation (3). In addition to the three tables for People's Names used in the first experiment, we create a simple Sports concept model, C(Sports Team Names), based on the team names for three professional sports, *NBA*, *NFL*, and *NHL*. If a test sentence, w_i, is not mapped to any of these four concept models, it is mapped to *other* in the EPG taxonomy, no further parsing applied. Table 6 contains the performance metric of the evaluation algorithms on the EPG-T test set.

Table 6. The performance metric for the EPG-T test set using the models created for 4 specific EPG concepts. All remaining EPG-T test sentences are classified as *other*.

Test Sentences	Test set	Model: PT			Model: EPG10		
		Mapped	Precision	Recall	Mapped	Precision	Recall
Program Titles	477,520	337,082	100%	70.6%	395,169	100%	82.8%

N=82,351	Test set	Rel %	Mapped	Precision	Recall	Models
Actors/Directors/Cast	9,311	11.3	9,311	100%	100%	C(People's Names)
Channel Names/#	2,390	2.9	2,390	100%	100%	C(Channel Names/Number)
Sports(NBA,NFL,NHL)	1,262	1.5	1,262	100%	100%	C(Sports Team Names)
other	69,359	84.2		Note: *No further classification*		

Rel %: *relative percentage* to the super set (N=82,351) created from those EPG-T test sentences that have been excluded from the test set for the Program Titles concept.

The test sets for the Names, Channel, and Sports concept models are created using the entries in their corresponding tables as a selector. This explains the high recall rate on the three EP concept models. But the algorithm relying on the strict table look-up is unable to map 14.1% (69,359) of all EPG-T test sentences.

5 Conclusions

This paper describes a new approach to corpus-based conceptual modeling of EPG content for purpose of building a high-precision NL-based search interface. By applying the Zipf-Mandelbrot law to the selection of n-gram texts based on their frequency ranking in the middle-tier up to $r=5000$, the resultant Program Titles concept model is able to achieve a 100% in precision. The vocabulary size for the Program Titles concept model is 22% smaller, 20,910 words for $C(PT)$ versus 26,386 for $C(EPG10)$, while the reduction of recall is less than 12.8% absolute (70.6% vs. 82.8%). The results of the two experiments with both spoken and typed search interface confirm the hypothesis discussed in Section 2. In ongoing work, further research is required for constructing *other* low-frequency concept models beyond the four concept models studied in this paper. The segmentation algorithm based on the Zipf-Mandelbrot law also needs to be tested on a much larger EPG corpus. Finally, a larger experiment in spoken interface with more users (N>30) is needed to confirm the language coverage produced by the same efficient EPG taxonomy as described in this paper.

References

1. Writtenburg, K., Lanning, T., Schwenke, D., Shubin, H., Vetro, A.: The Prospects for Unrestricted Speech Input for TV Content Search. In: Proceedings of the Working Conference on Advanced Visual Interfaces, pp. 352–359 (2006)
2. Johnston, M., D'Haro, L.-F., Levine, M., Renger, R.: A Multimodal Interface for Access to Content in the Home. In: ACL 2007, pp. 376–383 (2007)
3. Amores, J.G., Perez, G., Manchon, P.: A Multimodal and Multilingual Dialogue System for the Home Domain. In: ACL 2007, pp. 1–4 (2007)
4. Roark, B., Charniak, E.: Noun-phrase Co-Occurrence Statistics for Semi-Automatic Semantic Lexicon Construction. In: ACL 2007, pp. 1110–1116 (1998)
5. Widdows, D., Dorow, B.: A Graph Model for Unsupervised Lexical Acquisition. In: ACL 2002, pp. 1093–1099 (2002)
6. De Boni, M.: Automated Discovery of Telic Relationships for WordNet. In: Proc. of First International Word Net Conference (2002)
7. Widdows, D.: Unsupervised Methods for Developing Taxonomies by Combining Syntactic and Statistical Information. In: Proc. of HLT/NAACL 2003, pp. 197–204 (2003)
8. Rousu, J., Shawe-Taylor, J.: Efficient Computation of Gapped Substring Kernels on Large Alphabets. J. of Machine Learning Research 6, 1323–1344 (2005)
9. Kate, R.J., Mooney, R.J.: Using String-Kernels for Learning Semantic Parsers. In: Proc. of COLING/ACL 2006, pp. 913–920 (2006)
10. Beck, A., Borst, S., Ensor, B., Esteban, J.O., Hilt, V., Rimac, I., Walid, A.: New Challenges in Content Dissemination Networks. Bell Labs Technical Journal 13(3), 5–12 (2008)
11. Mandelbrot, B.B.: An Information Theory of the Statistical Structure of Language. In: Communication Theory, pp. 503–512. Academic Press, New York (1953)

Integration of Natural Language Dialogues into the Conceptual Model of Storyboard Design

Markus Berg[1], Antje Düsterhöft[1], and Bernhard Thalheim[2]

[1] Hochschule Wismar, Germany
Department of Electrical Engineering and Computer Science
{markus.berg,antje.duesterhoeft}@hs-wismar.de
[2] Christian-Albrechts-University Kiel, Germany
Department of Computer Science and Applied Mathematics
thalheim@is.informatik.uni-kiel.de

Abstract. Web information systems are growing with regard to complexity. Speech is a way of increasing usability and facilitating interaction. Moreover it nowadays is an important factor concerning customer needs and wishes. This paper focusses on the extension of an existing conceptual model called storyboard, to support the design of natural language dialogues. It includes a short description of storyboarding and natural language dialogues. Subsequently the outcomes are evaluated, condensed and then integrated into the storyboard model. The results of this work show that only little adaptions regarding the storyboard concept are necessary and the extension of the presentation layer with a channel-dependent renderer is sufficient to be able to model natural language dialogues.

1 Introduction

A web information system (WIS) [1] is a database-backed information system that is realised and distributed over the web with user access via web browsers. Information is made available via pages including a navigation structure between them. Furthermore, there should also be operations to retrieve data from the system or to update the underlying database. The design of web information systems requires detailed information about the users, their behaviour and the possibility to use different access channels. Storyboarding is a methodology which was created to address these problems by offering an abstract conceptual model. Because usability suffers along with the complexity of those systems, an interesting approach is the usage of natural language. As language is a very efficient way of exchanging information, it helps to simplify man-machine-interaction and thus increases usability. Since the modality *speech* hasn't been regarded in storyboarding yet, this paper will concentrate on the extension of the storyboard methodology by a possibility to design natural language dialogues.

After a short overview of related work, we will introduce the topics of storyboard design [1] and natural language dialogues. Then we will show how to integrate them into the conceptual model. Finally we will give a short conclusion which also gives an insight on benefits and future work.

C.J. Hopfe et al. (Eds.): NLDB 2010, LNCS 6177, pp. 196–203, 2010.

2   Related and Previous Work

The number of publications in the field of WIS is enourmous. The ARANEUS framework [2] determines that conceptual modeling of WIS consists of content, navigation and design aspects. This results in the modeling of databases, hypertext structures and page layout. Another similar approach is OOHDM [3] [4] which is completely object oriented. It also comprises three layers: object layer, hypermedia components and interface layer. Our own work on storyboarding has been reported in [5] and [6]. Moreover we have investigated the role and use of metaphors in storyboarding in [7]. Besides conceptual modeling also aspects of natural language dialogues have to be considered. The process of the development of speech interfaces is analysed in [8]. The work in [9] describes the conceptual base for the dialogue design process. In [10] the Wizard-of-Oz methodology is used for simulating interactive systems. Besides the dialogue description, also the user has to be modeled. This is done in [11]. The usage of different modalities when accessing the web is examined in [12].

3   Storyboard Design

Storyboarding is a methodology which was created for the design of large-scale data-intensive web information systems. It is based on the abstraction layer model (ALM) [13], which consists of the following layers. The strategic layer is used to describe the system in a general way concerning the intention. It is comparable to a mission statement. The business layer concretises this information by describing stories which symbolise paths through the system. The purpose of this layer is to anticipate the behaviour of the users. In the conceptual layer the scenes in the storyboard are analyzed and integrated. The design of abstract media types supports the scenes by providing a unit which combines content and functionality. The presentation layer associates presentation options to media types. In the implementation layer physical implementation aspects like setting up database schemata, page layout and the realisation of functionality by script languages are addressed. Each layer is associated with specific modeling tasks which allow the transition between the layers. To progress from the strategic to the business layer storyboarding and user profiling is required. To get from the business layer to the conceptual layer conceptual models have to be created, i.e. database modeling, operations modeling, view modeling and media type modeling. The transition to the presentation layer is characterised by the definition of presentation styles. In the implementation layer all implementation tasks have to be realised.

Storyboarding focusses on the business and the conceptual layer [13]. The business layer deals with user profiling and the design of the application story. The core of the story space can be expressed by a directed multi-graph, in which the vertices represent scenes and the edges actions by the users including navigation. If more details are added, application stories can be expressed by some form of process algebra. That is, we need atomic activities and constructors for sequencing, parallelism, choice, iteration, etc. to write stories. In the conceptual layer media types which support the scenes and operations which support the activities in the storyboard are modeled. Moreover hierarchical presentations, and the adaptivity to users, end-devices and channels are addressed in this layer.

WIS can be used by any web user. That's why the design of such systems requires anticipation of the user's behaviour. This problem is addressed by storyboarding. It describes the ways users may choose to interact with the system. A storyboard consists of three parts [13]: The *stories*, which are navigation paths through the system, the *actors* which comprise users with the same profile and *tasks* which link activities (resp. goals) of the actors with the story space which is a container for the description of the stories. Subgraphs of the story space are called *scenarios*. This enables a hierarchy and encapsulation of scenes. Every action can be equipped with pre- and postconditions or triggering events. This allows us to specify under which conditions an action can be executed.

SiteLang is a language which defines a story algebra and allows the formal representation of the theoretical storyboard model. The explanation of the SiteLang syntax is beyond the scope of this paper and can be read in [14].

4   Natural Language Dialogues

Natural language is the most effective form of human communication. People are used to it and don't need to learn it especially for interacting with a computer as they need it either way. That's why interacting with machines in a natural way is an ambitious and sensible goal. There is no need to learn new user interfaces when it is possible to interact with the machine like with man. There are many domains where speech applications increase usability (e.g. navigation systems), productivity (e.g. warehouse-systems) or raise life quality (e.g. for visually handicapped people). But along with all the advantages concerning ease of use there are of course drawbacks. Natural language is an unsharp, fuzzy, interpretable and context-sensitive way of expressing information. Apart from technical context information like physical states in the world model, also the mental model of the user has to be considered. Dependent on experience and emotions, different users have a different understanding of statements. We often learn a discrepancy between meaning and saying which often results in the phrase *"that's what I meant"*. This gap has to be filled by intelligent questions and contextual analysis.

A dialogue is described as the interaction between two people. Because the process of speaking does not base on abstract mathematical rules, models have to be found to describe a dialogue. To accomplish this, the parts of a dialogue have to be identified. For the exchange of information, outputs and inputs are needed. This leads to common action-response-systems which resemble the IPO principle as described in figure 1. The input symbolises a question or a command whereas the output is the answer. A sequence of questions Q and answers A leads to a dialogue D. The knowledge of recent dialogue steps is called discourse or dialogue context C.

$$D = (I + A)^* \ where \ I = Q + C_{(0,n-1)} \tag{1}$$

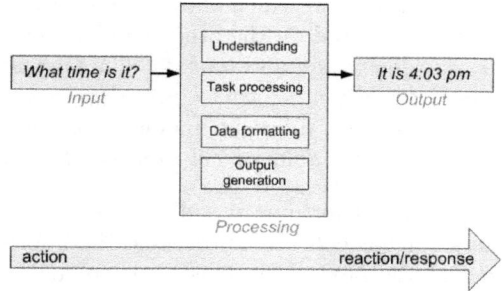

Fig. 1. IPO principle in natural language processing

In natural language it has to be considered that the dialogue flow and the initiative is not fixed. So we have to slightly adopt the model to allow succeeding questions or questions without answers.

Example 1.
A: "How many nights do you want to book in our hotel?"
B: "So, how much is one night?"

Example 2.
A: "How many nights do you want to book in our hotel?"
A: "I can't hear you. How long do you want to stay?"

As you can see, the sequence (QA)* has to be changed to Q*A*. Now we are flexible enough to model every dialogue flow. It has to be kept in mind that the input does not only occur in form of a question, but can also appear as a command: *"Please say the time"* vs. *"What time is it?"*. But regardless of the syntactic form, the meaning is the same and therefore questions and commands can be summarised as *inputs*. Outputs can be divided into questions (re-requests) and answers. This allows us to generalise the detailed view of dialogue parts to the IPO-principle mentioned at the beginning.

5   Integration of Natural Language Dialogues into Storyboards

As already mentioned, natural language is dynamic and unsharp, but natural and convenient. A very important requirement to create effective dialogues is the dialogue design. That's why we analyse the modeling of dialogues with storyboarding.

Modeling of WIS consists of different layers as mentioned in the first section. The integration of the dialogue logic has to be done in the business and conceptual layer as it is part of the storyboard design. The existing elements *scenario*, *scene* and *dialogue* can also be used for the design of natural language dialogues. The following example will explain the usage of these elements to model natural language dialogues with the method of storyboard design.

Example 3. Imagine an application which allows you on the one hand to inform yourself about touristic events and on the other hand to book these offers. This results in two scenarios: *information* and *booking*. The information-scenario consists of several scenes, e.g. *determine travel period* or *get number of persons*. These scenes can be divided into different dialogues: *get departure date* or *get arrival date* resp. *get number of adults* or *get number of children*.

Pre- and postconditions [15] determine if a dialogue can be accessed with the given information and if it has collected/processed the information which is required for proceeding to the next step. This leads to the following syntax: **on event if** `precondition` **do** `actions` **accept on** `postcondition`. The definition of a scene can be done as follows:

SCENE numberOfPersons	SCENE numberOfPersons
MediaObject: speechForm	MediaObject: submitForm
Actors: customer	Actors: customer
Context: channel = speech	Context: channel = web
Task: getMandatoryInformation	Task: getMandatoryInformation
Specification:	Specification:
on travelPeriodGathered	on linkClicked
if ∅	if ∅
doScene	doScene
adults=numberOfAd()	adults=numberOfAd()
children=numberOfCh()	children=numberOfCh()
accept on adults+children>0	accept on adults+children>0
else errordialogue	else errordialogue
nextScene: accommodationFeatures	nextScene: accommodationFeatures

Postconditions can be used to refer to error dialogues depending on the error type. The `nextScene` statement links to the subsequent part of the scene which only is followed if the postcondition holds. The references in `doScene` refer to dialogue objects, like the following `numberOfAdults` object, which are the most granular unit in the field of storyboarding.

DIALOGUEOBJECT **numberOfAdults**
EnabledActor: customer
if ∅
do adults=collect(numberOfAdults)
accept on adults>=0 && adults<10
else(nomatch) noMatchDialogue
else errorDialogueNoA

These examples are very simple and do not depend on each other. Of course there are situations, when succeeding dialog steps refer to previously entered information. Imagine the questions for credit card number and validation code which only make sense when having chosen credit card as payment method. This is realised with preconditions which only activate the scene when certain conditions are true.

In contrast to clicking on a link, the interaction via natural language is much more error prone. That's why error handling is crucial to natural language systems. This includes the need of helpful error messages, which allow the user to be able to correct misunderstandings and to proceed with the dialogue when a valid input has been given. Error messages can also be realized as dialogue steps.

As you can see, there is no relevant difference between the web and the speech channel, because the logic of an application is independent of the form of presentation. You still have the same dialogue flow, which is defined in the conceptual layer. The dialogue output itself is specified in the presentation layer. That's why the storyboard model (conceptual layer) can also be used for speech dialogues. You just have to change the presentation layer and provide the system with additional information, e.g. the words the user is allowed to say. With the help of this information it is possible to generate code in dependence of the channel or device. The transformation to Voice XML will be realised in the implementation layer:

DIALOGUEPRESENTATION
numberOfAdults(ID,dialogueID)
Channel: web
request: "Adults:"
input: ^{a}

_____

^{a} input validation is optional

DIALOGUEPRESENTATION
numberOfAdults(ID, dialogueID)
Channel: speech
request: random(
"Please say now the number of adults!",
"How many adults take part?")
input: builtin:digits(0..9)^a

DIALOGUEPRESENTATION
noMatch(ID, dialogueID)
Channel: speech
request: "I did not understand you."
input: $\emptyset$

_____

^{a} inherited from dialogue specification

```
<form name="ID" action="post">
Adults: <input type="text" />
<input type="submit"
  value="send" />
</form>
```

```
<form id="ID">
 <field name="answer">
 <nomatch>
   I did not understand you.
 </nomatch>
 <prompt>
  How many adults take part?
 </prompt>
 <grammar src="digits.xml#oneToNine"
   type="application/grammar+xml"/>
 </field>
</form>
```

Because the dialogue flow is specified in the conceptual layer whereas the real utterances are defined in the presentation layer, it is necessary to connect information from the different layers to create code. This is realised by unique identifiers. These link the application logic with the presentation style. As we are in the field of IVR-applications we use Voice XML as markup language. This requires the use of grammars. These grammars can be generated from the post-condition of the dialogue elements, as these declare the valid range of values. The usage of the web channel doesn't force a definition of what a user is able to say or key in. Nevertheless this description could be used in the frontend to realise an input checker (e.g. JavaScript). Although the definition of all input values is only mandatory for speech input, postconditions in the dialogue elements have to be set to guarantee a valid system state.

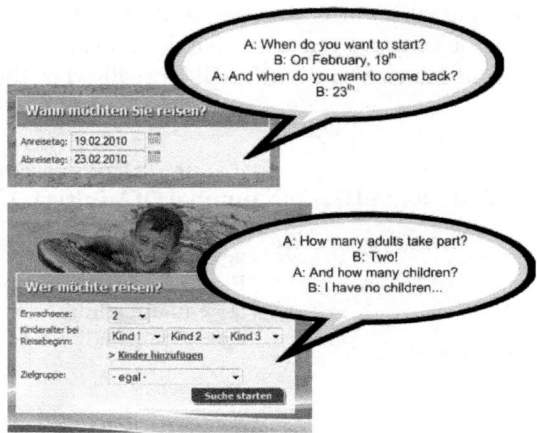

Fig. 2. Implementation Example

The concepts of storyboarding have been tested during the dialogue design for a project of our industry partner *Manet Marketing*. Figure 2 illustrates the comparison between the web and the speech channel at the example of a tourism portal.

6   Conclusion and Future Work

This paper has shown that graphical and system-initiative language-based dialogues have strong similarities. The application logic is not affected by the form of input. That's why the storyboard methodology can also be used for designing speech applications. Because of the higher error probability, a clear dialogue flow and a convenient error-handling are eminent for an efficient dialogue which is accepted by the users. For this reason, modeling of dialogues is the key for usability. The integration of new channels (like speech) takes part in the presentation layer of the ALM. In the implementation layer markup code which corresponds to the user's input device and channel is generated from information of higher-level-models.

The next step will be to increase naturalness with features like implicit correction. This requires the integration of further dialogue design strategies as well as the integration of more complex grammars and semantics. Moreover the mixing of channels has to be regarded as mobile devices offer more than one input channel. Another aspect is the extension of the model to support mixed-initiative dialogues.

Acknowledgements. This work is supported by the European Funds for Regional Development (EFRE).

References

1. Schewe, K.-D., Thalheim, B.: Conceptual modelling of web information systems. Data & Knowledge Engineering 54, 147–188 (2005)
2. Atzeni, P., Gupta, A., Sarawagi, S.: Design and maintenance of data-intensive websites. In: Schek, H.-J., Saltor, F., Ramos, I., Alonso, G. (eds.) EDBT 1998. LNCS, vol. 1377, pp. 436–450. Springer, Heidelberg (1998)
3. Rossi, G., Garrido, A., Schwabe, D.: Navigating between objects: Lessons from an object-oriented framework perspective. ACM Computing Surveys 32, 1 (2000)
4. Rossi, G., Schwabe, D., Lyardet, F.: Web application models are more than conceptual models. In: Kouloumdjian, J., Roddick, J., Chen, P.P., Embley, D.W., Liddle, S.W. (eds.) ER Workshops 1999. LNCS, vol. 1727, pp. 239–252. Springer, Heidelberg (1999)
5. Feyer, T., Thalheim, B.: E/R based scenario modeling for rapid prototyping of web information services. In: Kouloumdjian, J., Roddick, J., Chen, P.P., Embley, D.W., Liddle, S.W. (eds.) ER Workshops 1999. LNCS, vol. 1727, pp. 253–263. Springer, Heidelberg (1999)
6. Schewe, K.-D., Thalheim, B.: Integrating database and dialogue design. Knowledge and Information Systems 2 1, 1–32 (2000)
7. Thalheim, B., Düsterhöft, A.: The use of metaphorical structures for internet sites. Data & Knowledge Engineering 35, 161–180 (2000)
8. Cohen, M., et al.: Voice User Interface Design. Addison-Wesley, Reading (2004)
9. Harris, R.A.: Voice Interaction Design. In: Crafting the New Conversational Speech Systems, Morgan Kaufman Publ. Inc., San Francisco (2004)
10. Fraser, N., Gilbert, G.: Simulating Speech Systems. Computer, Speech and Language (1991)
11. Fischer, G.: User Modeling in Human-Computer Interaction. In: User Modeling and User-Adapted Interaction 11 (2001)
12. Wahlster, W.: SmartKom: Multimodal dialogues with Mobile Web Users. In: Proc. of the Cyber Assist International Symposium, pp. 33–34 (2001)
13. Schewe, K.-D., Thalheim, B.: Web Information Systems: Usage, Content, and Functionality Modelling. Technical Report (2005)
14. Thalheim, B., Düsterhöft, A.: SiteLang: conceptual modeling of internet sites. In: Kunii, H.S., Jajodia, S., Sølvberg, A. (eds.) ER 2001. LNCS, vol. 2224, pp. 179–192. Springer, Heidelberg (2001)
15. Schewe, K.-D., Thalheim, B.: Personalisation of web information systems – A term rewriting approach. Data & Knowledge Engineering 62, 101–117 (2007)

Autonomous Malicious Activity Inspector – AMAI

Umar Manzoor[1], Samia Nefti[1], and Yacine Rezgui[2]

[1] Department of Computer Science, School of Computing, Science and Engineering,
The University of Salford, Salford, Greater Manchester, United Kingdom
[2] Department of Computer Science, School of Engineering,
Cardiff University, Cardiff, United Kingdom
umarmanzoor@gmail.com, s.nefti-meziani@salford.ac.uk,
rezguiy@cardiff.ac.uk

Abstract. Computer networks today are far more complex and managing such networks is not more then a job of an expert. Monitoring systems helps network administrator in monitoring and protecting the network by not allowing the users to run illegal application or changing the configuration of the network node. In this paper, we have proposed Autonomous Malicious Activity Inspector – AMAI which uses ontology based knowledge base to predict unknown illegal applications based on known illegal application behaviors. AMAI is an Intelligent Multi Agent System used to detect known and unknown malicious activities carried out by the users over the network. We have compared ABSAMN and AMAI concurrently at the university campus having seven labs equipped with 20 to 300 number of PCs in various labs; results shows AMAI outperform ABSAMN in every aspect.

Keywords: Network Monitoring, Malicious Activity, Ontology, Cognitive Mobile Agent, Distributed Proxy Server, Collaborative Multi-Agent System.

1 Introduction

Network administrator is responsible for managing computer networks and it has become humanly impossible to manage these networks because of their increased complexity. The job of network administrator includes network monitoring (i.e. monitoring of malicious application on network nodes), network management (i.e. installation/un-installation applications on network nodes) and resource management (adding/removing resources on network). Many companies are trying to come up with automated tool(s) which can make life easier for network administrator. Monitoring system is one category among such products. Monitoring systems [9, 10, 11] helps network administrator in monitoring and protecting the network by not allowing the users to run illegal application or changing the configuration of the network node.

Monitoring systems are also used as Parental Control [15] which helps parents to monitor and restrict their children activities over internet. However, Parent Control tools are developed to monitor single node (PC), uses static rules and requires manual maintenance. Network administrator usually deploys Proxy Server on the network to monitor and restrict activities on the network nodes. Proxy Server is a software

C.J. Hopfe et al. (Eds.): NLDB 2010, LNCS 6177, pp. 204–215, 2010.
© Springer-Verlag Berlin Heidelberg 2010

program deployed on a server that acts as a mediator between a network node user and the Internet. Proxy Server has the following problems 1) It can easily be by-passed using tools available on the Internet [12, 18] 2) Filtering rules (keywords / URL etc) are defined by Network Administrator and requires manual update (i.e. static rules) 3) All the traffic is diverted to/from Proxy Server and it becomes bottleneck (i.e. centralized approach). 4) If Proxy Server crashes, Internet service will not be available on the whole network. 5) Proxy Server is unable to detect correlated websites.

The need of the time is to develop a tool which 1) is distributed and uses decentralized approach 2) does not require software installation on network node(s) 3) monitors the activities on the network node (i.e. client machine) 4) is so light and transparent that network user does not know that its activities are being monitored 5) uses application behavior to detect whether its malicious (legal) or non-malicious 6) has the capability of taking action (i.e. killing the malicious activity, disable user account etc) at runtime 7) is intelligent (dynamically update rules without intervention/interaction of network administrator).

Umar et al in [1] proposed "An Agent Based System for Activity Monitoring on Network – ABSAMN" for monitoring of resources over network, suitable for network of networks; commonly known as Campus Area Network. ABSAMN is a multi agent [7, 8] based system for the monitoring of illegal activities or applications running on the network nodes. An Agent is a software program which acts on the behalf of the user or other agent to perform a specific task in order to achieve its goal(s) [3, 2]. Agent based system have been used in many areas such as automated network installation, manufacturing, decision making, and business applications [19, 22, 20, 21].

ABSAMN has the following problems

- Rule set for malicious (illegal) applications are defined in the form of process name to action pair. Network administrator is responsible to define and update the rule set.
- Only process name is used to identify the malicious activity. If user changes the name of the illegal application, ABSAMN will never be able to monitor it.
- The rule set is static and maintained manually.
- The proposed framework cannot predict or capture malicious unknown applications.

In this paper we have modified existing An Agent Based System for Activity Monitoring on Network – ABSAMN architecture and proposed Autonomous Malicious Activity Inspector – AMAI which uses ontology based knowledge base to predict unknown malicious applications based on known application behaviors. AMAI is fully autonomous and once initialized it monitors whole network with the help of cognitive mobile agents. AMAI can dynamically update its Knowledge Base (KB) by adding new concepts and their relationships by detecting unknown illegal application.

The remainder of this paper is organized as follows. In Section 2, we present brief overview of ontology, knowledge base and AMAI knowledge base. Section 3 introduces the AMAI architecture, including the pre-processing, disambiguation-extrapolation-detection, and classification using ontology. In Section 4, we critically evaluate and compare the performance of AMAI with ABSAMN. Finally, the conclusion is drawn in Section 5.

2 Ontology, Agents and Knowledge Base

Ontology is explicit specification of abstract model consisting of facts and is used to construct the domain model [4, 6]. In Computer or Information Science, Ontology is a formal representation of a specific domain, containing concepts, their relationships and often represented in semantic network. The semantic network is internally represented in graph where nodes are concepts and links among nodes are relationships. There are two main types of ontology 1) Upper/top level ontology describe very general concepts and usually used to define common objects that are generally applicable across a wide range of domain ontologies, and 2) Domain ontology describes very specific concepts of a certain domain and usually these ontologies are incompatible with each other because each concept can have different meaning in different domain ontology [5].

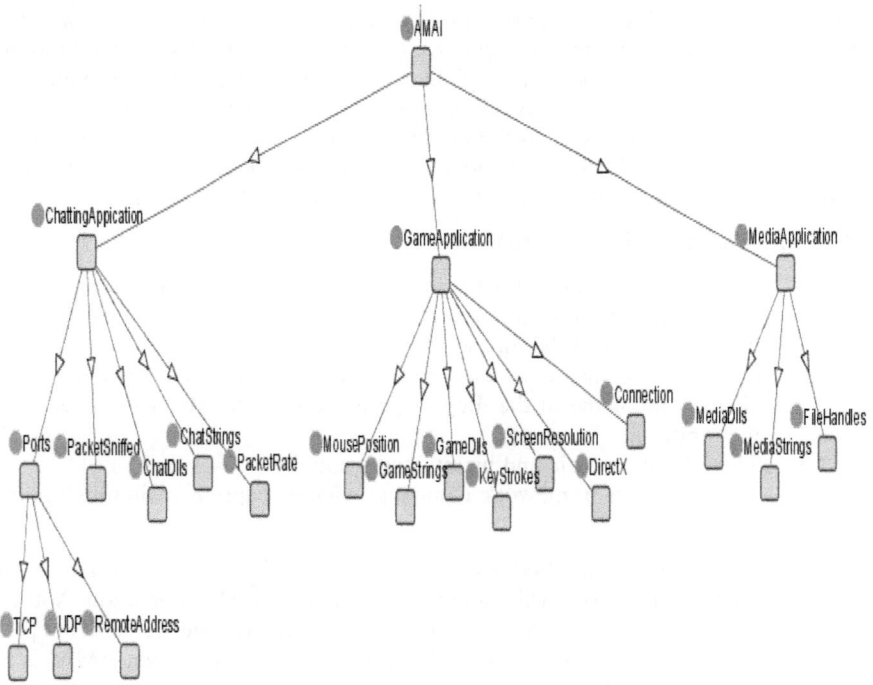

Fig. 1. Partial hierarchal tree of Autonomous Malicious Activity Inspector (AMAI)

Ontology building requires two steps 1) extracting knowledge from the domain using knowledge extracting techniques 2) representing the knowledge in a meaningful way. Ontology can be used to define the domain or do reasoning on the domain. Ontology plays an important in the development of intelligent distributed systems because it gives meaning and context to the data and can be dynamically updated (i.e. new concepts or relationship can be added) [6]. AMAI uses ontology based Knowledge base as shown in figure 2 to classify unknown application as malicious (illegal) or legal and is implemented in protégé 3.3.1 [14]. AMAI knowledge base is developed in a hierarchal way and contains the following classes.

- Application
- Application Type
- Resource

Malicious activity types (i.e. chatting application, illegal internet application etc) are defined in the application type. Resource class contains sub-classes which define all the resources that can be used by an application. Any unknown application can be malicious application if it uses any resources defined in the 'Resource Class'. An unknown application can be categorized in any of the 'Application Type' if it fulfills few properties defined against the application type. Camel notation is a common practice to describe properties and usually characteristics of the particular class are defined using 'has a' or "is a' property. Cognitive Ontology Agent is responsible to infer the unknown application into a defined 'Application Type' and update the ontology whenever new unknown illegal application is captured.

3 System Architecture

Autonomous Malicious Activity Inspector (AMAI) is an Intelligent Multi Agent System used to detect known and unknown malicious activities carried out by the users over the network. AMAI is implemented using Java Agent Development Framework (JADE) [13] and uses ontology based knowledge base to predict known / unknown application as malicious or not malicious by monitoring the application behavior. The architecture of the system consists of the following agents

- Knowledge Elicitation Agent (KEA)
- Sub-Network Agent (SNA)
- Cognitive Ontology Agent (COA)
- Action Agent

3.1 Knowledge Elicitation Agent (KEA)

Knowledge Elicitation Agent (KEA) is the core agent as it manages and initializes the system autonomously without user interaction. In initialization, KEA performs the following steps

1) KEA loads the network configuration from a pre-configured XML file. Network XML contains the information about the sub-networks of the network and which activities are allowed on these sub-networks.
2) KEA loads the ontology from Ontology based Knowledge Base.
3) KEA makes different Rule Profiles for the sub-networks.
4) KEA creates and initializes N number of Sub-Network Agents (SNA) where N depends on the number of sub-networks. KEA uses Sub-Network Agent to Sub-Network ratio of 1:1.

5) KEA passes Ontology, Rule Profile and Configuration of the sub-network as arguments to SNA.
6) Once SNA reaches its destination server, KEA transfers the compiled code of Cognitive Ontology Agent to Sub-Network Agent.

After initialization KEA waits for the updates from Sub-Network Agent and in case of any violation reported, it update its database with violation details.

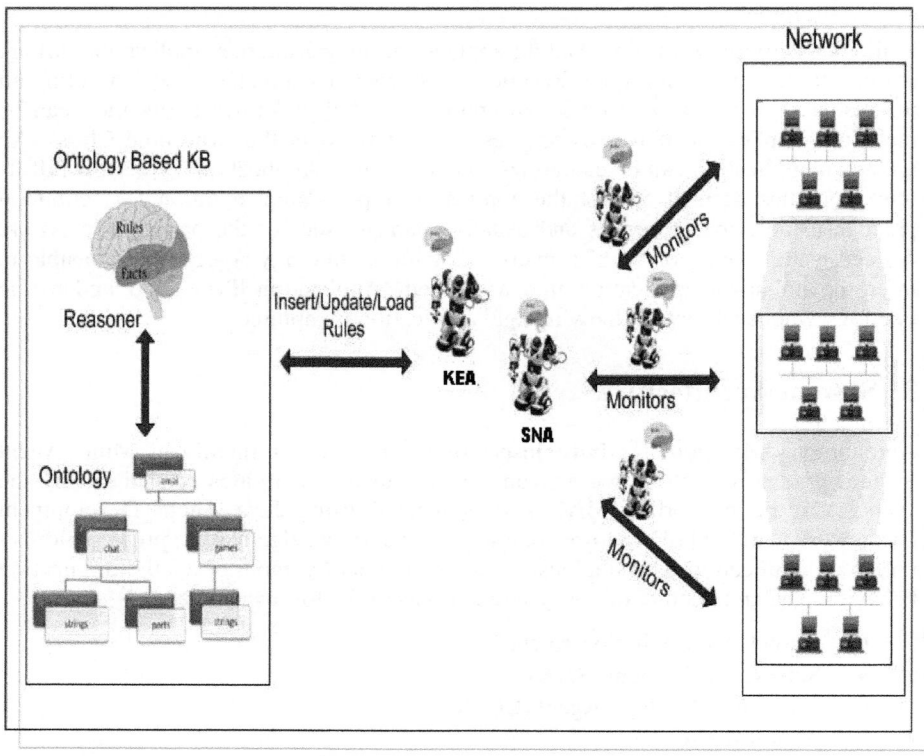

Fig. 2. Zero Level Architecture of Autonomous Malicious Activity Inspector (AMAI)

3.2 Sub-Network Agent (SNA)

After creation and initialization Sub-Network Agents (SNA) move to the assigned destination and performs the following steps

1) SNA loads ontology, network and rule profile configuration passed by the KEA.
2) SNA creates and initializes N number of Cognitive Ontology Agents (COA) where N depends on the number of network nodes present in the sub-network. SNA uses Cognitive Ontology Agent to Network nodes ratio of 1:5.
3) SNA passes Ontology, Rule Profile of the sub-network, and list of network nodes to monitor as arguments to COA.

4) If COA reports any malicious activity over the network; SNA creates Action Agent (AA) and initializes AA with network node name, Process ID, IP Address, and action to be performed on that node.

After creation and initialization of COAs, SNA waits for the updates from Cognitive Ontology Agents and in case of any violation reported, it updates its database with violation details.

3.3 Cognitive Ontology Agent (COA)

After initialization each Cognitive Ontology Agent moves to the nodes in the itinerary one by one and extracts all processes running on the network node and stores it in process vector set model V. The information includes name, size, path, company, built-in strings, ports (TCP, UDP, Remote-Address), keystrokes frequency, mouse clicks frequency, DLLs, process screen resolution, network connection, packet rate, packet contents, files opened, file extensions etc.

$$V = \{V_1, V_2, V_3, V_4, ..., V_m\} \tag{1}$$

A process can have few or all attributes mentioned above and the information is used by COA to classify the application. Cognitive Ontology Agent classification algorithm is similar to the approach proposed by Nefti in [16] but the difference is we are using WordNet [17] as knowledge base for text analysis. The classification algorithm is comprised of three steps: (I) Pre-processing (II) Concept Disambiguation, Extrapolation and Detection (III) Classification

Pre-processing: To generate the reduced process vector set P of each process, preprocessing will be performed on different attributes of the process vector set V.

- Built-in strings attribute V_i of process vector set V contains much useless strings, which needs to be removed. In Step 1, all standard stop-list / stemmer words like ("is", "the", "on", "and", "in", "with", "for", "by"...) and words including any non-alphabetic char(s) like ("%", "_", "#", "@", "-",...) are eliminated from built-in string vector attribute. In Step 2, homogeneous words like {("chat", "chatting", "chatted"), ("connected", "connecting", "connection")} are all substituted by the single word "chat" and "connect" respectively. In Step 3, the number of occurrences for each keyword is computed and multiple entries are eliminated from built-in string attribute.
- Keystroke frequency attribute v_j of process vector set V contains keystroke frequencies of all the keys, however we are interested in abnormal keystroke frequency. The keys with α (normal) key frequencies are removed from v_j where α is adjusted accordingly when an unknown illegal application is captured. Similarly, mouse click frequency attribute v_k is reduced using the same technique.
- Dynamic Link Library (DLL) attribute V_l of process vector set V contains all DLLs used by the process, however we are interested only in Operating System DLLs so all DLLs except Operating System DLLs are removed from V_l.

Concept Disambiguation, Extrapolation and Detection: Given the filtered process vector set P extracted from the previous step, built-in string and DLL attributes needs to be disambiguated in order to assign proper category to each of these attributes.

$$P = \{V_1, V_2, V_3, V_4, ..., V_m\} \tag{2}$$

Built-in string attribute $V_i = \{K_1, K_2, K_3, K_4, ..., K_m\}$ contains n keywords where each keyword K_i ($i=1,...,m$) has n possible meanings and $K_i S_q$ (q=1,...,n) represents the different meanings that K_i can express. In order to disambiguate and to assign the proper sense to each keyword, we compare all of the keyword senses by taking groups of two keywords at a time. For each of the words' senses, we select the most related senses that are semantically more related using the semantic relatedness between two words. The most related senses between two keywords are calculated by considering the number of nodes present in the paths that connect each of the two keywords' senses in a given taxonomy(WordNet), and selecting the two words' senses yielding the shortest path (smallest number of nodes) in the hierarchy.

$$Ws(x, y) = \{Min_N [Wsi(x, y)] : 0 \leq i \geq p\} \tag{3}$$

After identifying the most related senses for each pair of keywords, Concept set (C_S) is generated which contains either the Lowest Super Ordinate (LSO) for each pair of keywords or the individual concepts of keywords x and y. The selection depends on the parameter h which is used to control the level of generalization [34], if the LSO of each keyword(x, y) lies within or equal to h, LSO is added to the concept set else keywords x and y individual concepts are added to the set.

$$Cs_i = Opt \begin{Bmatrix} Lso(x, y) \mid lso_i(x, y) : 1 \leq i \geq h \\ Ck(x) \wedge Ck(y) \end{Bmatrix} \tag{4}$$

Malicious keyword concepts are loaded and stored in malicious set (M_k). Each concept in Concept set is looked up in WordNet and a list of synonym, hypernym synsets for each concept are stored in Synonym (S_c) and Hypernym (H_c) sets respectively. Each concept, its corresponding S_c and H_c are compared with the malicious concepts one by one and matches (if any) are stored in a separate resultant set (R_S) with labels C (Original Concept C_S), S (Concepts' Synonym) or H (Concepts' Hypernym). Each concept in R_S is assigned a score based on the assigned label weight multiply by the number of occurrences in the text; C is assigned maximum weight equal to 1, S is assigned a weight of 0.75 and H is assigned 0.50 respectively. The collective weight of the malicious content (M_C) found in string attribute is defined as the sum of the individual scores of all the concepts in R_S.

$$M_C = \sum_i Weight(Rs_i) * Occurrence(Rs_i) \tag{5}$$

where *Weight(Rs$_i$)* represents the individual weight of concept and and *Occurrence(Rs$_i$)* represents the frequency of appearance in text. If condition $Mc \geq \beta$ (where β is configurable and updated when new application text is assigned as malicious) is satisfied, 1) string attribute is assigned malicious tag and 2) New concepts (i.e. Synonyms and Hypernyms) are added to the knowledge base.

In order to disambiguate and assign proper category to DLL attribute, n Category Sets (C_n) are created where n represents the number of application categories in the ontology. Each DLL is compared with all category DLLs, if it coincides with one or more categories, the DLL is added to each of those category sets. Once all the DLLs are assigned, we calculate the score of each Category Set. The collective score of a Category Set (S_{Ci}) is defined as the sum of the individual scores of all the DLLs in C_i

$$S_{C_i} = \sum_i \text{Score}(Dc_i) \tag{6}$$

Where $Score(Dc_i)$ represents the individual score of each DLL which currently is uniform (i.e. 1) for all DLLs. DLL attribute of the process vector set is assigned category which has the maximum score. Similarly, keystroke and mouse-click attributes of process vector set are assigned abnormal or normal category based on the key frequencies.

Classification: Cognitive Ontology Agent (COA) uses the filtered and disambiguated process vector set W extracted from the previous step, to classify the process as malicious or un-malicious using the ontology.

$$W = \{V_1, V_2, V_3, V_4, ..., V_m\} \tag{7}$$

In order to assign malicious class to W, COA substitutes all attributes into ontology concepts and these concepts can belong to one or many ontology classes. In order to assign proper class to W, we assign each concept a weight. Once all the attributes are substituted by concepts, we calculate the weight of each ontology class O_i by summing up all the individual weights of concepts in O_i

$$W_{O_i} = \sum_i \text{Weight}(O_i) \tag{8}$$

where $Weight(O_i)$ represents individual weight of ontology class concepts which currently is either 0 or 1. Ontology class which has the maximum weight will be assigned to W only if the maximum weight ≥ 2 else Unknown class will be assigned.

3.4 Action Agent (AA)

After initialization each Action Agent moves to the network node assigned and perform the following tasks.

1) Get the handler of the Malicious Process P_i (passed in argument by SNA).
2) Perform the Action on P_i and report the result of action to SNA.
3) Wait for the response from SNA.
4) If SNA assigns a new task move to new destination and repeat step 1 else Kill itself.

4 Performance Analysis

We have compared An Agent Based System for Activity Monitoring on Network (ABSAMN) and Autonomous Malicious Activity Inspector (AMAI) concurrently at the university campus having seven labs equipped with 20 to 300 number of PCs in various labs.

Table 1. Showing comparison of ABSAMN and AMAI with respect to Average Time taken by Different Operations

Operation Description	Time	Operation Description	Time
ABSAMN		**AMAI**	
MCA Initialization	2.43	KEA Initialization	3.96
CA Creation & Initialization	2.96	SNA Creation & Initialization	4.07
Monitor Agent Creation & Initialization	2.17	COA Creation & Initialization	3.21
Monitor Agent Match Rule Activity	5.41	COA Match Rule Activity	8.79
Take an Action against a rule violation on the network using Action Agent	3.57	Take an Action against a rule violation on the network using Action Agent	3.57
Send a Message on the Network using Messaging Agent	1.03	Send a Message on the Network using Messaging Agent	1.03
Get Information of a System on the Network Using Information Agent	4.52	Get Information of a System on the Network Using Information Agent	4.89
Agents Movement on the Network	0.5	Agents Movement on the Network	0.89

ABSAMN and AMAI were compared on same configuration; results shows AMAI outperform ABSAMN in every aspect, some of the comparison is presented in this section. Table 1 shows the comparison of ABSAMN and AMAI with respect to different operations. AMAI different operations takes little more time then ABSAMN as it has to perform extra task (i.e. load/retrieve the ontology), however the time is ignorable because the difference between the scan time of one machine is only 3.38 seconds and the accuracy of AMAI is far more better than ABSAMN.

Fig 3 shows the comparison of malicious activities captured using ABSAMN and AMAI on the same configuration, AMAI captures more violations because it detects malicious unknown application based on the application behavior and does not rely on the process name. ABSAMN has no intelligence (i.e. rules are not updated dynamically, network administrator can add new rules or update existing rules which itself requires manual maintenance). AMAI learns from the environment and update its ontology based knowledge base by adding, updating or deleting concepts dynamically which enhances the efficiency of the system. Administrator can view the number of violations of any specific day, week or month or any specific date as shown in Fig 3.

Most importantly administrator has the option to view the number of violations per lab, using this graph administrator can easily track from which lab most violations are being performed. Fig 4 shows the number of violation per user captured using ABSAMN and AMAI, results shows that AMAI captures more malicious activities as compared to ABSAMN.

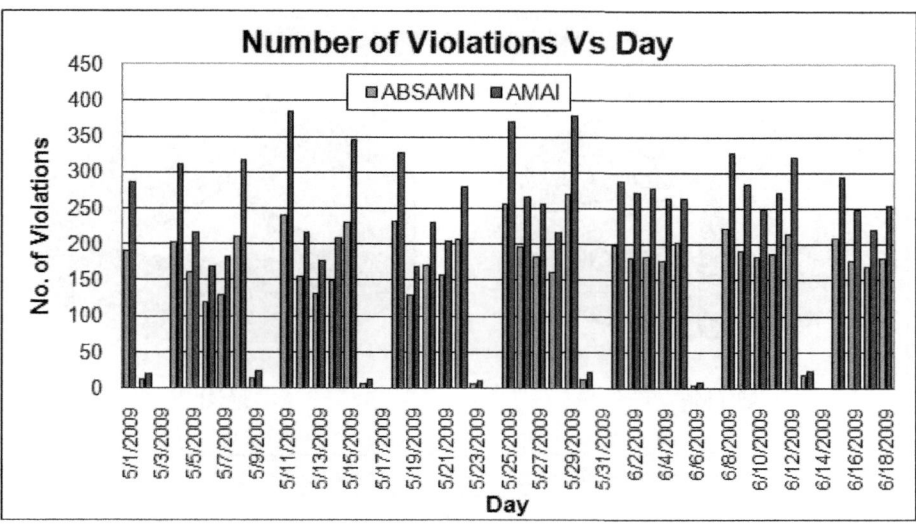

Fig. 3. Comparison of Number of Violations per Day captured using ABSAMN and AMAI

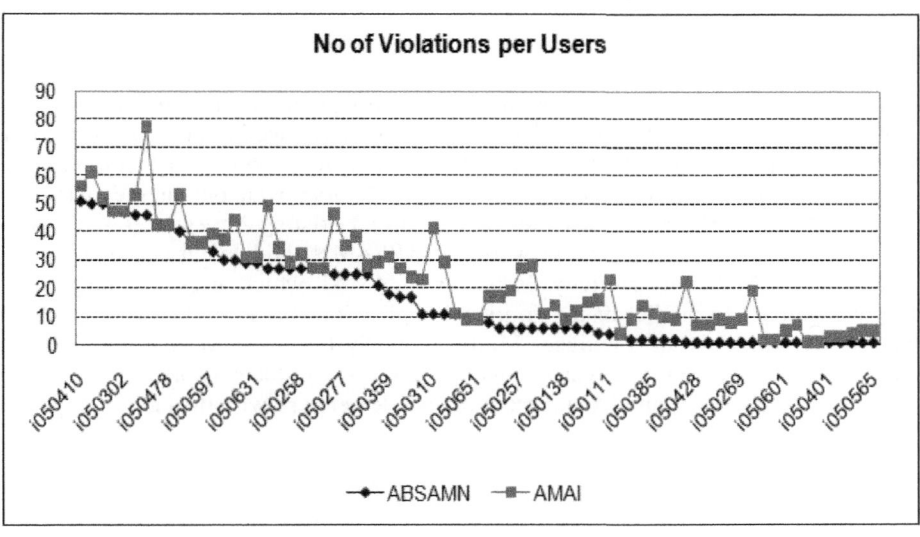

Fig. 4. No of Violations per Users captured using ABSAMN and AMAI

ABSAMN and AMAI both facilitate the administrator to view the top violated processes in pie chart or bar graph generated dynamically as shown in Fig 5. ABSAMN process share is not accurate because it is unable to capture all instances of illegal process (i.e. dependency on process name). On the other hand, AMAI captures all instance of the illegal process because it monitors the application behavior. These statistics helps the network administrator to track and un-install the illegal process from the network.

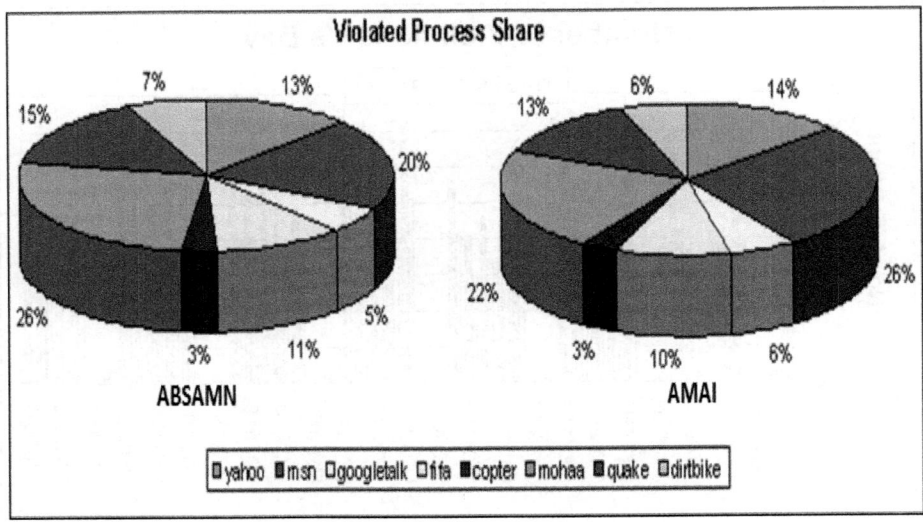

Fig. 5. Violated process share comparison of ABSAMN and AMAI

5 Conclusion

Today every organization big or small uses computers to manage their data / resources and usually share their data / resources by connecting computers by some sort of computer network. Computer networks have become very complex and usually network administrator uses monitoring tool(s) available in the market to monitor the network. In this paper we have modified existing An Agent Based System for Activity Monitoring on Network – ABSAMN architecture and proposed Autonomous Malicious Activity Inspector – AMAI which uses ontology based knowledge base to predict unknown illegal applications based on known application behaviors. AMAI is an Intelligent Multi Agent System used to detect known and unknown malicious activities carried out by the users over the network. AMAI is fully autonomous and once initialized it monitors whole network with the help of cognitive mobile agents.

References

1. Manzoor, U., Nefti, S.: An agent based system for activity monitoring on network – ABSAMN. Expert Systems with Applications 36(8), 10987–10994 (2009)
2. Ilarri, S., Mena, E., Illarramendi, A.: Using cooperative mobile agents to monitor distributed and dynamic environments. Information Sciences 178, 2105–2127 (2008)
3. Shah, N., Iqbal, R., James, A., Iqbal, K.: Exception representation and management in open multi-agent systems. Information Sciences 179(15), 2555–2561 (2009)
4. Lee, C., Jiang, C., Hsieh, T.: A genetic fuzzy agent using ontology model for meeting scheduling system. Information Sciences 176(9), 1131–1155 (2006)

5. Storey, V.C., Burton-Jones, A., Sugumaran, V., Purao, S.: A methodology for context-aware query processing on the World Wide Web. Information Systems Research 19(1), 3–25 (2008)
6. Huhns, M.N., Stephens, L.M.: Personal ontologies. IEEE Internet Computing 3(5), 85–87 (1999)
7. Weiss, G.: Multiagent Systems A Modern Approach to Distributed Artificial Intelligence, ch. 1-4. The MIT Press, Cambridge (1999)
8. Manzoor, U., Nefti, S.: Cognitive Agent for Automated Software Installation – CAASI. In: Lytras, M.D., Damiani, E., Carroll, J.M., Tennyson, R.D., Avison, D., Naeve, A., Dale, A., Lefrere, P., Tan, F., Sipior, J., Vossen, G. (eds.) WSKS 2009. LNCS, vol. 5736, pp. 543–552. Springer, Heidelberg (2009)
9. Paessler - PRTG Network Monitor (2009), http://www.paessler.com/prtg/
10. Network Monitoring Tools (2009), http://www.topology.org/comms/netmon.html
11. Nagios (2009), http://www.nagios.org/
12. YouHide (2009), http://www.youhide.com/
13. Java Agent Development Framework – JADE (2009), http://jade.tilab.com/
14. Protégé (2009), http://protege.stanford.edu/
15. Çankaya, S., Odabaşıa, H.F.: Parental controls on children's computer and Internet use. Procedia - Social and Behavioral Sciences 1(1), 1105–1109 (2009)
16. Nefti, S., Oussalah, M., Rezgui, Y.: A modified fuzzy clustering for documents retrieval: application to document categorization. Journal of the Operational Research Society 60(3), 384–394 (2009)
17. WordNet (2009), http://wordnet.princeton.edu/
18. Anonymous Proxy Server - Browser9 (2009), http://www.browser9.com/
19. Manzoor, U., Nefti, S.: QUIET: A Methodology for Autonomous Software Deployment using Mobile Agents. Journal of Network and Computer Applications (2010), http://dx.doi.org/10.1016/j.jnca.2010.03.015
20. Chen, S.-H.: Computationally intelligent agents in economics and finance. Information Sciences 177(5), 1153–1168 (2007)
21. Rajiv, K., Hong, Z., Ramesh, R.: Enterprise integration using the agent paradigm: foundations of multi-agent-based integrative business information systems. Decision Support Systems 42(1), 48–78 (2006)
22. Guo, Q., Zhang, M.: A novel approach for multi-agent-based Intelligent Manufacturing System. Information Sciences 179(18), 3079–3090 (2009)

Towards Geographic Databases Enrichment

Khaoula Mahmoudi[1] and Sami Faïz[2]

[1] Laboratoire URISA – SUPCOM, Cité Technologique des Communications,
Route de Raoued Km 3.5-2083 El Ghazala, Ariana, Tunisie
Khaoula.mahmoudi@supcom.rnu.tn
[2] Laboratoire LTSIRS – ENIT, Campus Universitaire El Manar, Tunis, Tunisie
Sami.Faiz@insat.rnu.tn

Abstract. The geographic database (GDB) is the backbone of the geographic information system (GIS). Indeed, all kinds of data managements are based and strongly affected by the type, relevancy and scope of the stored data. Nevertheless, this dataset is sometimes insufficient to make the adequate decision. Then, it is of primary interest to provide other data sources to complement the inherent GDB. In this context, we propose a four staged semantic data enrichment approach consisting of: text segmentation, theme identification, delegation and text filtering. Besides, a refinement is eventually executed to enhance the data enrichment results.

Keywords: GIS, segmentation, theme identification, delegation, text filtering.

1   Introduction

Increasingly used, the geographic data are related to the description and the localization of natural or man made objects, activities and phenomena on the earth. Information systems managing these geographic data are geographic information systems (GIS) [1]. GIS is a vital tool for decision making in business, environmental protection, crime prevention and so on. The soundness of the decisions to make depends closely upon the data stored into the GDB. The latter commonly do not propose all the data that can support different users with different preoccupations. This is due mainly to the cost entailed by the data acquisition and the incapacity to account for all the valuable information since the world state changes continuously. Indeed, it is common to integrate in the same problem different data describing different entities that cohabit or coexist in the same environment and which features can not be captured via unique values of different types. For instance, to keep track of a disease spreading (the case of the bubonic plague detected recently in Libya), the decision makers have to cope with a huge amount of data. This is related mainly to the social, economic, environmental information, citizen habits, garbage management, the interaction of the inhabitants with other natural or artificial entities, etc. All these data are commonly not stored in the same GDB due to the cost of data processing and maintenance. Besides, by referring to the data abstraction, we loose information. Some preoccupations need a meaningful description of the current state usually conveyed by well formed texts. For the rural zones, it is worth informing to know

C.J. Hopfe et al. (Eds.): NLDB 2010, LNCS 6177, pp. 216–223, 2010.
© Springer-Verlag Berlin Heidelberg 2010

where the animals are placed and how the inhabitants behave regarding the animals waste, the water and food nature they provide, etc. In front to such situations, the data enrichment was asserted as a solution to complement the GDB by providing additional valuable information about the geographic entities. A review of the literature has revealed a set of works towards semantic data enrichment, namely, MetaCarta [2], PIV[3], GeoNode [4] and Perseus [5]. They operate the enrichment by linking textual documents to the corresponding geographic entities. Although, they do not propose (or a slight) sophisticated means to manage the textual contents. To relieve the GIS user from steady, time consuming and fastidious works to execute to reach the target information, we propose a new way to enrich the semantic data. We go beyond merely linking the documents to mine the textual data to extract the essence of information. Given a corpus of documents related to the geographic entities at hand, our approach [6] proceeds to: (i) segmenting each document into homogeneous text segments, (ii) annotating each segment by its main topic and (iii) filtering the texts which are thematically related to the same topic to keep the essence. Besides, our approach offers a supplementary stage that can be triggered once the previous process is fruitless. It deals with a refinement exploiting the spatial relations to locate more accurately the solicited geographic entities in order to reach the relevant documents. To model our approach we comply with the multi-agent paradigm [7]. We adopted a society of agents sheltering: an Interface agent (I_{ag}), Task agents (T_{ag}) and eventually Geographic agents (G_{ag}). Generally, the system agents fall into two categories: (i) the agents (T_{ag}) with the highest granularity since they undergo almost all the data enrichment processing and (ii) the agents considered as supervisors (I_{ag} and G_{ag}) intervening during all the data enrichment stages to support the T_{ag} agents. All these agents communicate via peer-to-peer message passing. Once the user handles one geographic entity, only the I_{ag} agent and the T_{ag} agents are involved in the problem resolution. If the user is looking for information about a zone, besides these two classes of agents we involve the G_{ag} agents. Each G_{ag} agent will be considered as a supervisor of a subset of T_{ag} agents relative to a given geographic entity. Whatever the case, it is the I_{ag} agent which triggers the overall data enrichment process. This process is to be detailed in the rest of the paper.

2   Our Semantic Data Enrichment Approach

2.1   Text Boundaries Delimitation

To segment a document, each T_{ag} agent applies our modified version of the well known and established TextTiling algorithm [8]. The native version of this algorithm consists in fragmenting the text into blocks of fixed size of token sequences. Between each pair of adjacent blocks, the similarity is computed according to the cosine method. The high similarity scores are requisite to merge the adjacent blocks to form a text segment. This algorithm exhibits a deficiency that may be observed at the first stage when the blocks are set up. Because this stage is not guided heuristically, it is common to gather token sequences that are thematically heterogeneous in the same block. This may entail a high similarity score among the said block and the adjacent ones. So the segment to be generated (from the merging of the similar blocks) will be also heterogeneous. As a remedy, we propose to intervene during the block creation stage. To purify each block, we apply C99 [9] which is the best tailored for short texts. C99 operates by building a similarity matrix issued from the similarity scores (using the cosine function) between every pair of adjacent sentences. These values serve to stand for a rank matrix. A rank refers to the number of neighboring elements with similarity values lower than the value at hand. To delimit the text boundaries, C99 adopts a maximization algorithm to maximize the measure of internal segments cohesion. By applying C99 to a block, we look for further decomposition if any. If it is the case, the resulting sub-blocks situated at the frontier of the associated block will undergo the rest of TextTiling processing (similarity computation and segment boundary determination). The sub-blocks not at the frontiers are considered definitively as text segments. At the end of the segmentation stage, each T_{ag} agent holds a fragmented document.

2.2 Thematic Annotation

A T_{ag} agent annotates each segment by the theme it tackles. The thematic annotation is based on the term frequency. Roughly two cases are distinguished: (i) the frequencies which are heterogeneously distributed along the segment and (ii) the frequencies which are homogeneously distributed. In the first case, the word with the highest frequency is assigned as a text topic. The second case is observed once no word is sharply reproduced. Hence, the topic is the word that is thematically referenced by the most of the text words. The semantic relations are depicted by exploring the WordNet thesaurus [10]. Then, the frequencies are adjusted to the number of links (we consider all the semantic relations), a noun maintains with the other segment nouns. For the sake of clarity, we consider the following text: *"The state run Tunisian Radio and Television Establishment operates two national TV channels and several radio networks. Until 2003, the state had a monopoly on radio broadcasting"*. None of the nouns is highly occurring and by the way we can not deduce the text theme. We exploit the WordNet thesaurus to depict the semantic relations among the nouns. The main relations are reproduced in what follows: *Hypernymy(TV)=broadcasting, Hypernymy(radio)=broadcasting, Synonym(TV)=television, Topic-domain(network)= broadcasting*. Then we infer that *broadcasting* is the noun that reveals the main idea of the text. Once this stage is over, each T_{ag} agent maintains a fragmented annotated document. To allow a straightforward processing of the text segments topically related, the latter are gathered in the same container that we call *generated document*. The latter is to be processed by one delegate agent affected by the supervisor via a delegation process.

2.3 Delegation

The objective underpinning this stage is to distribute equitably the topics among different T_{ag} agents. This stage is achieved by the supervisor agent (I_{ag} or every G_{ag} agent). While affecting the T_{ag} agents as delegates, the supervisor tries to distribute with equity the work overload meanwhile it looks to minimize the communication overhead. For the workload, it is computed by recording the size of the segments (in terms of words length) which are already affected to the T_{ag} agent. For the communication, it is set to 0 whether the delegate is responsible of a topic that it tackles within its document. We do so to penalize every assignment of one T_{ag} agent to a topic that it does not tackle in its document. In such case, the communication is set equal to the number of the acquaintances (the T_{ag} agents maintaining the same topics) of the delegate. At the end of the delegation stage, each delegate agent solicits its acquaintances to send the segments related to the topics it holds. Whether the delegate agent is responsible of some topic that it does not tackle in its document, it has to communicate with the supervisor (which knows all the agents) to determine the T_{ag} agents maintaining the required segments. By collecting the segments, each delegate agent maintains a so-called *generated document* for each topic under its jurisdiction.

2.4   Text Filtering

Each delegate agent has to analyse each of its *generated documents* to pick out the essential material and to discard all subsidiary information. This analysis is performed according to the rhetorical structure theory (RST) [11]. The latter considers a text as a set of textual units linked to each others by rhetorical relations via cue phrases (like: for instance and because). RST makes the distinction among the satellite and the nucleus units. By removing the satellite, the text remains still comprehensible while removing the nucleus yields to an incomprehensible text. Based on these concepts and the extensions proposed by [16], the text may be represented as a binary tree whose leaves are the textual units. By exploiting this theory, a T_{ag} agent builds a binary tree to each segment under its jurisdiction to maintain the essential textual units. For the sake of clarity we consider the following text fragment: *"[Although the workers are working hard,[1]][the building is not yet achieved.[2]]"*. By detecting the cue *although* and by referring to the database, the RST tree is a two leaf nodes; one considered as a nucleus (unit 2) and one as a satellite (unit 1). Then, at the inner node (representing the whole text span [1-2]) we detect *concession* as a rhetorical relation. Then, by retaining only the most important unit (2) we give a concise view of the text. It deals with the fragment: *the building is not yet achieved.* Indeed, some adjacent sentence units are not linked by cue phrases. So, in such cases the delegate agent computes the similarity among these units to be able to infer the nucleus unit. The computation is performed using the cosine similarity. For the dissimilar units, we keep the two units which are considered of equal importance (multi-nuclear). For the similar units, we maintain the one situated at the left side since naturally the authors announce the fact and then proceed to its development. Besides, because a *generated document* is a set of segments that are thematically identical, we may obtain redundant text units. To bypass such redundancy, we maintain certain interactivity with the end user who is able to choose: (i) one segment from the *generated document,* (ii) keep the whole *generated document* (iii) or the whole *generated document* with the possibility to discard the segments that are likely to report redundant information. The latter facility relies upon a similarity computation (using the cosine function) to maintain one instance of the similar units. At the end of this text filtering stage, each delegate agent maintains a condensed view of each *generated document* under its jurisdiction. These results are sent to the supervisor to be dispatched towards the GIS user.

2.5   Refinement of the Results

This stage may be triggered once the user feels unsatisfied by the results of the process described above. This may be entailed from the ambiguity, the lack of the required data, etc. So, in an attempt to reach the desirable information, we look to describe more accurately the geographic entity at hand. Hence, we examine the vicinity of the geographic entity to disclose those that have some relationships with it. Our argument is the first law of geography stated as follows: *"everything is related to everything else, but near things are more related than distant things"*[12]. Concretely, we exploit the following spatial relationships [18]: adjacency, proximity, connectivity and the nearest neighbor. Besides, we exploit the overlaying operation. One illustration of the refinement exploitation, concerns a situation where a user is looking for information about a river contamination state. Whenever, the user fails to found such information in the documents relative to the river, it will be interesting to examine the connected rivers. The state of the connected branches may be worth informing about the river state. Hence, we enlarge our investigation space to look for information through documents relative to the related entities according to the connectivity topological relation. By the way, the GIS user can deduce the decision about the river state or the measurement to undertake in order to prevent eventual contamination. The ambiguity is another problem that confronts the user during the data enrichment process. For instance, when looking for information about Carthage (the historical city of Tunisia) an information retrieval session returns about 9870000 documents. A large subset of the resulting documents is not relative to our geographic entity. For instance, some documents are relative to Carthage city located to the U.S.A. (www.carthagetx.com). Indeed, even we mention Tunisia with Carthage in the query we have a non negligible subset of non relevant documents. This occurs because sometimes the term Tunisia is mentioned only to highlight the origin of the term Carthage. To deal with such situations, we can exploit one of the spatial relations mentioned above to describe the geographic vicinity of the city. The surrounding geographic environments are no way to be the same. The references of the detected geographic entities are to be used to enhance and refine the results. This is achieved by adopting the attributes of these entities as key components to reach the target information. Another example is shown in figure 1.b. In this situation, a user is looking for information about Uganda and the relationships maintained with its neighbors that he ignores. Such information are not stored in our GDB relative to the world countries. By exploiting the adjacency relation, the system highlights graphically the neighboring countries meanwhile it provides the attributes that will serve as part of the search key. Besides, to guide more the GIS user, we report the statistics relative to the number of web documents reporting information of the association of each neighbor with Uganda. So, the user will be informed about the amount of data to handle. By the way we assist him in choosing the attributes of the geographic entities that are likely to allow reaching the relevant documents.

3   Simulation and Evaluation

To allow the exploitation of the devised semantic data enrichment approach, we developed the SDET (for Semantic Data Enrichment Tool) tool. The SDET functionalities have been integrated into a GIS, namely, the OpenJUMP (see figure 1.a and 1.b). The implementation was performed using the Java programming language and this to comply with our distributed architecture.

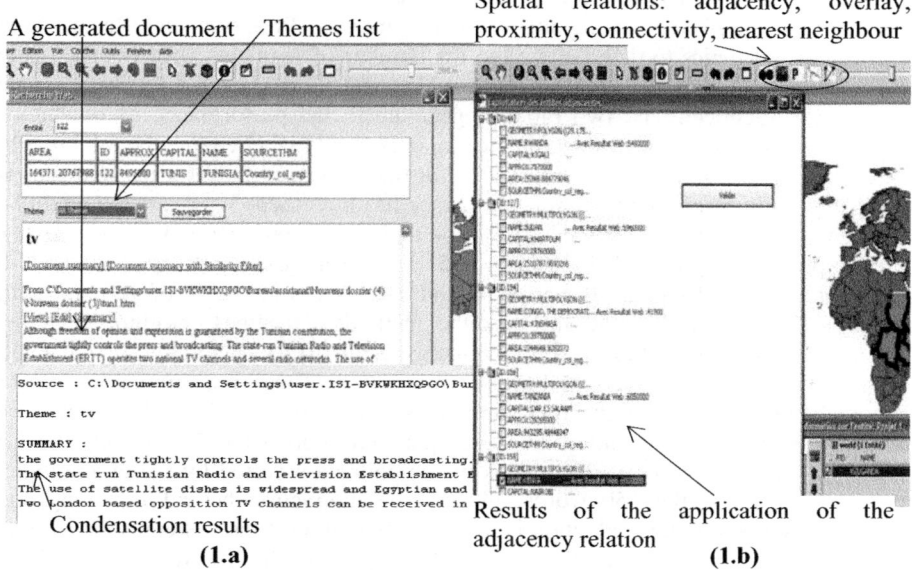

Spatial relations: adjacency, overlay, proximity, connectivity, nearest neighbour

A generated document   Themes list

Condensation results

(1.a)

Results of the application of the adjacency relation

(1.b)

Fig. 1. (a) A SDET session. (b) A Refinement of the data enrichment process via the adjacency relation.

To assess the performance of our SDET system, we evaluate its major subsystems, namely, the segmentation, the theme identification and the text filtering. To this end, we stand up an annotated corpus by harvesting 50 web documents (we followed Hearst [8] and others [13] in such a way that the small number of documents was compensated by the high number of judges) of different formats relative to various geographic entities. We have confided the documents to 10 judges who are linguistics with different qualification degrees. The documents are processed manually by first segmenting the texts, second annotating the resulting segments and lastly generating a condensed representation of each set of segments topically related (*the generated documents*). To evaluate the segmentation, we compare C99, TextTiling (T_{native}) and our adapted version of TextTiling ($T_{adapted}$) by using P_k^1 [14] and *WindowDiff* [14]. We obtain the following (P_k, *WindowDiff*) scores: (0.215, 0.245), (0.193, 0.201) and (0.127, 0.139) for C99, T_{native} and $T_{adapted}$, respectively. $T_{adapted}$ outperforms C99 and T_{native}. Also, we observe that P_k scores are lower than *WindowDiff* ones. This is due to the fact that P_k is a more lenient evaluation score than *WindowDiff*. The high scores of *WindowDiff* in comparison to P_k ones, show that errors are due essentially to imprecision in positioning the boundaries. For the thematic annotation, it is evaluated according to two strategies: *Exact* where the machine theme has to be identical to the

---

[1] Computes the distance (in terms of sentences) between a found boundary and the right boundary to find.

[2] Computes the difference between the number of boundaries in the interval ranging from position i to i+k in both the reference and system segmentation.

manual one and *Relaxed* where we accept the themes which are synonymous. The evaluation is preformed using the following indices: Precision and Recall. Since the number of topics to detect is tied to the number of segments already established, the number of topics detected by the machine and those detected by the reference system are identical. The two ratios degenerate to the performance ratio which is defined as the ratio among the number of topic correctly labelled and the total number of topics. The results are as follows: 0.75 for the *Exact* and 0.89 for the *Relaxed*. By adopting a relaxation, the system performance has increased. To assess the quality of the generated condensed texts relative to the topics scattered among the documents of the corpus, the ROUGE-BE package [15] was used. Especially, we adopted the commonly used Rouge-2, Rouge-SU4 and BE. We runt each of the following systems, SDET, MEAD and the baseline methods (baseline1, baseline2) and we compare them against the human judgments using ROUGE toolkit. In the DUC' fashion, we have defined two baseline references. These are summaries created automatically in terms of the following rules: one baseline selects 150 first words from the most recent document in the corpus and another baseline selects the first sentences within the (1...n) documents of the collection sorted by the chronological order till reaching the 150 words. Indeed, the motivation behind involving such baseline methods in the evaluation is to ensure that at least the system to be assessed does not perform worse than those methods. Regarding the MEAD system [17] we have supplied as input different *generated documents* relative to the detected topics instead of the documents of the corpus. Our system is neither a generic nor a query-based in a strict sense. Hence, we do so, to insure a fair comparison. The resulting (Rouge-2, Rouge-SU4, BE) scores are as follows: SDET (0.4685, 0.2956, 0.20024), MEAD (0.2969, 0.1868, 0.0997), baseline1 (0.2001, 0.10321, 0.08796) and baseline2 (0.16034, 0.07906, 0.04965). The relevancy of a system is measured by the highness of ROUGE scores. SDET is classified at the first position with the best evaluation scores. The two baselines bring the lower scores. In fact, while the hypothesis behind the two baselines is always or quasi-always valid, it is not sufficient for a multi-document summarization task which needs a special care of the similarities and differences through out the documents of the corpus.

4   Conclusion

In this paper, we have detailed our semantic data enrichment approach devised to complement the semantic GDB data. Our approach exploits extensively different text management techniques to provide end users with the required information. The main text processing consists in: the fragmentation of textual documents into segments, the identification of the topic of each segment and the condensation to retain the essence of text relative to the segments that are thematically identical. The overall approach was modeled in a distributed fashion with respect to the multi-agent paradigm. Besides, our approach exploits the spatial relationships maintained among the geographic entities to enhance the results. Our system was evaluated by following the standard evaluation schemes in the corresponding text management fields. The results were satisfactory.

References

1. Pornon, H.: Systèmes d'information géographique, pouvoir et organisations, Géomatique et stratégies d'acteurs, p. 254. Harmattan, Paris (2001)
2. MetaCarta: MetaCarta Document Templates Integrator's Guide March 13 (2008)
3. Lesbegueries, J., Gaio, M., Loustau, P.: Geographical information access for non-structured data. In: ACM Symposium on Applied Computing (SAC), Dijon, France, pp. 83–89 (2006)
4. Hyland, R., Clifton, C., Holland, R.: GeoNODE: Visualizing News in Geospatial Environments. In: Proceedings of the Federal Data Mining Symposium and Exhibition 1999, AFCEA, Washington D.C (1999)
5. Crane, G.: The Perseus Project and the Problems of Digital Humanities. In: Standards und Methoden der Volltextdigitalisierung Trier. Mainz Academy, Germany (2001)
6. Mahmoudi, K., Faïz, S.: Un processus d'acquisition d'information pour les besoins d'enrichissement des BDG. Revue des Nouvelles Technologies de l'Information (RNTI), pp. 691–696. Editions Cepaduès (2008)
7. Ferber, J.: Les Systèmes Multi-Agents vers une intelligence collective. Edition InterEditions, p. 522 (1997)
8. Hearst, M.: Texttiling: Segmenting text into multi-paragraph subtopic passages. Computational Linguistics 23(1), 33–64 (1997)
9. Choi, F.Y.Y.: Advances in independent linear text segmentation. In: Proceedings of the 1st Meeting of the North American Chapter of the Association for Computational Linguistics (ANLP-NAACL 2000), pp. 26–33 (2000)
10. Fellbaum, C.: WordNet: An Electronic Database. MIT Press, Cambridge (1998)
11. Mann, W., Thompson, S.A.: Rhetorical structure theory: Toward a functional theory of text organization. Text 8(3), 243–281 (1988)
12. Tobler, W.R.: A Computer Movie Simulating Urban Growth in Detroit Region. Economic Geography 46, 234–240 (1970)
13. Bestgen, Y., Piérard, S.: Comment évaluer les algorithmes de segmentation automatique? Essai de construction d'un matériel de référence. In: TALN 2006, Leuven, pp. 10–13 (2006)
14. Pevzner, L., Hearst, M.: A critique and improvement of an evaluation metric for text segmentation. Computational Linguistics 28(1), 19–36 (2002)
15. Hovy, E., Lin, C.Y., Zhou, L.: Automated summarization evaluation with basic elements. In: Proc. of the 5th International Conference on Language Resources and Evaluation (2006)
16. Marcu, D.: Discourse trees are good indicators of importance in text. In: Mani, I., Maybury, M. (eds.) Advances in Automatic Text Summarization, pp. 123–136 (1999)
17. Radev, D., Otterbacher, J., Qi, H., Tam: MEAD reducs: Michigan at DUC 2003, pp. 160–167. ACL, Edmonton (2003)
18. Larson, R.R.: Geographical Information Retrieval and Spatial Browsing. In: Geographical Information Systems and Libraries: Patrons, Maps, and Spatial Information, pp. 81–124 (1996)

On-Demand Extraction of Domain Concepts and Relationships from Social Tagging Websites

Vijayan Sugumaran[1,2], Sandeep Purao[3,4], Veda C. Storey[5], and Jordi Conesa[6]

[1] School of Business Administration, Oakland University
Rochester, MI 48309
sugumara@oakland.edu
[2] Department of Service Systems Management and Engineering, Sogang University,
#1 Shinsoo-Dong, Mapo-Gu, Seoul 121-742, South Korea
[3] Visiting Scholar, School of Information, University of Washington, Seattle, WA
[4] College of Information Sciences & Technology, The Pennsylvania State University,
University Park State College, PA 16802
spurao@ist.psu.edu
[5] Department of Computer Information Systems, J. Mark Robinson Collage of Business
Georgia State University, Box 4015 Atlanta, GA 30302
vstorey@cis.gsu.edu
[6] Estudis d'Informatica i Multimedia, Universitat Oberta de Catalunya,
Rambla del Poblenou, 156, E-08018, Barcelona, Spain
jconesac@uoc.edu

Abstract. Much content on the World Wide Web is becoming tagged with simple words or phrases in natural language as web citizens create tags that organize information primarily to facilitate their personal retrieval and use. These tags represent, often incomplete, pieces of knowledge about concepts in a domain. Aggregated across a large number of contributors, these tags provide the potential to identify, in a bottom-up manner, key constructs in a domain. This research develops a set of heuristics that aggregate and analyze tags contributed by individual users on the web to extract and generate domain-level constructs. The heuristics infer the existence of constructs, and distinguish entities, attributes, and relationships.

1 Introduction

With increasing amounts of information available on the World Wide Web, web citizens often turn to tagging web sites for convenient retrieval of information that is of importance to them. This large, and growing, layer of user-contributed annotations – consisting of natural language words and phrases – is being used to describe and link information on the web. The user-directed and free-form nature of these annotations is a direct response to the importance of searching the mass of data on the web. They appear as tags in folksonomy sites, such as delicious (www.delicious.com). They, in essence, capture some notion of semantics about the underlying resources [3]. Aggregated, they have the potential to create a bottom-up definition of a domain, thus contributing the meta-data structures needed to understand a domain [7]. For example, a web citizen

C.J. Hopfe et al. (Eds.): NLDB 2010, LNCS 6177, pp. 224–232, 2010.
© Springer-Verlag Berlin Heidelberg 2010

might tag a website that captures information on green technology as "sustainability;" for example, *green technology contributes-to sustainability.* This exposure of the concepts "green technology" and "sustainability" could provide useful input to a domain model for designing an information system dealing with environmental initiatives.

The objective of this research is to: *analyze tags contributed by web citizens to semi-automatically extract useful domain-level constructs and to distinguish them as entities, attributes and relationships so they can be used to generate a domain model.* The contribution of this work is a set of heuristics that use natural language manipulation to infer domain-level concepts useful for the creation of domain models.

2 Related Work

Effectively using information on the World Wide Web requires simultaneously addressing the difficult challenges of: 1) scale and 2) semantics. Tagging allows dealing with *scale* because it facilitates participation from a large number of web citizens. The layer of annotations contributed can be exploited to understand the *semantics* of the underlying data layer.

2.1 Identifying Domain-Level Constructs

Understanding and representing the main concepts in a domain involves abstracting from the real world and representing the concepts in a form that captures essential elements in the domain. It is usually carried out by a designer in conjunction with user(s). In spite of significant prior efforts aimed at automated conceptual modeling, this remains a challenge. Domain-level constructs represent entities, attributes and relationships in a chosen universe of discourse. Consider an example from the hotel business. Key constructs in this domain could be Guest, Reservation, and Room (entities), Date and Number (attributes), *Guest-makes-Reservation, Maid prepares Room* and others. Identifying these constructs is difficult. Successfully identifying them, however, could contribute to building an effective domain model.

2.2 Tagging, Wisdom of the Crowd, and Text Processing

The "wisdom of the crowd" reflects the collective opinion of a group of individuals [6] with independent, diverse, opinions. Without a mechanism to turn these diverse and independent opinions into an aggregate, these opinions remain isolated. The notion of folksonomies captures this combination of input from a large number of individuals (folks) that results in a taxonomy [1]. Mechanisms to convert independent, diverse opinions into a taxonomy can benefit from simple, as well as sophisticated, text-processing mechanisms. For example, tags are likely to be semantically related to each other if they tag the same resource [8]. They may also be semantically related if they are used to tag related resources many times [8, 2]. They may be semantically related if multiple users use the same tag. Although these are not traditional text processing mechanisms, they take into account authorship, frequency of occurrence and co-occurrence of tags similar to text processing algorithms elsewhere. These tagging practices can be examined to identify patterns [5] which, in turn, can lead to identification of concepts.

3 Identifying Domain-Level Constructs

This paper proposes a set of heuristics that use text-processing mechanisms for identifying domain level constructs. The heuristics use the following notations.

Resources and Tags

$s,t \in T$: Tags; $u,v \in V$: Users; $q,r \in R$: Resources; $title_r$: Title of resource r
tag_{rt} Resource r is tagged with Tag t $\{0,1\}$
$tagged_{rtu}$ Resource r is tagged with Tag t by User u $\{0,1\}$
$num (tag_{rt})$ Number of Resources tagged with Tag t, $num (tag_{rt}) = \sum (tag_{rt}) \ \forall \ r \in R$
$date (tagged_{rtu})$ Date on which Resource r was tagged with Tag t by User u
$freq (tag_{rt})$ Frequency with which Tag t is used to describe Resource r
$co\text{-}occur_{str}$ Tags s and t co-occur for Resource r
$freq (co\text{-}occur_{str})$ Frequency with Tag s and t co-occur for Resource r
$co\text{-}exist_{st}$ Tags s and t co-exist $\{0,1\}$, $co\text{-}exist_{st} = 1$ iff $\exists \ r \in R$ s.t. $freq (co\text{-}occur_{str}) > 0$
$freq (co\text{-}exist_{st})$ Frequency with which Tag s and t co-exist,
                      $freq (co\text{-}exist_{st}) = \sum (co\text{-}occur_{str}) \ \forall \ r \in R$, $co\text{-}exist_{st} = 1$ iff $freq (co\text{-}exist_{st}) > 0$

Identifying Domain-level Constructs

$candidate_t$ Probability that Tag t is a candidate domain-level construct $[0,1]$
$connected_{st}$ Probability that Tags s and t are connected $[0,1]$
$entity_t$ Probability that Tag t represents an Entity $[0,1]$
$attribute_t$ Probability that Tag t represents an Attribute $[0,1]$
$relationship_t$ Probability that Tag t represents an Attribute $[0,1]$
$instance_t$ Probability that Tag t represents an Instance of an Entity $[0,1]$
$value_t$ Probability that Tag t represents a Value of an Attribute $[0,1]$
$subsume_{st}$ Tag s subsumes Tag t $\{0,1\}$
$synonym_{st}$ Tags s and t are synonyms $\{0,1\}$
$hypernym_{st}$ Tag s is a hypernym of Tag t $\{0,1\}$
$dominant\text{-}sense_t$ Dominant wordsense for a tag t $\{noun, verb, adjective\}$
                     For the purpose of this research, others are considered irrelevant.
m_n Probability indicator, $m > 0$, $m_n > m_{n-1}$, $n=\{0,1,2,\dots\}$

3.1 Heuristics to Identify Concepts and Connectedness

Five heuristics assess the probability (m_n) that a Tag is a *Candidate* domain-level construct and identifying pairwise *Candidate* connections. The heuristics to identify concepts from user-contributed tags provide indicators that progressively increase the probability that a tag represents a domain-level construct.

Heuristic 1 - Frequency of Authored Tag: The first heuristic leverages the number of users who use a tag, regardless of the resources they use to describe the tag. The heuristic follows the rationale that the larger the number of users who use the tag, the greater the agreement that the tag represents a meta-level concept.

$$\sum (tag_{rtu}) \geq \text{Threshold-U}, \ \forall \ u \in U, r \in R$$
$$\Rightarrow candidate_t = m_n \mid n=1, m_1 > 0 \tag{1}$$

Example: Searching delicious.com for "air travel" yields resources and tags used to describe them. Depending upon the threshold, tags are marked as candidates for domain-level constructs. For example, the tags Travel, Flights and Airfare was used by 7953, 4993, and 2892 users respectively.

Heuristic 2 - Number of Resources tagged with a Tag: The second heuristic suggests that the number of resources tagged with a given tag indicates the importance of that tag for the domain. This heuristic, in essence, increases the probability that the candidate tag is a domain-level construct.

$$\text{num } (tag_{rt}) \geq \text{Threshold-R} \mid candidate_t = m_n \; \forall \; u \in U, r \in R$$
$$\Rightarrow candidate_t = m_{n+1} \tag{2}$$

Example: For the air travel domain, the number of resources tagged are obtained from delicious.com. For example, the tags Travel, Flights and Airfare were associated with 842103, 36605, and 13307 resources respectively.

Heuristic 3 - Frequency of Tag Authorship: The third heuristic examines the frequency of tags and uses the rationale that the greater the frequency, the more important the tag. Because this heuristic examines only resources generated from the previous heuristic, it effectively bounds the search, and results in a scoping decision.

$$\sum \text{freq } (tag_{rt}) \geq \text{Threshold-F}, \; \forall \; r \in R \mid ((res_{rt}) = 1 \; \forall \; t \in T \mid candidate_t = m_n)$$
$$\Rightarrow candidate_t = m_{n+1} \tag{3}$$

Example: Tags applied to resources identified in the previous heuristic are examined. Table 1 shows an example.

Table 1. Sample Resources and Tag Frequencies

Resource	Number of People tagging	Top tags for the resource - Tag(Frequency)
Kayak.com	16,114	Travel(12055), flights(4514), airfare(4390), search(2784), tickets(2286), airline(1652), hotel(1643), shopping(1085), cheap(698), tools(562), vacation(526)
Farecast.com	13,338	Travel(9436), airfare(5539), flights(3900), tickets(3152), airline(2314), shopping(2004), tools(658), deals(593)
Sidestep.com	4,378	Travel(3190), airfare(1117), flights(1109), search(672), tickets(655), airline(618)

The heuristics, in effect, move from users to tags to resources, in each step, increasing the confidence in the assessment that a Tag is a domain-level construct.

Heuristic 4 - Connectedness of Tags: This heuristic only considers tags that have a reasonable probability of being domain-level constructs, i.e., the universe of tags identified by the heuristics thus far.

$$\text{If freq } (co\text{-}exist_{st} > \text{Threshold-T}) \; \forall \; s,t \in T \mid candidate_s = m_n \wedge candidate_t = m_n$$
$$\Rightarrow connected_{st} = m_{n+1} \tag{4}$$

Example: The tags *Travel* and *Flights*, identified by the previous heuristics, co-occur 31 times in the first 100 results returned.

The result of this first set of heuristics is a set of tags that are likely candidates for domain-level constructs, and potential connections among these constructs.

3.2   Heuristics to Distinguish Entities and Relationships

Seven heuristics, drawing on lexical sources [4] support this phase to improve the probability that a Tag, identified as a *Candidate*, represents an *Entity* or a *Relationship*. If greater support is found for the interpretation that a tag is an Entity, it improves the probability that the tag represents an Entity, and simultaneously lowers the probability of a Relationship. The set of candidates identified in the final heuristic in the previous sub-section {candidate$_t$} | candidate$_t$ = m$_n$ constitute the starting point. The probability m$_n$ provides the initial probability that a tag represents both, an Entity and a Relationship. The heuristics uses Wordnet [4].

Heuristic 5 - No Match in a Lexicon: A tag that is a Proper Noun is not likely to be an Entity, but may indicate the need for an Entity (with the Tag an instance). If the Tag is not present in lexicon, Wordnet, it is considered to represent a Proper Noun. The inference from the proper noun to the Entity is left for the user. The inference is that the tag is an instance of an Entity, and *not* a relationship.

$$\text{If dominant-sense}_t = \text{Null} \wedge \text{candidate}_t = m_n$$
$$\Rightarrow \text{instance}_t = m_{n+1} \wedge \text{relationship}_t = m_{n-1} \tag{5}$$

Example: The tags "Europe", "Japan", "London" represent instances of country and city. For the example, manual examination shows that no tags are proper nouns.

Heuristic 6 - Word-sense=Noun: This heuristic argues that if the dominant word-sense is Noun, the tag is a likely to be an Entity (or Attribute) but not a relationship.

$$\text{If dominant-sense}_t = \text{'noun'} \wedge \text{candidate}_t = m_n$$
$$\Rightarrow \text{entity}_t = m_{n+1} \wedge \text{attribute}_t = m_{n+1} \wedge \text{relationship}_t = m_{n-1} \tag{6}$$

Example: The tags *Flight* and *Airfare* are identified as nouns (dominant word-sense frequency from WordNet are 6 and 1 as a noun, respectively). The probability that each is either an Entity or Attribute is increased.

Heuristic 7 - Word-sense=Adjective: If the dominant word-sense is Adjective, the tag is a likely to be a Value of an Attribute, not a relationship.

$$\text{If dominant-sense}_t = \text{'adjective'} \wedge \text{candidate}_t = m_n$$
$$\Rightarrow \text{value}_t = m_{n+1} \wedge \text{relationship}_t = m_{n-1} \tag{7}$$

Example: The tags *Cheap* and *LowCost* (low-cost) are identified as adjectives by WordNet and are potential attributes of entities, not relationships.

Heuristic 8 - Word-sense=Verb: If the dominant word-sense for a Tag is Verb, it is likely to be a Relationship but not an Entity.

$$\text{If dominant-sense}_t = \text{'verb'} \wedge \text{candidate}_t = m_n$$
$$\Rightarrow \text{relationship}_t = m_{n+1} \wedge \text{entity}_t = m_{n-1} \tag{8}$$

Example: The tag *Flying* is identified by WordNet as a verb and, hence, likely to represent a relationship.

Heuristic 9 - (Non-)Synonyms: This heuristic uses connectedness of tags, and a check of whether they are synonyms. If the tags are not synonyms, the heuristic increases the probability that the tags represent separate entities. If they represent synonyms, it combines the tags and ascribes the increased probability to the combined tag, removing one of the tags arbitrarily from consideration.

$$\text{If connected}_{st} = m_n \wedge \text{entity}_s = m_n \wedge \text{entity}_t = m_n \wedge \text{synonym}_{st} \neq 1$$
$$\Rightarrow \text{entity}_s = m_{n+1} \wedge \text{entity}_t = m_{n+1} \tag{9a}$$

$$\text{If connected}_{st} = m_n \wedge \text{entity}_s = m_n \wedge \text{entity}_t = m_n \wedge \text{synonym}_{st} = 1$$
$$\Rightarrow \text{subsumes}_{st} = 1 \wedge \text{entity}_t = m_1 \wedge \text{entity}_t = m_{n+1} \tag{9b}$$

Example: *Flight* and *Airline* are connected but are not synonyms; hence the probability that these two tags represent entities is high. *Transport* and *Transportation* are synonyms (in same synset in WordNet); one subsumes the other; one is removed from further consideration.

Heuristic 10 - (Non-)Synonyms-2: This uses the connectedness of tags to improve the probability of a tag as an Entity and a Relationship if they are connected.

$$\text{If connected}_{st} = m_n \wedge \text{entity}_s = m_n \wedge \text{relationship}_t = m_n \wedge \text{synonym}_{st} \neq 1$$
$$\Rightarrow \text{entity}_s = m_{n+1} \wedge \text{relationship}_t = m_{n+1} \tag{10}$$

Example: If the tags *Flight* and *Search* are connected, from WordNet, it can be shown that these two tags are not synonyms, so the probability they represent an entity and a relationship is high.

Heuristic 11 - Need for a Relationship: This heuristic leverages the connectedness of tags to identify the need for a relationship. If a needed relationship is inferred because two entities are connected, it introduces a new tag.

$$\text{If connected}_{st} = m_n \wedge \text{synonym}_{st} \neq 1$$
$$\Rightarrow \text{candidate}_{t'} = 1 \wedge \text{relationship}_{t'} = m_n \tag{11}$$
$$\text{where } t' \text{ (t-prime) is a new, un-labeled, tag}$$

Example: The tags *Airline* and *Flight* are related though their connectedness property, but there may not be an explicit relationship tag between them. This warrants the addition of a new relationship tag (e.g., *owns* or *has*).

The result of this second set of heuristics is improved probabilities for tags that are likely to be domain-level Entities or Relationships.

3.3 Heuristics to Identify Attributes and Properties

Three heuristics distinguish *Entities* and *Attributes*, and identify additional *Attributes* based on the Candidate Tags available as Concepts and their Connectedness.

Heuristic 12 - Word-Sense Combination, Noun-Null: If (a) two Tags are connected, (b) the dominant word-sense for one Tag is Null, suggesting a proper noun (see (5) above), and (c) the other is a Noun (see (6) above), then the first tag is likely to be an Entity and not an Attribute.

$$\text{If connected}_{st} = m_n \wedge$$
$$\text{If dominant-sense}_s = \text{Null} \wedge \text{candidate}_s = m_n \Rightarrow \text{instance}_s = m_{n+1} \text{ (from 5)} \wedge$$
$$\text{If dominant-sense}_t = \text{'noun'} \wedge \text{candidate}_t = m_n \Rightarrow \text{entity}_t = m_{n+1} \text{ (from 6)} \tag{12}$$
$$\Rightarrow \text{entity}_t = m_{n+2}$$

Example: If the tags *London* and *Tour* are connected (based on connectedness), then the *Tour* is likely to be an entity and the *London* an instance.

Heuristic 13 - Word-Sense Combination, Noun-Adjective: The second heuristic argues that if: (a) two Tags are connected, (b) the dominant word-sense for one Tag is a Noun (see (6) above), and the (c) other is an Adjective (see (7) above), then the first tag is likely to be an Attribute, not an Entity.

$$\text{If connected}_{st} = m_n \wedge$$
$$\text{If dominant-sense}_t = \text{'noun'} \wedge \text{candidate}_t = m_n \Rightarrow \text{instance}_t = m_{n+1} \text{ (from 6)} \wedge$$
$$\text{If dominant-sense}_t = \text{'adjective'} \wedge \text{candidate}_t = m_n \Rightarrow \text{attribute}_t = m_{n+1} \text{ (from 7)} \tag{13}$$
$$\Rightarrow \text{attribute}_t = m_{n+2}$$

Example: The tags *Cheap* and *Flight*, are connected to each other. The dominant word sense for *Cheap* is an adjective and for *Flight* is a noun (both determined from WordNet). Then, *Cheap* is a potential attribute and the tag *Flight* is an entity.

4 Implementation

A prototype environment is currently under development. The prototype will contain modules for automatically collecting tags and processing them to extract concepts and relationships. The modules are being implemented using Java and Perl and the databases with MYSQL. A *Tagged Resources Identifier* module will provide the interface for the user to specify the search phrase in natural language and will parse the input to generate the search terms by removing stop words, etc. This module will be responsible for interfacing with social bookmarking websites. A *Tag Extractor* module will extract tags associated with the resources that result from the search. Along with the tags, additional information will be culled such as frequency, number of users etc. The extracted tags will be time-stamped and stored in a *Tag Database*. A *Concept-Relationship Extractor* module will implement the heuristics. It will help the user in a systematic way and elicit feedback. It will interface with *WordNet*.

5 An Application

The approach was applied to delicious.com, which provides a facility for searching the resources that have been tagged. Figure 1 shows the results found by searching for the label 'corporate sustainability.' Figure 2 shows all tags for an article.

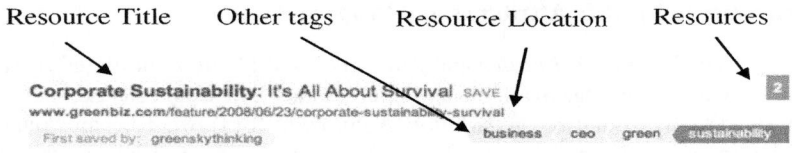

Fig. 1. An Example of Starting with a Core Tag

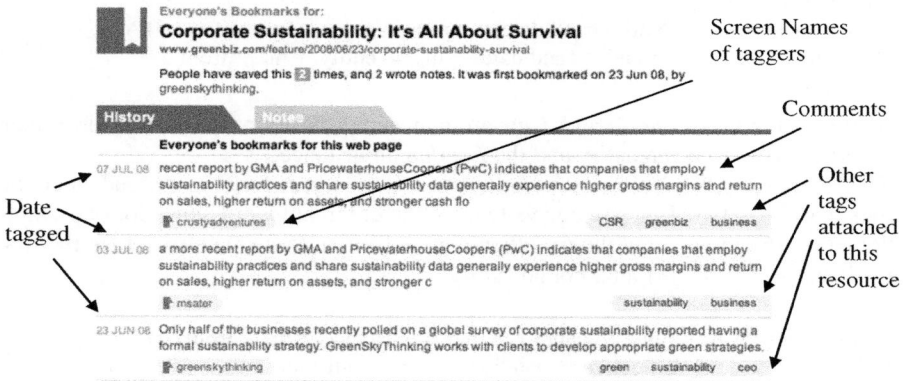

Fig. 2. Multiple Tags associated with a Resource

The concepts and relationships identified with the help of heuristics can be represented as local graphs (see Figure 3a). They are integrated to generate a larger graph (see Figure 3b). Word senses and synonyms obtained from WordNet are used to find common nodes for integration.

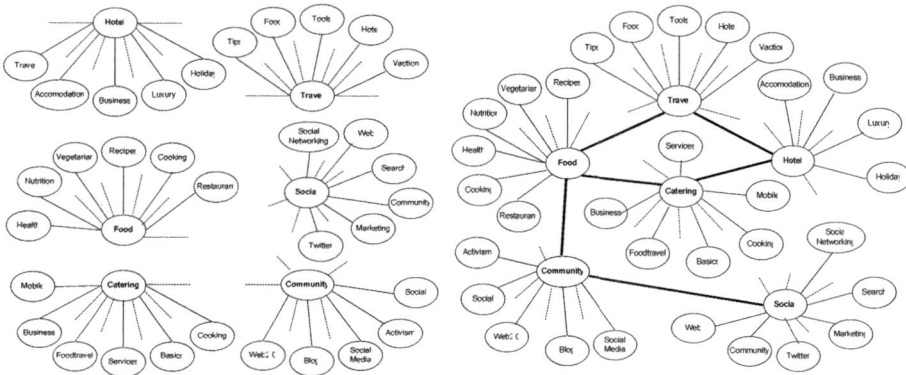

Fig. 3a. Local Graphs for Hospitality Domain (Partial)   **Fig. 3b.** Merged Graph for Hospitality Domain (Partial)

6   Conclusion

A set of heuristics based on simple text-processing mechanisms to extract and represent concepts for domain models have been developed. To illustrate feasibility and provide proof-of-concept, the heuristics were applied across several domains. The results are encouraging and illustrate that simple text processing mechanisms with the tags layer can help identify and organize concepts in a domain. Further research will refine the heuristics, complete the implementation and evaluate the outcomes.

Acknowledgment. This work has been supported by Sogang Business School's World Class University Program (R31-20002) funded by Korea Research Foundation, the Georgia State University, and the Spanish Ministry MICINN (TIN2008-00444).

References

1. Angeletou, S., Sabou, M., Specia, L., Motta, E.: Bridging the Gap Between Folksonomies and the Semantic Web: An Experience Report. In: Workshop Bridging the Gap between Semantic Web & Web 2.0 at 4th European Sem. Web Conf., Innsbruck (2007)
2. Campbell, D.G.: A Phenomenological Framework for the Relationship Between the Semantic Web and User-Centered Tagging Systems. In: Proceedings of the 17th ASIS&T SIG/CR Classification Research Workshop, Austin, TX, USA (2006)
3. Dye, J.: Folksonomy: A Game of High-tech (and High-stakes) Tag. EContent 29(3), 38–43 (2006)
4. Fellbaum, C. (ed.): Wordnet: An electronic lexical database. MIT Press, Cambridge (1998)

5. Kipp, M., Campbell, D.: Patterns and Inconsistencies in Collaborative Tagging Systems: an Examination of Tagging Practices. In: Prs of the ASIST 2006, Austin, TX, USA (2006)
6. Sunstein, C.: Infotopia: How Many Minds Produce Knowledge. Oxford Univ. Press, Oxford (2006)
7. Weikum, G., Kasneci, G., Ramanath, M., Suchanek, F.: Database and information-retrieval methods for knowledge discovery. C ACM 52(4), 56–64 (2009)
8. Wu, X., Zhang, L., Yu, Y.: Exploring social annotations for the semantic web. In: Proceedings of the 15th Int. Conf. on WWW, Edinburgh, Scotland, pp. 417–426 (2006)

Analysis of Definitions of Verbs
in an Explanatory Dictionary
for Automatic Extraction of Actants
Based on Detection of Patterns

Noé Alejandro Castro-Sánchez and Grigori Sidorov

Natural Language and Text Processing Laboratory,
Center for Research in Computer Science (CIC),
National Polytechnic (Technical) Institute (IPN),
Av. Juan Dios Batiz, s/n, Zacatenco, 07738, Mexico City,
Mexico
noe.acastro@gmail.com, sidorov@cic.ipn.mx

Abstract. Due to the importance that verbs have in language an identification of their actants (obligatory complements) is important for understanding of the meaning of sentences. Usually, the solution of this problem in natural language processing is based on machine learning approaches, which are trained on large sets of tagged texts. We show that it is possible to work with other kind of sources, i.e., explanatory dictionaries. Dictionary definitions have patterns that provide enough information for identifying actants. We develop a heuristic approach in order to obtain this information and developed an algorithm for detection of actants in texts.

Keywords: actants extraction, definition pattern, explanatory dictionary.

1 Introduction

The verb is considered as a linguistic center of a sentence. It expresses states, actions or processes that involve some other entities. These entities are known as *actants,* a term proposed by L. Tesnière in 1959. This term was motivated by an analogy with the theatre play, containing the action (verb), the actors (actants) and the decoration (circumstances). The actants are defined as obligatory expressions demanded by verbs: their absence would produce ungrammatical sentences. The circumstances can be omitted without harming a sentence being their principal function to broaden its meaning [6].

The correct identification of these entities is crucial for various linguistic theories and is widely used in various tasks of natural language processing. Usually they are extracted from a corpus using the concept of subcategorization frames on the basis of their grammar category, number of actants, and their syntactic position. Majority of these methods are developed for English ([1, 4, 5, 10, 17]), also there are works related to Spanish ([14]), Hungarian ([16]), Czech ([15]), Bulgarian ([13]), Italian ([9]), etc.

C.J. Hopfe et al. (Eds.): NLDB 2010, LNCS 6177, pp. 233–239, 2010.

Other data sources also have been used for this task, such as the Web ([10, 18]), multilingual resources and texts ([1]), bilingual dictionaries, as in [5], where the dictionaries are used to extend the number of entries in a dictionary of valences, assuming that verbs with similar translations should have the same valence structure.

In this work, we use a large explanatory dictionary of Spanish to identify the actants required by verbs. We assume that definitions of verbs in dictionaries provide enough information to identify the actants, also providing some selectional restrictions that would help to identify them in sentences.

2   Patterns in Definitions

We used for our experiments the dictionary of Spanish Royal Academy which contains 162,362 definitions (senses) grouped in 89,799 lexical entries. From these, 12,008 lexical entries correspond to verbs, which contain 27,668 definitions (senses). The distribution of number of definitions (senses) among verbs corresponds to the well-known Zipf law, where there are few verbs with many senses, while the majority of verbs have only one sense.

A typical dictionary definition presents two terms: *genus* or hyperonym, and *differentia*, which shows the characteristics that distinguish the lexical entry from other items grouped within the same genus. This last term contains some elements named *pattern in definition*, which are not strictly related with the semantic content of the lexical unit, but are included to help the correct use of the defined term, providing some contextual restrictions ([5, 19]). We use this information to identify the verb actants. This can be appreciated carrying out the exchange task, a process that allows extraction of a pattern from the definition. For example:

Sostener: Sustentar, mantener firme algo
(Hold: To take something firmly).

Suppose that we have the following sentence:

Juan sostiene los libros
(John holds the books).

The exchange is carried out by replacing the defined term by its definition, as it is shown in the following sentence:

Juan sustenta, mantiene firme los libros
(John takes the books firmly).

We can see in the last sentence, that it was necessary to remove the element *"algo"* (*"something"*) contained in the definition, because it is replaced by the noun phrase *"los libros"* (*"the books"*). This situation is called *pattern in definition*, and it exposes the actant that is required by the verb *sostener* (*to hold*).

3   Processing of Definitions

Verb meanings are extracted from the dictionary and grammatically tagged with the FreeLing parser, an open source text analysis tool for various languages including Spanish [11].

We carry out processing of definitions based on the following algorithm, which we will explain in detail in the following sections:

- Identify the patterns in definitions.
- Extract the verbs that have any pattern in their first sense.
- Identify the senses of other verbs where the verbs obtained at the previous step are genus of the definition.

3.1 Extraction of Patterns from Definitions

First, the patterns in definition are identified. In order to do that we consider that nouns and indefinite pronouns are the only parts of speech that can represent patterns. We measure their frequency in definitions, see Table 1.

Table 1. Frequencies of the usage of nouns and indefinite pronouns in definitions

Word	Frequency
Algo (something)	3,462
Alguien (somebody)	2,154
Cosa (thing)	1,594
Persona (person)	1,175
Lugar (place)	408
Cuerpo (body, section)	349
Animal (animal)	300
Agua, acción, tierra, fuerza, ... (water, action, land, force, ...)	< 300

It can be seen that the most frequent elements represent very general classes of words. For example, *algo (something)* can be matched with any word that refers to any non-animated thing, *alguien (someone)* matches with any word that refers to a person, *cosa (thing)* behaves in the same way as *algo (something)*, etc. For our experiment, we select from the list above eight most frequently used elements; let us name them *general patterns*. It can be seen that some of these elements are synonyms, i.e., in the same context it is possible to exchange them, say, to change *alguien (somebody)* with *persona (person)*, or *algo (something)* with *cosa (thing)*, etc.

In the majority of cases the patterns observed in the definition do not correspond to the complete list of valences (subcategorization frame). For example,

Conducir: Llevar, transportar de una parte a otra.
(Drive: move, transport from one part to another).

This definition does not contain information about the direct object that is necessary. Our preliminary evaluation is that more than 80% of definitions are incomplete in this sense. Nevertheless, each definition contains information about some valences and the proposed algorithm allows their extraction by making substitutions. Again, our preliminary evaluation is that in more than 60% of cases the absent actants are restored using this methodology.

Considering that usually the first sense of a lexical entry is the most frequently used, we extracted verbs that have general patterns in their first sense. The number of these verbs in the dictionary is 2,748.

3.2 Processing of Genus

Almost all definitions included in the dictionary follow the typical formula represented by *genus + differentia* (see Section 2). The predictable position of these elements allowed us to identify them in an automatic way.

From 12,008 verbs that we considered, 3,751 are used as genus. Usually, verbs that have general patterns in their first sense are used as genus. In the following table we show the frequency of the five most frequent verbs.

Table 2. Verbs with highest frequency in definitions as genus

Verb	Frequency
Hacer (to do)	1,462
Dar (to give)	969
Poner (to put)	836
Quitar (to remove)	503
Echar (to throw)	271

We got 5,987 definitions that have a genus that is part of the verbs that have general pattern in their first sense (see above).

After this processing we got the information of actants of dictionary verbs that have patterns.

4   Mapping between Patterns and Actants in Sentences

After this processing we detected patterns in definitions of verbs. The next step is to apply them for identifying the actants in the input sentences. One manner of identifying actants is usage of lexical relations (hyperonim/hiponym) that are detected using ontology like WordNet, for example. However, we considered that the same dictionary used at the first stage can also help us to carry out this task.

The mapping is based on the following algorithm. We maintain the list of the processed words for avoiding possible circles.

1. Identify nouns and indefinite pronouns in sentences. They are the candidate actants.
2. Start with one of the candidates getting its meaning from the dictionary. Clear the list of the processed words.

3. Identify the genus of the definition.
4. Compare this genus with the verb patterns possibly present in the sentence.
    4.1. If genus is different from all patterns, verify that it is not present in the list of the already processed genus and go to step 2, taking the new genus as the candidate for replacing and add the new genus to the list of the already processed words. If it is already processed then go to step 5.
    4.2. Otherwise mark the found actant and proceed to step 5.
5. Determine if there are more candidates to process.
    5.1. If that is the case, go to step 1, with the next candidate.
    5.2. Otherwise finish.

Let us apply the algorithm to the following sentence:

> *María regaló una trenca a su esposo*
> *(Mary gave a duffle-coat to her husband).*

In this example, we find as candidates the words "*trenca*" ("*duffle-coat*") and "*esposo*" ("*husband*"). The next step is to get the meaning of the verb:

> *Regalar: Dar a alguien, sin recibir nada a cambio, algo…*
> *(Give a present: Give something to someone, without expecting compensation …).*

The patterns in this definition are "*alguien*" ("*someone*") and "*algo*" ("*something*"). We execute the previously presented algorithm for the first candidate:

1. **Trenca (duffle-coat)**. *Abrigo corto, con capucha y piezas alargadas (a coat which has toggle buttons and usually has a hood).*
2. Genus: *Abrigo* (*coat*).
3. Compare genus with patterns: *abrigo = alguien* or *abrigo = algo* (*coat = someone* or *coat = something*)
    3.1. Words are different. Go to step 1 of the algorithm with genus "*abrigo*" ("*coat*").

Now we repeat the algorithm with the new word.

1. **Abrigo (coat)**. *Cosa que abriga (something to wrap up).*
2. Genus: *Cosa* (*something*).
3. Compare genus with pattern *cosa = alguien* or *cosa = algo* (*something = someone* or *something = something*)?
    3.1. *Cosa* (*something*) is equal to *algo* (*something*) (see sub-section 3.1). We found the actant here.
4. Go to step 1 and continue with other candidates.

In this example we found the actant at the second iteration.

5 Conclusions

The aim of this work was to explore new resources and methodologies for automatic identification of the actants of verbs. Traditional approaches are based on the usage of texts corpora as training sources for automatic learning systems. Our proposal is based on the recursive analysis of dictionary definitions.

More or less regular structure that lexicographers use while developing definitions allows applying of simple heuristics for identifying the definitions terms and analyzing their structure. A typical definition of a verb contains the actants that can be iteratively reduced to words with very abstract meaning providing semantic restrictions that can be used for identifying the actants in sentences (like *something, someone, place*, etc.). We found that definitions of many verbs in Spanish can be transformed into definitions with abstract words that help in identifying their actants. Our preliminary evaluation is that at least in 70% of cases the absent actants are restored using this methodology.

We showed that the proposed method can be applied for detection of actants in texts.

As a future work we can mention more extended experiments and evaluation, analysis if verbs with the same genus keep the same number and type of actants.

Acknowledgements. Work done under partial support of Mexican Government (CONACYT projects 50206-H and 83270, SNI) and National Polytechnic Institute, Mexico (projects SIP 20080787, 20091587, 20090772, 20100773, 20100668; COFAA, PIFI).

References

1. Alcina, A., Valero, E.: Análisis de las definiciones del diccionario cerámico científico-práctico, sugerencias para la elaboración de patrones de definición. Debate Terminológico, 4 (2008)
2. Aone, C., MacKee, D.: Acquiring Predicate-Argument Mapping Information from Multilingual Texts. In: Corpus processing for lexical acquisition, pp. 191–202. MIT Press, Cambridge (1996)
3. Brent, M.: Automatic acquisition of subcategorization frames from untagged text. In: 29th Annual Meeting of the Association for Computational Linguistics, Berkeley, CA, pp. 209–214 (1991)
4. Brent, M.: From grammar to lexicon: unsupervised learning of lexical syntax. Computational Linguistics 19(3), 243–262 (1993)
5. Cordero, S.: Diccionario De la Lengua Española. Secundaria (Diles): Planta para su elaboración con algunos apuntes básicos de metalexicografía. Káñina. Revista Artes y Letras XXXI, 167–195 (2007)
6. Frías, X.: Introducción a la semántica de la oración del español. Revista Philologica Romanica (2001)
7. Fujita, S., Bond, F.: An Automatic Method of Creating Valency Entries using Plain Bilingual Dictionaries
8. Gahl, S.: Automatic extraction of subcorpora based on subcategorization frames from a part-of-speech tagged corpus. In: 36th Annual Meeting of the Association for Computational Linguistics and 17th International Conference on Computational Linguistics, pp. 428–432. Association for Computational Linguistics, Morristown (1998)
9. Ienco, D., Villata, S., Bosco, C.: Automatic Extraction of Subcategorization Frames for Italian. In: International Conference on Language Resources and Evaluation IREC (2008)
10. Kawahara, D., Kurohashi, S.: Case frame compilation from the web using high-performance computing. In: International Conference on Language Resources and Evaluation, pp. 1344–1347 (2006)

11. Atserias, J., Casas, B., Comelles, E., Gonzáles, M., Padró, L., Padró, M.: FreeLing 1.3: Syntactic and Semantic Services in an Open-Source NLP Library. In: Fifth international conference on Language Resources and Evaluation, Genoa, Italy nlp/freeling (2006), http://www.lsi.upc.edu/nlp/freeling
12. Manning, C.: Automatic acquisition of a large subcategorization dictionary from corpora. In: 31st Annual Meeting of the Association for Computational Linguistics, pp. 235–242. Association for Computational Linguistics, Columbus (1993)
13. Marinov, S., Hamming, C.: Automatic Extraction of Subcategorization Frames from the Bulgarian Tree Bank (2004)
14. Monedero, J., et al.: Obtención automática de marcos de subcategorización verbal a partir de texto etiquetado: el sistema SOAMAS. Procesamiento del lenguaje natural 17, 241–254 (1995)
15. Sarkar, A., Zeman, D.: Automatic Extraction of Subcategorization Frames for Czech
16. Séreny, A., Simon, E., Babarczy, A.: Automatic Acquisition of Hungarian Subcategorization Frames. In: 9th International Symposium of Hungarian Researchers on Computational Intelligence and Informatics, Budapest, Hungary (2008)
17. Ushioda, A., Evans, D., Gibson, T., Waibel, A.: The automatic acquisition of frequencies of verb subcategorization frames from tagged corpora. In: Workshop on the Acquisition of Lexical Knowledge from Text, pp. 95–106. Association for Computational Linguistics, Morristown (1993)
18. Usun, E., et al.: Web-based Acquisition of Subcategorization Frames for Turkish. In: 9th International Conference on Artificial Intelligence and Soft Computing, IEEE Computational Intelligence Society, Los Alamitos (2008)
19. Wortjak, G.: Reflexiones acerca de construcciones verbo-nominales funcionales. Revista de Estudos Linguísticos da Universidade do Porto 1, 257–280 (1998)

An Automatic Definition Extraction in Arabic Language

Omar Trigui[1], Lamia Hadrich Belguith[1], and Paolo Rosso[2]

[1] ANLP Research Group- MIRACL Laboratory, University of Sfax, Tunisia
[2] Natural Language Engineering Lab. - ELiRF, Universidad Politécnica de Valencia, Spain
{omar.trigui,l.belguith}@fsegs.rnu.tn, prosso@dsic.upv.es

Abstract. During the last few years, a lot of researches have focused on automatic definition extraction in the context of question answering systems. Although, these researches have been conducted for different languages, no research has been proposed for Arabic. In this paper, we tackle the automatic definition extraction in the context of Question Answering systems. We propose a method based on patterns to automatically identify a definition answer to a definition question. The proposed method is implemented in an Arabic definitional question answering system. We experimented this system using a set of 50 definition questions, and a corpus of 2000 snippets collected from the Web. The obtained results are very encouraging: 94% of the definition questions have complete definitions among their first 5 answers.

Keywords: Question answering system, pattern-based approach, Arabic language, definition extraction.

1 Introduction

Automatic definition extraction is an important task for many research fields such as question answering (QA) systems, automatic building of glossaries or dictionaries and development of ontologies.

This paper explores definition extraction in the context of QA systems. Definition questions in QA systems ask for interesting information about a particular person e.g. 'Who is Mohamed ElBaradei?' or thing such as 'What is the League of Arab States?'.

During the last decade, many researches have focused on the problematic of definition question in different evaluation conferences such as TREC[1] and CLEF[2].

In this paper, we focus on a core component of a definitional Arabic QA system namely 'Definition Extraction', where we apply our method to identify the definition answers. The paper is structured in six sections. Section 2 presents a brief overview of the state of the art. In Section 3, we propose our method to identify the candidate definitions (*CDs*). In Section 4, we discuss the integration of the proposed method into a definitional QA system. In Section 5, we show the impact of this method in definition QA systems. In Section 6, we present the conclusions and some future directions for our research work.

---

[1] Text Retrieval Conference http://trec.nist.gov/
[2] Cross-Language Evaluation Forum http://clef-campaign.org/

C.J. Hopfe et al. (Eds.): NLDB 2010, LNCS 6177, pp. 240–247, 2010.
© Springer-Verlag Berlin Heidelberg 2010

2 Related Works

Extracting definitions is a very important process in definitional QA systems. Many tasks have been dedicated to this process by TREC since 2003 and also by CLEF in the recent years.

Initial researches in this area were related to pattern-based approaches [1][2][3]. There are three main methods based on this approach. The first method consists of building manually the patterns. This method is the most used in QA systems (e.g. Xu et al. [4] used 23 manual distinct patterns). In spite of their effectiveness, the manually constructed patterns require linguistic experts and a lot of manual labors [5]. The second method consists of building the patterns automatically. This method was developed by Ravichandran and Hovy [6]. It performed better for factual patterns than for definition patterns. The third method consists of generating soft patterns (probabilistic lexico-syntactic patterns) by using unsupervised learning [5]. Two main methods are used to process pattern matching. Hard matching is the simplest one. It is based on a rigid match and does not deal with the language variation. The second method is a soft pattern matching (probabilistic lexico-syntactic pattern matching) developed by Cui et al. [5] who proposed two soft pattern matching models to identify the definition sentences. This method outperformed significantly the hard matching patterns in a definitional QA system [5].

Many statistical methods and techniques have been proposed to rank the *CDs*. Yang et al. [7] used the centroid based ranking method to rank their *CDs*. They built for each topic a centroid vector (i.e., vector of highly relevant words to a topic) basing on definitions in the external knowledge resource. Then, they select as answers the *CDs* which have the best similarity with this centroid vector. Cui et al. [5], Xu et al. [4] and Zhang et al. [8] used mining external definitions method to rank the definitions. This method consists in calculating the similarity between the *CDs* and other definitions from external web knowledge resources (i.e., Encyclopedia[3], Wikipedia[4] and a biography dictionary[5]). Joho and Sanderson [2] proposed to rank the *CDs* according to three scores: the score of the pattern that matches the candidate definition (*CD*), the position of the sentence containing the *CD* in its document, and the percentage of the *CD* words which are occurred across *CDs*.

Later, other researches have combined pattern-based approach with machine learning [9][10][11][12]. This combination aims at filtering the incorrect answers.

There are other researches which have not used any pattern-based approach. Blair-Goldensohn et al. [13] proposed to combine data-driven statistical method and machine learning rules to generate definitions. Prager et al. [14] proposed to decompose a definition question into a series of factoid questions: from the answers to the factoid questions, they deduce answers to the definition question. Zhang et al. [8] identified and selected definition sentences from document collection by using Web knowledge bases. Their system was ranked second (with an f-measure score of 0.404) among all participating systems in TREC 2004.

---

[3] http://www.encyclopedia.com/

[4] http://www.wikipedia.org/

[5] http://www.s9.com/

3 The Proposed Method

In this section, we propose an original method to identify the definition answers of definition questions written in Arabic. The method consists of two main steps: (i) identify the candidate definitions by using a pattern matching process, and (ii) rank the correct definitions by using a statistical approach.

The originality of our idea consists basically of combining two methods for the pattern matching process. The first method called hard matching is well known and has been used in various researches (e.g. Cui et al. [5] and Xu et al. [4]). The second method that we propose is called tolerant matching and it is applied when the HM fails to identify a definition in a snippet.

3.1 Building Lexical Patterns

We manually constructed a set of lexical patterns based on a newspaper articles collection that we have extracted from BBCArabic[6] and Euronews[7] web sites. This collection contains more than 2,000 snippets. For each definition sentence of this collection, we annotated manually the topic, the definition, and their connectors. On the basis of these annotated definition sentences, we constructed a set of patterns to identify the definitions. We associated a score to each pattern according to its precision (*pr*). Where the precision for each pattern is calculated as:

$$pr = \frac{number\ of\ complete\ definitions\ identified\ by\ this\ pattern}{number\ of\ CDs\ identified\ by\ this\ pattern} \tag{1}$$

3.2 Identifying Candidate Definitions

The identification of the *CDs* is based on a pattern matching process. We propose to use two methods for this process. The first method consists of a hard matching (HM) where the alignment between a snippet and a lexical pattern is done slot-by-slot without any tolerance. The second method which we called a tolerant matching (TM) is applied when the HM fails to identify a definition in a snippet. TM aims to exploit the snippets containing definitions missed by HM. It handles tokens that are not expected in the lexical pattern. An example of TM is showed in (Fig. 1). Table 1 presents the tokens of the snippet and the pattern related to the example of (Fig. 1).

Tolerant Matching process. We propose the TM process as a mechanism to overcome some problems of strict slot-by-slot matching which often misses definitions due to minor variations that are not expected in the lexical patterns. This mechanism operates as follows: the topic is identified inside the patterns and the snippets, then the right and left parts of each pair of a pattern and a snippet are aligned according to the delimited topic position. Note that both left and right parts of the snippet and the pattern are matched separately (Fig. 1).

---

[6] http://www.bbc.co.uk/arabic/index.shtml
[7] http://arabic.euronews.net

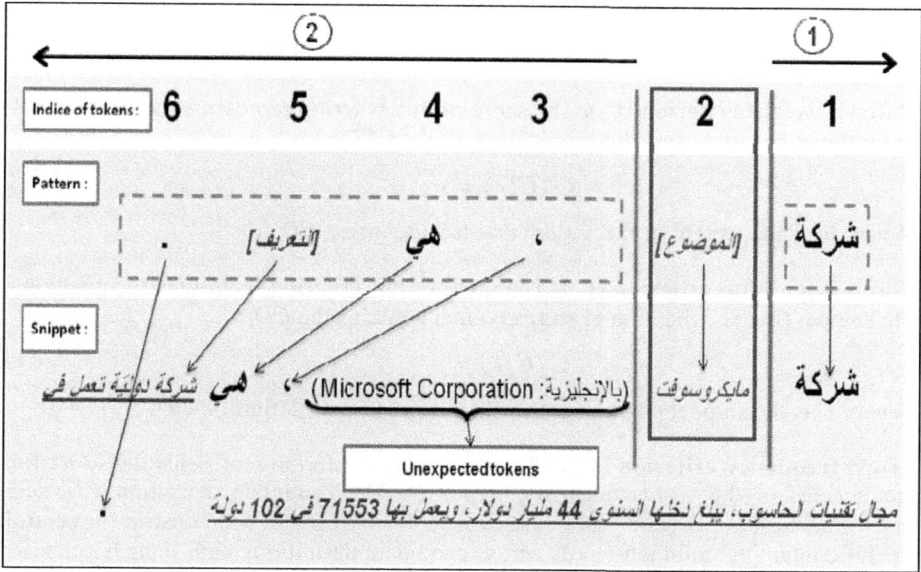

Fig. 1. Example of a TM process

Thus, the matching process of the pattern left part and the snippet left part, starts from $token_{h+1}$ to $token_n$ where h is the topic position and n is the pattern token number. Consequently, the matching process of the pattern right part and the snippet right part starts from $token_{h-1}$ to $token_1$. The right/left part matching process succeeds, if all tokens of the relative part are aligned. It is important to note that the alignment process does not consider the unexpected tokens to the lexical pattern. TM of a pair of pattern and snippet succeeds if the matching processes of both right and left parts succeed. The result of this step is a set of *CDs*.

Table 1. The snippet and pattern tokens

Token (in English)	Token (in Arabic)
Company	شركة
topic	الموضوع
Microsoft	مايكروسوفت
She	هي
(Named: Microsoft Corporation)	(بالإنجليزية:Microsoft Corporation)
Definition	التعريف
An international company working in the field of computer technology, with an annual income of 44 billion \$, and employs 71,553 in 102 countries	شركة دوليّة تعمل في مجال تقنيات الحاسوب، يبلغ دخلها السنوي 44 مليار دولار، ويعمل بها 71553 في 102 دولة

3.3 Ranking Correct Definitions

To rank the candidate definitions (*CDs*), we use the three following criteria:

Pattern weight criterion (C_1). The score of this criterion represents the weight of the pattern that has identified the candidate definition CD_i :

$$C_1(CD_i) = w_i \qquad (2)$$

Where w_i is the weight of the pattern that has identified CD_i

Snippets position criterion (C_2). The score of this criterion represents the position of the snippet (in the collection of snippets) that contains the CD_i :

$$C_2(CD_i) = p_i \qquad (3)$$

Where p_i is the snippet position containing the candidate definition CD_i.

Word frequency criterion (C_3). The score of this criterion represents the sum of the frequencies of the words occurring in a *CD*. The candidate definition CD_i score according to this criterion is calculated as follows. Firstly, we construct a centroid vector containing common words across candidate definitions with their frequencies, beyond stopwords. Secondly, we calculate the frequency sum of the words recurring in both CD_i and centroid vector as indicated by the following formulate:

$$C_3(CD_i) = \sum_{k=1}^{n} f_{ik} \qquad (4)$$

Where n is the number of words which occur in the centroid vector and in the candidate definition CD_i, $1 \leq k \leq n$ and f_{ik} is the frequency of word$_k$.

Criteria aggregation. In order to aggregate the three criteria described above, we first proceed to the normalization of the score of each criterion by dividing it by the maximum score as follows:

$$C'_{i,j} = {C_{i,j}} \Big/ {MaxC_i} \qquad (5)$$

Where i indicates the candidate definition and j the criterion.

Then, we combine the three normalized scores in order to obtain the global score *GS* of the candidate definition CD_i. This global score is done by:

$$GS(CD_i) = \sum_{j=1}^{3} C'_{i,j} \qquad (6)$$

4 Definitional Question Answering System

In this section, we detail the integration of the proposed method in our definitional QA system called *DefArabicQA* [15]. Figure 2 shows the general architecture of this system. *DefArabicQA* starts by analyzing a given definition question in the module "Question Analysis". As a result, the topic is identified and the question type is specified. In the module "Passage Retrieval", the topic is reformulated in two queries for Google[8] and Wikipedia Arabic version[9]. Then, from the retrieved results, the system selects the top 20 snippets returned by each Web resource. Therefore, it considers 40 snippets for each question. In the module "Definition Extraction", the lexical patterns are used to identify the *CDs* from the selected snippets (i.e., we have manually constructed a set of 20 lexical patterns). Finally the retained *CDs* are ranked according to their global scores computed according to the three criteria described in Section 3.3.

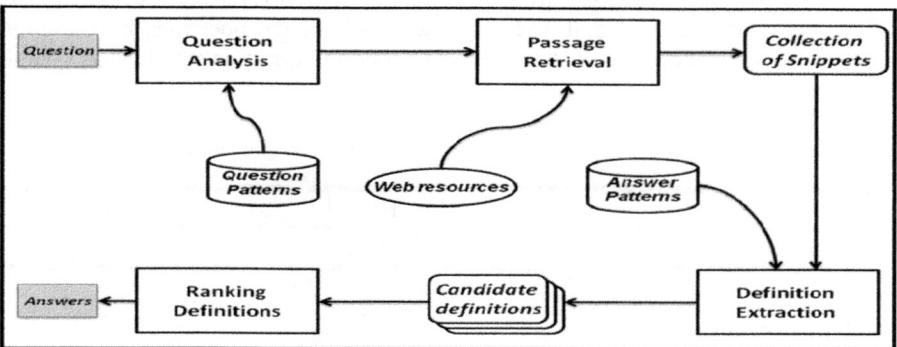

Fig. 2. Architecture of the Arabic definition QA system *DefArabicQA*

5 Experiments

We evaluated our definitional QA system using 50 organization definition questions, and 40 snippets for each question collected from Google and Wikipedia[10].

5.1 Repartition of the Candidate Definitions

Table 2 shows the contribution of hard matching process (HM) and tolerant matching process (TM) to the identification of the candidate definitions (*CDs*). Thus, HM has identified 1670 *CDs* (87% of the total of *CDs*) and TM has identified 244 *CDs* (13% of the total *CDs*). Table 2 shows also the distribution of the collection of *CDs* according to the Web resources. 61% of the *CDs* were identified from the Wikipedia snippets, and the rest (39%) are identified by Google snippets.

---

[8] http://www.google.com/intl/ar/
[9] http://ar.wikipedia.org/w/index.php?
   title=خاص:بحث&search=&go=اذهب
[10] Resources available for research purpose at
   http://sites.google.com/site/omartrigui/downloads

Table 2. *CDs* distribution according to the matching processes and the Web resources

Candidate definitions (*CDs*)	Wikipedia	Google	Total
HM	990 (84%)	680 (92%)	1670 **(87%)**
TM	184 (16%)	60 (8%)	244 **(13%)**
Both (HM+TM)	1174 (61%)	740 (39%)	**1914** (100%)

5.2 Final Results

Table 3 shows the success rate of the identification of complete definition by HM and TM processes. 92% of the questions have complete definitions using only HM process and 50% of the questions have complete definitions using only TM process. When we used both HM and TM processes, 94% of the questions have complete definitions. We deduce that TM process increased the question rate having complete definition by 2%.

Table 3. The percentage of successful questions for each type of matching process

	HM	TM	Both (HM+TM)
Total questions have complete answers	**92%**	50%	**94%**

6 Conclusions

In this paper we have proposed a novel method to identify the definition answers in Arabic definition QA systems. The novelty of the method consists of a two-step approach based on identifying the candidate definitions by using a pattern matching process; and ranking the correct definitions by using a statistical mechanism.

We have proposed also, in the identification phase of candidate definitions, a new method called tolerant matching which is able to deal with various languages in the matching process. This method has increased the rate of questions with complete definitions by 2%.

We have experimented our work in the *DefArabicQA* definitional QA system using Web resources and 50 organization definition questions. The evaluation shows that obtained results are very encouraging. Our system succeeded in generating pertinent definitions for 47 out of 50 definition questions (94% of the total questions) in the top-five answers.

As perspectives we plan to employ machine learning techniques such as Naïve Bayes to classify the candidate definitions before ranking them. We also plan to experiment our method with different type of corpus.

Acknowledgments. This research work started thanks to the bilateral Spain-Tunisia research project on "Answer Extraction for Definition Questions in Arabic" (AECID-PCI B/017961/08). The work of the third author was carried out in the framework of the AECID-PCI C/026728/09 and the MICINN TIN2009-13391-C04-03 (Plan I+D+i) research projects.

References

1. Soubbotin, M.M., Soubbotin, S.M.: Patterns of Potential Answer Expressions as Clues to the Right Answers. In: 10th Text Retrieval Conference (TREC 2001), Gaithersburg, MD, pp. 293–302 (2001)
2. Joho, H., Sanderson, M.: Retrieving Descriptive Phrases from Large Amounts of Free Text. In: 9th ACM CIKM Conference, McLean, VA, November 2000, pp. 180–186 (2000)
3. Joho, H., Liu, Y.K., Sanderson, M.: Large scale testing of a descriptive phrase finder. In: 1st Human Language Technology Conference, San Diego, CA, pp. 219–221 (2001)
4. Xu, J., Licuanan, A., Weischedel, R.M.: TREC 2003 QA at BBN: Answering definitional questions. In: 12th TREC, Washington, DC, pp. 98–106 (2003)
5. Cui, H., Kan, M.Y., Chua, T.S.: Soft Pattern Matching Models for Definitional Question Answering. ACM Transactions on Information Systems (TOIS) 25(2) (2007)
6. Ravichandran, D., Hovy, E.H.: Learning surface text patterns for a Question Answering system. In: ACL 2002, Philadelphia, PA, pp. 41–47 (2002)
7. Yang, H., Cui, H., Maslennikov, M., Qiu, L., Kan, M.Y., Chua, T.S.: QUALIFIER in TREC-12 QA main task. In: TREC 2003, Gaithersburg, MD, pp. 480–488 (2003)
8. Zhang, Z., Zhou, Y., Huang, X., Wu, L.: Answering definition questions using web knowledge bases. In: Dale, R., Wong, K.-F., Su, J., Kwong, O.Y. (eds.) IJCNLP 2005. LNCS (LNAI), vol. 3651, pp. 498–506. Springer, Heidelberg (2005)
9. Westerhout, E.: Extraction of definitions using grammar-enhanced machine learning. In: Student Research Workshop, EACL 2009, Athens, Greece, pp. 88–96 (2009)
10. Borg, C., Rosner, M., Pace, G.: Evolutionary Algorithms for Definition Extraction. In: The 1st Workshop on Definition Extraction 2009, RANLP, Borovets, Bulgaria (2009)
11. Fahmi, I., Bouma, G.: Learning to Identify Definitions using Syntactic Features. In: Workshop of Learning Structured Information in Natural Language Applications, EACL, Trento, Italy (2006)
12. Miliaraki, S., Androutsopoulos, I.: Learning to identify single-snippet answers to definition questions. In: COLING 2004, Geneva, Switzerland, pp. 1360–1366 (2004)
13. Blair-Goldensohn, S., McKeown, K., Schlaikjer, A.H.: Answering Definitional Questions: A Hybrid Approach. In: New Directions in Question Answering 2004, pp. 47–58 (2004)
14. Prager, J.M., Chu-Carroll, J., Czuba, K., Welty, C., Ittycheiach, A., Mahindru, R.: IBM's PIQUANT in TREC 2003. In: 12th Text REtreival Conference, Gathersburg, MD, pp. 283–292 (2003)
15. Trigui, O., Belguith, H.L., Rosso, P.: DefArabicQA: Arabic Definition Question Answering System. In: Workshop on Language Resources and Human Language Technologies for Semitic Languages, 7th LREC, Valletta, Malta (2010)

Automatic Term Extraction
Using Log-Likelihood Based Comparison
with General Reference Corpus

Alexander Gelbukh[1], Grigori Sidorov[1],
Eduardo Lavin-Villa[1], and Liliana Chanona-Hernandez[2]

[1] Center for Computing Research (CIC),
National Polytechnic Institute (IPN),
Av. Juan de Dios Bátiz, Zacatenco,
Mexico DF, 07738, Mexico
[2] Engineering faculty (ESIME),
National Polytechnic Institute (IPN),
Av. Juan de Dios Bátiz, Zacatenco,
Mexico DF, 07738, Mexico
www.gelbukh.com, www.g-sidorov.org

Abstract. In the paper we present a method that allows an extraction of single-word terms for a specific domain. At the next stage these terms can be used as candidates for multi-word term extraction. The proposed method is based on comparison with general reference corpus using log-likelihood similarity. We also perform clustering of the extracted terms using k-means algorithm and cosine similarity measure. We made experiments using texts of the domain of computer science. The obtained term list is analyzed in detail.

Keywords: Single-word term extraction, log-likelihood, reference corpus, term clustering.

1 Introduction

Automatic term extraction is an important task in the field of natural language processing [1]. Even preliminary term extraction with certain degree of errors for further manual processing is very useful. Extracted terms can be used, for example, in ontology construction, information retrieval, etc.

Manual term extraction is possible but it relies on human knowledge of an expert, so, this process is expensive and very slow. Another important consideration is that the extracted terms would be subjective [9], i.e. the experts would have different opinions, while the automatic processing is more objective though it depends on the availability of the corpus data.

Traditionally, the investigations on term extraction are focused on extraction of multi-word terms. POS tagging and various parsers, as well as statistical methods are used in this task [6]. In this case, the main purpose of the statistical methods is the evaluation of the strength of connection between words in multi-word terms.

C.J. Hopfe et al. (Eds.): NLDB 2010, LNCS 6177, pp. 248–255, 2010.

In our opinion, this task should be separated into two steps. At the first step, we detect most probable single words that are candidates for being terms in a specific domain. At the second step, we can apply techniques for multi-word term extraction for obtained single-word terms.

In this paper, we present a method that corresponds to the first step of term extraction (single-word terms extraction) and corresponding experiments for Spanish language. We also perform automatic clustering of terms.

Further in the paper we first describe the method, then present detailed results of our experiments and discuss them, and finally conclusions are drawn.

2 Description of the Term Extraction Method

The proposed method is a modified version of the method presented in [8] for Chinese language. The input data are texts from a specific domain: in our experiments we used texts of the domain of computer science. Also, some general reference corpus should be used for comparison. It is expected that the reference corpus is rather big because otherwise we will not be able to filter out general words of the domain corpus.

The general idea is related to comparison of weighted frequencies in the two corpora: if a word appears much more frequently in the domain corpus, it is a probable term. Note that in [8] it is stated that log-likelihood comparison gives better results than more traditional *tf-idf* based comparison.

There are two main stages during the whole processing: preprocessing and proper term extraction using log-likelihood.

We modified the method [8] in the following aspects: it is applied to Indo-European language (Spanish) with corresponding changes in preprocessing; we do not use any enrichment with additional resources (for example, WordNet) that is important part of the original method; we changed the formula for calculation of the log-likelihood similarity: instead of using more traditional log-likelihood test, we calculate the log-likelihood based distance [7], [2]. Note that this distance measure was precisely developed for comparison of corpora. We also was obliged to add an additional step in the calculations that distinguishes the domain terms as compared to possible words from the general corpus with the same properties, see Formula 4.

Several operations are performed at the preprocessing step. First we tokenize the texts. Then all words are changed to the unified register. We ignore punctuation marks, special symbols, and numbers; all words are lemmatized using freely available lemmatizer for Spanish developed in our laboratory. We filter out auxiliary words: prepositions, articles, auxiliary words, etc.

For calculation of weights of words we used the following formula [7].

$$
G = 2 * \left(\left(freq_{domain} * \log\left(\frac{freq_{domain}}{freq\_Expected_{domain}} \right) \right) + \left(freq_{general} * \log\left(\frac{freq_{general}}{freq\_Expected_{general}} \right) \right) \right)
\tag{1}
$$

where $freq_{domain}$ and $freq_{general}$. are real frequencies in the domain corpus and in the reference corpus.

$freq\_Expected_{domain}$ and $freq\_Expected_{general}$ are expected frequencies in the domain corpus and in the reference corpus. They are calculated according to the following formulae:

$$freq\_Expected_{domain} = size_{domain} * \frac{freq_{domain} + freq_{general}}{size_{domain} + size_{general}} \qquad (2)$$

$$freq\_Expected_{general} = size_{general} * \frac{freq_{domain} + freq_{general}}{size_{domain} + size_{general}} \qquad (3)$$

As the result of application of this formula, all words in the domain corpus are assigned weights.

Another important step in the algorithm consists in the following. Note that Formula 1 cannot distinguish to which of the two corpora the term belongs, i.e. Formula 1 is symmetrical for both corpus. We should somehow correct this situation because we are searching terms specifically in the domain corpus, and not all words with similar properties. So, we add an additional condition: we take into account only terms that satisfy Formula 4, i.e. their relative frequency is bigger in the domain corpus than in the reference corpus. If this condition is false, then we discard the word as a possible term: in our case, we multiply its weight by -1.

$$\frac{freq_{domain}}{size_{domain}} > \frac{freq_{general}}{size_{general}} \qquad (4)$$

After application of Formula 4, some words of the domain corpus will be discarded as possible terms: their weight will be negative. See an example given in Table 1.

Table 1. Examples of the calculated data

Word	$freq$ domain	$freq$ general	$freq\_Expected$ domain	$freq\_Expected$ general	G
... socket	1	0	0.010	0.989	9.153
sofisticado (sofisticated)	5	169	1.789	172.210	3.912
soft	1	12	0.133	12.866	2.351
software	430	831	12.971	1248.028	2334.961
sofware[1]	2	2	0.041	3.958	12.803
sol (sun)	2	933	9.618	925.381	-9.016
solamente (just) ...	20	1714	17.837	1716.162	0.254

---

[1] This is a spelling error in the corpus, the correct form is *software*.

It can be seen that the word *software* has very high weight. At the same time, though the word *sol* (*sun*) also has rather high weight, it is discarded because its relative frequency in the reference corpus is greater than in the domain corpus.

After application of this processing we have a list of words from the domain corpus ordered by their weight. Still, the question remains what is the value of the threshold for selection of the upper part of this list. For the moment we use the empirical value for this threshold, see below.

3 Experiments and Discussion

In our experiments we used the following data (in Spanish). We used issues of the *Excelsior* newspaper (Mexico, 1990s) as a general reference corpus, total 1,365,991 running words. We used texts related to computer science loaded from Wikipedia as a domain corpus, for example, articles about *informatics, software, programming*, etc. Totally we used 26 articles that contain 44,495 words.

After several experiments, we decided empirically to use threshold of 270 elements for the selection of the terms with the highest weight. In the paper [8], threshold of 216 terms was selected, but they also used additional analysis of relations in Chinese analogue of WordNet.

After extraction of terms and selection of the set of high scored terms, we cluster them using standard *k-means* algorithm. Note that *k-means* algorithm needs a manual selection of number of classes for clustering. For calculating of the similarity during clustering we used the standard cosine measure calculated over *tf-idf* values. We used an empirical threshold of 19 classes for this algorithm.

An interesting question arises if verbs should be part of the list of terms because in technical writing verbs are often lexical functions to the corresponding nouns. For example, for a noun *program* the lexical function $Oper_1$ will be *design (a program)* or *develop (a program),* see [5]. Similar question arises for nouns derived from the corresponding verbs, e.g., *development*. For the moment, we decided to exclude verbs for evaluation of results, but leave the derived nouns.

Results of one of the experiments are presented in **Table 2**. The words are clustered using *k-means* algorithm. The first word in the cell is the centre of the class. Some words were extracted in English, like, *for, to, DAQ*, etc., i.e. in the form they were represented in the source texts. The extracted words belong to various fields of computer science: programming, bioinformatics, electronics, etc.

In the table, we stroke out the words that are clearly errors of the term detection algorithm. Some of these errors can be easily corrected, like, *to* or *etc*. We also underlined verbs, which we do not take into account.

If we have a look at the obtained list, it can be easily seen that it has many words that are terms of science in general, like *analysis, model, science, theory*, etc. We marked these words with italic. It is not clear if these words are errors of the method or not. Since we make comparison with general reference corpus, we do not have the possibility to distinguish them from more specific terms. On the other hand, it seems very plausible that we can detect them if we also make a comparison with a domain corpus of other scientific field.

Table 2. Obtained classes of extracted single word terms

Classes of detected terms (Spanish)	Classes of detected terms (translated)
algoritmo, for, *implementación*, array, <u>implementar</u>, árbol	algorithm, for, *implementation*, array, <u>implement</u>, tree (search)
analógica, voltaje, binario	analog, voltage, binary
as, if, int, integer, pseudocódigo, return, vtemp, *diagrama, descripción*, Turing, end	as, if, int, integer (number), pseudocode, return, vtemp, *diagram, description*, Turing, end
b2b, *business*, hosting, cliente, servidor, internet, ~~to~~, electrónico, <u>consistir</u>	B2B, *business*, hosting, client, server, Internet, ~~to~~, electronic, <u>consist</u>
biología, bioinformática, ADN, alineamiento, clustalw, fago, gen, genoma, genomas, genome, genómica, génica, homología, *human*, microarrays, *modelado*, nucleótidos, *predicción*, proteína, proteína-proteína, sanger, secuenciación, evolutivo, *secuencia, biológico*, computacional, protocolo, *variedad, análisis, técnica, estructura, interacción*, <u>completar</u>, *montaje, herramienta*, ~~menudo~~, <u>usar</u>, <u>talar</u>, software, <u>visualizar</u>, *cuantificación, modelo*, <u>automatizar</u>, búsqueda	*biology*, bioinformatics, DNA, alignment, ClustalW, fag, gene, genome, genomes, genomics, genetic, homology, *human*, microarrays, *modeling*, nucleotides, *prediction*, protein, protein- protein, sanger, sequencing, evolutionary, *sequence, biological*, computational, protocol, *variety, analysis, technical, structure, interaction* <u>complement</u>, assembly, tool, ~~often~~, <u>use, fell</u>, software, <u>visualize</u>, *quantification, model*, <u>automate</u>, search
componente, transistor, tubo, <u>funcionar</u>, conexión, dispositivo, ~~etc~~, *tecnología*, digitales, microprocesadores, *velocidad, lógica*, <u>soler</u>, altavoz	component, transistor, tube, <u>function</u>, connection, device, ~~etc.~~, *technology*, digital, microprocessor, *speed, logic*, <u>happens</u>, speaker
computación, *ciencia, constable, científica*, cómputo, *disciplina, matemática*, ~~usualmente~~, *teoría*, computacionales, *ingeniería*, <u>estudiar</u>, artificial, *matemático*, informática, paralelo, programación	computer, science, *constable, scientific*, computing, *discipline, mathematics*, ~~usually~~, *theory*, computing, *engineering*, <u>study</u>, artificial, *mathematical*, informatics, parallel, programming
conjunto, *notación, problema*, finito, binaria, complejidad, np, np-completo, *número, tamaño, elemento*, coste, lineal, ~~comúnmente~~, ~~montículo~~	set, *notation, problem*, finite, binary, complexity, np, np-complete, *number, size, item*, cost, linear, ~~commonly~~, ~~mound~~
código, compilador, compiladores, lenguaje, máquina, programa, ~~compuesto~~	code, compiler, compilers, language, machine, program, ~~consisting~~

Table 2. *(continued)*

descifrar, criptografía, cifrar, *método*, texto, denominar	decode, cryptography, encrypt, *method*, text, name
dimensión, cubo, *espacial*, almacén, marts, metadato, middleware, warehouse, data, olap, tabla, operacional, variable, *definición*, especificar, usuario, poseer, almacenar, dato, colección, arquitectura, registro	*dimension*, cube, *space*, warehouse, marts, metadata, middleware, warehouse, data, olap, chart, operational, variable, *definition*, specify, user, possess, store, data, collection, architecture, register
diseñar, diseñador, objeto, funcional, procesar, proceso	*design, designer*, object, functional, process, process
formato, avi, compresión, *especificación*, formatos, mov, archivar, vídeo, audio, archivo, informático, codificar, *estándar*	format, avi, compression, *specification* formats, mov, archive, video, audio, file, computer, code, *standard*
potencia, válvula, analógicos, semiconductor, corriente, alternar, analizador, electrónica, conmutación, eléctrico, sonido, pila, supercomputadoras	power, valve, analog, semiconductor, power, switch, analyzer, electronic, switching, electrical, audio, battery, supercomputers
red, *principal, artículo*, permitir, utilizar, ~~vario~~, aplicación, información, ~~través~~, ~~tipo~~, *sistema*, ~~ejemplo~~, característica, interfaz, *forma*, gestión, operativo, acceder, ~~diferente~~, base, contener, operación, función, clasificar, ordenador, ejecutar, programador, cálculo, modelar, relacionales, interfaces, objeto, relacional	network, *main, article*, allow, use, ~~various~~, application, information, ~~through~~, ~~type~~, *system*, ~~example~~, feature, interface, *form*, management, operating, access, ~~different~~, base, contain, operation, function, classify, computer, execute, programmer, calculation, model, relational, interfaces, object, relational
~~rápido~~, acceso, ~~sencillo~~, soportar, web, ~~específico~~, ~~central~~, fiabilidad, paralelismo	~~fast~~, access, ~~easy~~, support, web, ~~specific~~, ~~central~~, reliability, parallelism
señal, transductores, transductor, impedancia, filtrar, conversión, acondicionamiento, convertidor, daq, *adquisición*, analógico, conectar, adaptación, frecuencia, medir, tensión, sensores, digital, cable, control, ~~física~~, entrada, medición, ~~físico~~, salida, ~~normalmente~~, bus, dato	signal, transducers, transducer, impedance, filter, converting, packaging, converter, daq, *acquisition*, analogue, connect, adapt, frequency, measure, voltage, sensors, digital, cable, control, ~~physics~~, input, measurement, ~~physical~~, output, ~~normally~~, bus, data
térmico, ci, cápsula, integration, scale, chip, circuito, chips, integrar, híbrido, silicio, reproductor, amplificador, ~~fabricación~~	heat, ci, capsule, integration, scale, chip, circuit, chips, integrated, hybrid, silicon, player, amplifier, ~~manufacturing~~

Some words that were absent in the dictionary of the system of the morphological analysis appear as two morphological forms, e.g. *genoma* and *genomas* (*genome, genomes*). Note that usually they belong to the same cluster that is an additional proof of the correct functioning of the clustering algorithm.

The next step is evaluation of the obtained results. As usual in case of term detection and classification, evaluation is not trivial since there are no gold standards. This is an objective situation due to the fact that term extraction is really subjective.

In the paper [8], after manual evaluation of results, it is reported around 70% of precision.

We evaluate our results using the following calculations. Totally we obtained 270 terms, form these terms there are 31 verbs (underlined), i.e., there are 239 left. There are 19 errors that clearly are not terms of the domain (stroke out). Thus, we obtain precision of (239-19)/239=92.5%. If we consider terms of general science (48 terms) as errors, then we get the following precision (239-(19+48))/239 = 72%. Calculation of recall would be even more subjective; some domain dictionary might be used for this; still, the compilation of this dictionary keeps being subjective and does not guarantee its completeness.

4   Conclusions

In the paper we presented the method that allows for extraction of single-word terms. We consider that this can be the first step for multi-word term extraction. The proposed method is based on comparison with general reference corpus using log-likelihood similarity that is used for corpus comparison. In addition, we perform term clustering using *k-means* algorithm and cosine similarity measure.

We made experiments using texts of the domain of computer science. We analyzed in detail the obtained term list. Our results show precision of 92.5% using manual evaluation if we consider general scientific terms as correct ones and 72% if we consider them as errors.

As the main direction of future work we would like to mention:

- Comparison with various corpora for filtering terms that belong to domain of general science.
- Comparison of various log-likelihood measures.
- Extraction of multi-word terms using the extracted single-word terms as the candidates.

Acknowledgments. Work done under partial support of Mexican Government (CONACYT, SNI) and National Polytechnic Institute, Mexico (SIP, COFAA, PIFI), projects 20100668 and 20100772. We thank anonymous reviewers for their important comments.

References

1. Cimiano, P.: Ontology learning and population from text, algorithms, evaluation and applications. Springer, New York (2006)
2. Dunning, T.: Accurate methods for the statistics of surprise and coincidence. Computational Linguistics 19(1), 61–74 (1993)

3. Gómez-Pérez, A., Fernandez-López, M., Corcho, O.: Ontological Engineering. Springer, London (2004)
4. Maedche, A., Staab, S.: Discovering conceptual relations from text. In: Proceedings of ECAI 2000 (2000)
5. Melchuk, I.A.: Lexical Functions in Lexicographic Description. In: Proceedings of VIII Annual Meeting of the Berkeley Linguistic Society, Berkeley, UCB, pp. 427–444 (1982)
6. Punuru, J.: Knowledge-based methods for automatic extraction of domain-specific ontologies. PhD thesis (2007)
7. Rayson, P., Berridge, D., Francis, B.: Extending the Cochran rule for the comparison of word frequencies between corpora. In: Purnelle, G., Fairon, C., Dister, A. (eds.) Le poids des mots: Proceedings of the 7th International Conference on Statistical analysis of textual data (JADT 2004), Louvain-la-Neuve, Belgium, March 10-12. Presses universitaires de Louvain, vol. II, pp.926–936. Presses universitaires de Louvain (2004)
8. He, T., Zhang, X., Xinghuo, Y.: An Approach to Automatically Constructing Domain Ontology. In: PACLIC 2006, Wuhan, China, November 1-3, pp. 150–157 (2006)
9. Uschold, M., Grunninger, M.: Ontologies: Principles, Methods and Applications. Knowledge Egineering Review (1996)

Weighted Vote Based Classifier Ensemble Selection Using Genetic Algorithm for Named Entity Recognition

Asif Ekbal[1] and Sriparna Saha[2,*]

[1] Department of Computational Linguistics, University of Heidelberg,
Heidelberg-69120, Germany
ekbal@cl.uni-heidelberg.de, asif.ekbal@gmail.com
[2] Interdisciplinary Center for Scientific Computing (IWR), University of Heidelberg,
Heidelberg-69120, Germany
sriparna.saha@iwr.uni-heidelberg.de, sriparna.saha@gmail.com

Abstract. In this paper, we report the search capability of genetic algorithm (GA) to construct a weighted vote based classifier ensemble for Named Entity Recognition (NER). Our underlying assumption is that the reliability of predictions of each classifier differs among the various named entity (NE) classes. Weights of voting should be high for the NE classes for which the classifier is most reliable and low for the NE classes for which the classifier is not at all reliable. Here, an attempt is made to quantify the amount of voting for each class in each classifier using GA. We use Maximum Entropy (ME) framework to build a number of classifiers depending upon the various representations of a set of features, language independent in nature. The proposed technique is evaluated with two resource-constrained languages, namely Bengali and Hindi. Evaluation results yield the recall, precision and F-measure values of 73.81%, 84.92% and 78.98%, respectively for Bengali and 65.12%, 82.03% and 72.60%, respectively for Hindi. Results also show that the proposed weighted vote based classifier ensemble identified by GA outperforms all the individual classifiers and three conventional *baseline* ensemble techniques for both the languages.

1   Introduction

Named Entity Recognition (NER) is an important tool in almost all Natural Language Processing (NLP) application areas such as Information Retrieval, Information Extraction, Machine Translation, Question Answering and Automatic Summarization etc. The objective of NER is to identify and classify every word/term in a document into some predefined categories like person name, location name, organization name, miscellaneous name (date, time, percentage and monetary expressions etc.) and "none-of-the-above". A lot of works in the area of NER has already been carried out using various techniques. The languages covered include English, most of the European languages and some of

---

* Authors equally contributed to this work.

C.J. Hopfe et al. (Eds.): NLDB 2010, LNCS 6177, pp. 256–267, 2010.

the Asian languages like Chinese, Japanese and Korean. India is a multilingual country with great linguistic and cultural diversities. People speak in 22 different official languages that are derived from almost all the dominant linguistic families in the world. However, the works related to NER in Indian languages have started to emerge only very recently. Named Entity (NE) identification in Indian languages is more difficult and challenging compared to others due to the (i). lack of capitalization information, (ii). appearance of NEs in the dictionary as common nouns, (iii). free word order nature of the languages and (iv). resource-constrained environment, i.e., non-availability of corpus, annotated corpus, name dictionaries, morphological analyzers, part of speech (POS) taggers etc. Some of the works related to Indian languages can be found in [1,2,3] for Bengali and in [4] for Hindi.

Rather than selecting the best-fitting feature set, ensembling several NER systems where each one is based on different feature representation can be more effective in order to achieve a reasonably high accurate system. Ensembling of classifiers is done to increase the generalization accuracy. This generalization accuracy of an ensemble classifier depends on the diversity of each individual classifier as well as on their individual performance. But, selecting the appropriate subset of classifiers that could participate in constructing an ensemble remains a difficult problem. Moreover, all the classifiers are not good to detect all types of classes. Some classifiers are good to detect *person names* whereas some are good to detect *location names*. For ensembling the outputs of all classifiers, either majority voting or weighted voting is used. In case of weighted voting, weights should vary among the various NE classes in each classifier. In a voted system, the weight of a particular classifier should be high for that particular NE class for which it performs good. Otherwise, weights should be low for NE classes for which it's outputs are not very reliable. So, it is very crucial to select the appropriate weights of votes per classifier. In the present work, we propose a single objective optimization based technique that uses GA [5] in order to find a suitable classifier ensemble using weighted voting.

Genetic algorithms [5] are randomized search and optimization techniques guided by the principles of evolution and genetics, having a large amount of implicit parallelism. GAs perform search in complex, large and multimodal landscapes, and provide near-optimal solutions for objective or fitness function of an optimization problem. In GAs, the parameters of the search space are encoded in the form of strings (called *chromosomes*). A collection of such strings is called a *population*. Initially, a random population is created, which represents different points in the search space. An *objective* or a *fitness* function is associated with each string that represents the degree of *goodness* of the string. Based on the principle of survival of the fittest, a few of the strings are selected and each is assigned a number of copies that go into the mating pool. Biologically inspired operators like *crossover* and *mutation* are applied on these strings to yield a new generation of strings. The process of selection, crossover and mutation continues for a fixed number of generations or till a termination condition is satisfied.

We use ME framework as a base classifier. Depending on the different available feature combinations, different versions of this classifier are built. One most interesting and important characteristics of these features is that these are language independent in nature; and can be easily derived for almost all the languages with a very little effort. Thereafter, GA is used to search for the appropriate classifier selection using weighted voting. Our assumption is that the reliability of each classifier differs among the various NE classes. The main focus is given to investigate appropriate weights for voting rather than searching for individual classifiers. Weights for voting of each classifier for every NE classes are encoded in a chromosome. Mutation and crossover operators are modified accordingly. A new mutation operator is used to handle real encoding. Adaptive mutation and crossover operators are used to accelerate the convergence of GA. We also use elitism.

We evaluate our proposed approach for two different resource-constrained languages, namely Bengali and Hindi. In terms of native speakers, Bengali ranks fifth in the world and second in India. It is also the national language in Bangladesh. Hindi is the third most spoken language in the world and the national language in India. Evaluation results show the effectiveness of the proposed approach with the overall recall, precision and F-measure values of 73.81%, 84.92% and 78.98%, respectively for Bengali and 65.12%, 82.03% and 72.60%, respectively for Hindi. A voted NER system for Bengali was reported in [6] by combining three different classifiers, namely ME, Conditional Random Field (CRF) and Support Vector Machine (SVM). But, they made use of more complex set of features, gazetteers, post-processing techniques and unlabeled data to improve the performance in each of the classifiers as well as the overall system. The main contributions of our work are listed below:

- GA is used for selecting best weights to form a classifier ensemble. We tried to establish that such ensembling is capable to increase the classification quality by a considerable margin compared to the conventional ensembling methods.
- ME is used as a test classifier due to it's less computational overhead. However, the proposed method will work for any set of classifiers, i.e., either homogeneous or heterogeneous. The proposed technique is a very general approach and it's performance may further improve depending upon the choice and/or the number of classifiers as well as the use of more complex features.
- The proposed technique is language independent that can be replicated for any resource-poor language very easily.
- The proposed framework is applicable for any type of classification problems like NER, POS-tagging, question-answering etc. To the best of our knowledge, use of GA to select appropriate weights for voting is a novel contribution.
- Note, that our work proposes a novel way of ensembling the available classifiers. Thus, existing ensembling works (e.g., [6] etc.) can be further improved with our framework.

2   Named Entity Features

We use the following features for constructing the various classifiers based on the ME framework.

1. **Context words:** These are the preceding and succeeding words of the current word. This feature is considered with the observation that surrounding words carry effective information for identification of NEs.
2. **Word suffix and prefix:** Fixed length (say, n) word suffixes and prefixes are very effective to identify NEs and work well for the highly inflective Indian languages like Bengali and Hindi. Actually, these are the fixed length character sequences stripped from either the rightmost or leftmost positions of the words.
3. **First word:** This is a binary valued feature that checks whether the current token is the first word of the sentence or not. We consider this feature with the observation that the first word of the sentence is most likely a NE.
4. **Length of the word:** This binary valued feature checks whether the length of the token is less than a predetermined threshold value and based on the observation that very short words are most probably not the NEs. In the present work, the threshold value is set to 5 for both the languages.
5. **Infrequent word:** A cut off frequency is chosen to consider the infrequent words in the training corpus with the observation that very frequent words are rarely NEs. Here, the cut off frequencies are set to 7 and 10 for Bengali and Hindi, respectively. A binary valued feature 'INFRQ' fires if the current word appears in this list.
6. **Part of Speech (POS) information:** POS information of the current and/or the surrounding word(s) are effective for NE identification. We use a SVM based POS tagger [7]. In the present work, we evaluate this POS tagger with a coarse-grained tagset of three tags, namely Nominal, PREP (Postpositions) and Other. The coarse-grained POS tagger is found to perform better compared to a fine-grained one.
7. **Position of the word:** This binary valued feature checks the position of the word in the sentence. Sometimes, position of the word in a sentence acts as a good indicator for NE identification. In the present work, this feature is used with the observation that verbs generally appear in the last position of the sentence.
8. **Digit features:** Several digit features are defined depending upon the presence and/or the number of digits and/or symbols in a token. These features are digitComma (token contains digit and comma), digitPercentage (token contains digit and percentage), digitPeriod (token contains digit and period), digitSlash (token contains digit and slash), digitHyphen (token contains digit and hyphen) and digitFour (token consists of four digits only). These features are helpful to identify miscellaneous NEs.

```
Begin
    1. t = 0
    2. initialize population P(t)  /* Popsize = |P| */
    3. for i = 1 to Popsize
          compute fitness P(t)
    4. t = t + 1
    5. if termination criterion achieved go to step 10
    6. select (P)
    7. crossover (P)
    8. mutate (P)
    9. go to step 3
    10. output best chromosome and stop
End
```

Fig. 1. Basic Steps of GA

3   Proposed Approach

The proposed single objective GA based ensemble selection method is described below. The basic steps of the proposed approach closely follow those of the conventional GA as shown in Figure 1.

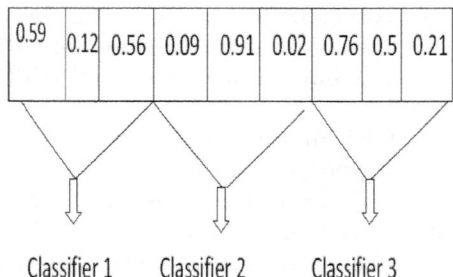

Fig. 2. Chromosome Representation

3.1   String Representation and Population Initialization

If the total number of available classifiers is M and total number of output tags (or, classes) is O, then the length of the chromosome is $M \times O$ (each chromosome encodes the weights of votes for possible O tags for each classifier). As an example, the encoding of a particular chromosome is represented in Figure 2. Here, $M = 3$ and $O = 3$ (i.e., total 9 votes can be possible). The chromosome represents the following voting ensemble:

The weights of votes for 3 different output tags in classifier 1 are 0.59, 0.12 and 0.56, respectively. Similarly, weights of votes for 3 different output tags are 0.09, 0.91 and 0.02, respectively in classifier 2 and 0.76, 0.5 and 0.21, respectively in classifier 3.

In the present work, we use real encoding. The entries of each chromosome are randomly initialized to a real value (r) between 0 and 1. Here, r is an uniformly distributed random number between 0 and 1. If the population size is P then all the P number of chromosomes of this population are initialized in the above way.

3.2 Fitness Computation

Initially, the F-measure values of all the ME based classifiers are calculated using 3-fold cross validation on the available training data. Each of these classifiers is built using various representations of the set of language independent features that can be easily derived for almost all the languages. We execute the following steps to compute the fitness value of each chromosome.

1. Suppose, there are total M number of classifiers. Let, the overall F-measure values of these M classifiers be F_i, $i = 1 \ldots M$.
2. Initially, the training data is divided into 3 parts. Each classifier is trained using 2/3 of the training data and tested with the remaining 1/3 part. Now, for the ensemble classifier the output label for each word in the 1/3 training data is determined using the weighted voting of these M classifiers' outputs. The weight of the NE tag provided by the mth classifier is equal to $I(m, i) \times F_m$. Here, $I(m, i)$ is the entry of the chromosome corresponding to m^{th} classifier and i^{th} NE class. The combined score of a particular class for a particular word w is:

$$f(c_i) = \sum I(m, i) \times F_m \quad \forall m = 1 \text{ to } M \text{ and } op(w, m) = c_i \qquad (1)$$

Here, $op(w, m)$ denotes the output label (or, tag) provided by the mth classifier for the word w. The class receiving the maximum combined score is selected as the joint decision.
3. Now, the overall F-measure value of this voted ensemble for the 1/3 training data is calculated.
4. Steps 2 and 3 are repeated 3 times to perform 3-fold cross validation.
5. The average F-measure value of this 3-fold cross validation is used as the fitness value of the particular chromosome. The fitness function is denoted by $fit = $ F-measure$_{avg}$, which is maximized using the search capability of GA.

3.3 Selection

Roulette wheel selection is used to implement the proportional selection strategy.

3.4 Crossover

Here, we use the normal single point crossover [8]. As an example, let the two chromosomes are :

P1: 0.24 0.16 0.54 0.87 0.66 0.76 0.01 0.88 0.21
P2: 0.12 0.09 0.89 0.71 0.65 0.82 0.69 0.43 0.15
At first a crossover point has to be selected uniformly random between 1 to 9 (length of the chromosome) by generating some random number between 1 and 9. Let the crossover point, here, be 4. Then after crossover, the offsprings are:
O1: 0.24 0.16 0.54 0.87 0.65 0.82 0.69 0.43 0.15 (taking the first 4 positions from P1 and rest from P2)
O2: 0.12 0.09 0.89 0.71 0.66 0.76 0.01 0.88 0.21 (taking the first 4 positions from P2 and rest from P1)

Crossover probability is selected adaptively as in [9]. The expressions for crossover probabilities are computed as follows:

Let f_{max} be the maximum fitness value of the current population, $\overline{f}$ be the average fitness value of the population and f' be the larger of the fitness values of the solutions to be crossed. Then the probability of crossover, μ_c, is calculated as:

$$\mu_c = \begin{cases} k_1 \times \frac{(f_{max} - f')}{(f_{max} - \overline{f})} & \text{if } f' > \overline{f} \\ k_3 & \text{otherwise} \end{cases} \qquad (2)$$

Here, as in [9], the values of k_1 and k_3 are kept equal to 1.0. Note that, when $f_{max} = \overline{f}$, then $f' = f_{max}$ and μ_c will be equal to k_3. The aim behind this adaptation is to achieve a trade-off between exploration and exploitation in a different manner. The value of μ_c is increased when the better of the two chromosomes to be crossed is itself quite poor. In contrast, when it is a good solution, μ_c is low so as to reduce the likelihood of disrupting a good solution by crossover.

3.5   Mutation

Each chromosome undergoes mutation with a probability μ_m. The mutation probability is also selected adaptively for each chromosome as in [9]. The expression for mutation probability, μ_m, is given below:

$$\mu_m = \begin{cases} k_2 \times \frac{(f_{max} - f)}{(f_{max} - \overline{f})} & \text{if } f > \overline{f} \\ k_4 & \text{otherwise} \end{cases} \qquad (3)$$

Here, values of k_2 and k_4 are kept equal to 0.5. This adaptive mutation helps GA to come out of local optimum. When GA converges to a local optimum, i.e., when $f_{max} - \overline{f}$ decreases, μ_c and μ_m both will be increased. As a result the GA may come out of this. It will also happen for the global optimum and may result in disruption of the near-optimal solutions. As a result, GA will never converge to the global optimum. The μ_c and μ_m will get lower values for high fitness solutions and get higher values for low fitness solutions. While the high fitness solutions aid in the convergence of GA, the low fitness solutions prevent the GA from getting stuck at a local optimum. The use of elitism will also keep the best solution intact. For a solution with the maximum fitness value, μ_c and μ_m are both zero. The best solution in a population is transferred undisrupted into the next generation. Together with the selection mechanism, this may lead to an exponential growth of the solution in the population and may cause premature convergence.

Here, each position in a chromosome is mutated with probability μ_m in the following way. The value is replaced with a random variable drawn from a Laplacian distribution, $p(\epsilon) \propto e^{-\frac{|\epsilon-\mu|}{\delta}}$, where the scaling factor δ sets the magnitude of perturbation. Here, μ is the value at the position which is to be perturbed. The scaling factor δ is chosen equal to 0.1. The old value at the position is replaced with the newly generated value. By generating a random variable using Laplacian distribution, there is a non-zero probability of generating any valid position from any other valid position while probability of generating a value near the old value is more.

3.6   Termination Condition

In this approach, the processes of fitness computation, selection, crossover, and mutation are executed for a maximum number of generations. The best string seen up to the last generation provides the solution to the above classifier ensemble problem. Elitism is implemented at each generation by preserving the best string seen up to that generation in a location outside the population. Thus on termination, this location contains the best classifier ensemble.

4   Experimental Results and Discussions

We use the OpenNLP Java based ME package [1]. Model parameters are computed with 200 iterations without feature frequency cutoff. For GA based ensemble, the following parameter values are used: population size=50, number of generations=40. The mutation and crossover probabilities are selected adaptively. We define three different *baseline* ensemble systems as below:

- *Baseline 1*: This is based on the majority voting among the classifiers.
- *Baseline 2*: This is a weighted voting approach. In each classifier, weights are calculated based on the average F-measure value of the 3-fold cross validation on the training data.
- *Baseline 3*: For each classifier, the average F-measure value of each class is computed from the 3-fold cross validation on the training data. The weight of any classifier is set to the average F-measure value of the corresponding class that it assigns to a word.

4.1   Datasets for NER

Indian languages are resource-constrained in nature. For NER, we use a Bengali news corpus [10], developed from the archive of a leading Bengali newspaper available in the web. A portion of this corpus containing approximately 250K

---

[1] http://maxent.sourceforge.net/

wordforms is manually annotated with a coarse-grained NE tagset of four tags namely, PER (*Person name*), LOC (*Location name*), ORG (*Organization name*) and MISC (*Miscellaneous name*). The miscellaneous name includes date, time, number, percentages, monetary expressions and measurement expressions. The data is collected mostly from the *National, States, Sports* domains and the various sub-domains of *District* of the particular newspaper. This annotation was carried out by one of the authors and verified by an expert. We also use the IJCNLP-08 NER on South and South East Asian Languages (NERSSEAL)[2] Shared Task data of around 100K wordforms that were originally annotated with a fine-grained tagset of twelve tags. This data is mostly from the agriculture and scientific domains. For Hindi, we use the dataset of approximately 502,913 wordforms obtained from the NERSSEAL shared task. An appropriate mapping is defined to convert the fine-grained NE annotated data to the desired forms, i.e., tagged with a coarse-grained tasget of four tags. In order to report the evaluation results, we randomly partition each dataset into training and test sets. The training and test sets statistics are presented in Table 1. In order to properly denote the boundaries of NEs, four basic NE tags are further divided into the format I-TYPE (TYPE→PER/LOC/ORG/MISC) which means that the word is inside a NE of type TYPE. Only if two NEs of the same type immediately follow each other, the first word of the second NE will have tag B-TYPE to show that it starts a new NE. This is the standard IOB format that was followed in the CoNLL-2003 shared task [11].

Table 1. Statistics of the datasets

Language	# words in training	#NEs in training	#words in test	#NEs in test	Unknown NEs (in %)
Bengali	312,947	37,009	37,053	4,413	35.1
Hindi	444,231	43,021	58,682	3,005	40

4.2 Discussion of Results

We build a number of different ME models by considering the various combinations of the available NE features that are language independent in nature. We consider the (i). context of size five (previous two and the next two words), (ii). word suffixes and prefixes of length up to three (3+3 different features) or four (4+4 different features) characters of the current token, (iii). POS information of the current token, (iv). first word of the sentence, (v). length of the word, (vi). position of the word in the sentence and (vii). several digit features.

Varying these features, we construct 19 different classifiers as shown in Table 2 for Bengali. The best individual classifier shows the recall, precision and F-measure values of 69.76%, 81.75% and 75.28%, respectively. The overall performance of the best individual classifier, three different *baseline* ensembles and the weighted vote based classifier ensemble identified by our proposed GA based technique are presented in Table 3 for Bengali. Results show that the overall

---

[2] http://ltrc.iiit.ac.in/ner-ssea-08

Table 2. Feature types used for training different ME based classifiers for Bengali. Here, the following abbreviations are used: 'CW':Context words, 'Pre-size': Size of the prefix, 'Suf-size': Size of the suffix, 'WL': Word length, 'IW': Infrequent word, 'PW': Position of the word, 'FW':First word, DI: 'Digit-Information', X: Denotes the presence of the corresponding feature.

Classifier	CW	FW	PRE-SIZE	SUF-SIZE	WL	IW	PW	DI	POS	recall (in %)	precision (in %)	F-measure (in %)
M_1	X	X						X	X	35.59	62.74	45.42
M_2	X	X	3					X	X	63.12	78.61	70.02
M_3	X	X	3	3				X	X	68.81	81.34	74.55
M_4	X	X	3	3	X			X	X	68.65	81.57	74.55
M_5	X	X	3	3	X	X		X	X	69.35	81.37	74.88
M_6	X	X	3	3	X	X	X	X	X	69.15	81.53	74.83
M_7	X	X	4					X	X	65.45	79.43	71.76
M_8	X	X	4	3				X	X	68.42	81.58	74.42
M_9	X	X	3	4				X	X	69.39	81.66	75.03
M_{10}	X	X	4	4				X	X	68.65	81.13	74.37
M_{11}	X	X	4	3	X			X	X	67.81	81.53	74.04
M_{12}	X	X	3	4	X			X	X	69.39	82.02	75.18
M_{13}	X	X	4	4	X			X	X	68.01	81.00	73.94
M_{14}	X	X	4	3	X	X		X	X	68.69	81.46	74.53
M_{15}	X	X	3	4	X	X		X	X	69.76	81.75	75.28
M_{16}	X	X	4	4	X	X		X	X	68.87	80.89	74.40
M_{17}	X	X	4	3	X	X	X	X	X	68.58	81.64	74.54
M_{18}	X	X	3	4	X	X	X	X	X	69.67	81.85	75.27
M_{19}	X	X	4	4	X	X	X	X	X	68.51	81.01	74.24

Table 3. Overall results for Bengali

Model	recall(in %)	precision (in %)	F-measure (in %)
Best individual classifier	69.76	81.75	75.28
Baseline 1	69.83	82.90	75.81
Baseline 2	70.25	82.97	76.08
Baseline 3	70.97	83.34	76.65
GA based ensemble	73.81	84.92	78.98

performance attained by the classifier ensemble determined by the proposed algorithm outperforms (an improvement in recall, precision and F-measure values by 4.05%, 3.17% and 3.70%, respectively) the best individual classifier. The proposed GA based ensemble technique also performs reasonably better than the three *baseline* models. It shows improvement of 3.98% recall, 2.02% precision and 3.17% F-measure values over the majority vote based ensemble, i.e., *Baseline 1*. We also observe that our proposed algorithm performs superior compared to the other two weighted vote based *baseline* ensemble techniques by 2.90% (*Baseline 2*) and 2.33% (*Baseline 3*) F-measure values, respectively.

Thereafter, the proposed system is evaluated for Hindi data. The various classifier combinations are reported in Table 4 for Hindi. Each of the classifiers is trained with Hindi training data and with the same set of features as Bengali. The best individual classifier yields the recall, precision and F-measure values as 62.12%, 79.68% and 69.82%, respectively. Overall evaluation results are presented in Table 5. The GA based classifier ensemble technique shows the overall recall, precision and F-measure values as 65.12%, 82.03% and 72.60%, respectively. The proposed technique performs better with more than 2.78%, 1.65%, 1.03% and 0.63% F-measure values over the best individual classifier, *Baseline 1*, *Baseline 2* and *Baseline 3*, respectively. The GA based ensemble yields

Table 4. Feature types used for training different ME based classifiers for Hindi. Here, the abbreviations are same as Bengali.

Classifier	CW	FW	PRE-SIZE	SUF-SIZE	WL	IW	PW	POS	DI	recall (in %)	precision (in %)	F-measure (in %)
M_1	X	X						X	X	29.36	69.30	41.25
M_2	X	X	3					X	X	58.79	79.68	67.66
M_3	X	X	3	3				X	X	62.09	79.57	69.75
M_4	X	X	3	3	X			X	X	61.85	79.63	69.63
M_5	X	X	3	3	X	X		X	X	62.16	79.52	69.78
M_6	X	X	3	3	X	X	X	X	X	62.12	79.68	69.82
M_7	X	X	4					X	X	50.54	78.83	61.59
M_8	X	X	4	3				X	X	54.08	80.04	64.55
M_9	X	X	3	4				X	X	60.10	79.05	68.29
M_{10}	X	X	4	4				X	X	52.33	79.18	63.01
M_{11}	X	X	4	3	X			X	X	59.73	79.04	68.04
M_{12}	X	X	3	4	X			X	X	53.57	79.75	64.09
M_{13}	X	X	4	4	X			X	X	51.62	79.65	62.64
M_{14}	X	X	4	3	X	X		X	X	59.97	78.70	68.07
M_{15}	X	X	3	4	X	X		X	X	54.18	79.61	64.48
M_{16}	X	X	4	4	X	X		X	X	52.49	79.36	63.19
M_{17}	X	X	4	3	X	X	X	X	X	59.94	78.97	68.15
M_{18}	X	X	3	4	X	X	X	X	X	54.01	79.83	64.43
M_{19}	X	X	4	4	X	X	X	X	X	52.29	79.73	63.16

Table 5. Overall results for Hindi

Model	recall (in %)	precision (in %)	F-measure (in %)
Best individual classifier	62.12	79.68	69.82
Baseline 1	63.57	80.28	70.95
Baseline 2	64.12	80.97	71.57
Baseline 3	64.58	81.28	71.97
GA based ensemble	65.12	82.03	72.60

better performance for Bengali compared to Hindi. The possible reasons may be (i) higher ratio between non-NEs and NEs in Hindi (9.33:1) than Bengali (7.46:1) and (ii). presence of more unknown NEs in the Hindi test data than Bengali.

5   Conclusion

In this paper, we propose a classifier ensemble technique based on GA for NER. We used ME classifier as the base classifier. The most important characteristic part of our system is that it makes use of only language independent features that can be easily derived for almost all the languages without any prior knowledge. Experiments with two resource poor languages like Bengali and Hindi demonstrated the effectiveness of our proposed approach over all the individual classifiers as well as the three different conventional *baseline* ensembles. In future we would like to add some more language independent as well as some language specific features to generate more classifiers. In this work, we considered only ME as the underlying classification technique. Future works include the development of weighted vote based classifier ensembles using some other well-known classifiers like Conditional Random Fields and Support Vector Machines.

References

1. Ekbal, A., Bandyopadhyay, S.: Lexical Pattern Learning from Corpus Data for Named Entity Recognition. In: Proceedings of the 5th International Conference on Natural Language Processing (ICON), India, pp. 123–128 (2007)
2. Ekbal, A., Naskar, S., Bandyopadhyay, S.: Named Entity Recognition and Transliteration in Bengali. Named Entities: Recognition, Classification and Use, Special Issue of Lingvisticae Investigationes Journal 30(1), 95–114 (2007)
3. Ekbal, A., Haque, R., Bandyopadhyay, S.: Named Entity Recognition in Bengali: A Conditional Random Field Approach . In: Proceedings of the 3rd International Joint Conference on Natural Language Processing (IJCNLP 2008), pp. 589–594 (2008)
4. Li, W., McCallum, A.: Rapid Development of Hindi Named Entity Recognition using Conditional Random Fields and Feature Induction. ACM Transactions on Asian Languages Information Processing 2(3), 290–294 (2004)
5. Goldberg, D.E.: Genetic Algorithms in Search, Optimization and Machine Learning. Addison-Wesley, New York (1989)
6. Ekbal, A., Bandyopadhyay, S.: Voted NER System using Appropriate Unlabeled Data. In: Proceedings of the 2009 Named Entities Workshop: Shared Task on Transliteration (NEWS 2009), ACL-IJCNLP, pp. 202–210 (2009)
7. Ekbal, A., Bandyopadhyay, S.: Web-based Bengali News Corpus for Lexicon Development and POS Tagging. POLIBITS 37, 20–29 (2008)
8. Holland, J.H.: Adaptation in Natural and Artificial Systems. The University of Michigan Press, AnnArbor (1975)
9. Srinivas, M., Patnaik, L.M.: Adaptive Probabilities of Crossover and Mutation in Genetic Algorithms. IEEE Transactions on Systems, Man and Cybernatics 24(4), 656–667 (1994)
10. Ekbal, A., Bandyopadhyay, S.: A Web-based Bengali News Corpus for Named Entity Recognition. Language Resources and Evaluation Journal 42(2), 173–182 (2008)
11. Tjong Kim Sang, E.F., De Meulder, F.: Introduction to the Conll-2003 Shared Task: Language Independent Named Entity Recognition. In: Proceedings of the Seventh Conference on Natural Language Learning at HLT-NAACL 2003, pp. 142–147 (2003)

Refactoring of Process Model Activity Labels

Henrik Leopold[1], Sergey Smirnov[2], and Jan Mendling[1]

[1] Humboldt-Universität zu Berlin, Germany
{henrik.leopold,jan.mendling}@wiwi.hu-berlin.de
[2] Hasso Plattner Institute, Potsdam, Germany
sergey.smirnov@hpi.uni-potsdam.de

Abstract. Recently many companies have expanded their business process modeling projects such that thousands of process models are designed and maintained. Activity labels of these models are related to different styles according to their grammatical structure. There are several guidelines that suggest using a verb-object labeling style. Meanwhile, real-world process models often include labels that do not follow this style. In this paper we investigate the potential to improve the label quality automatically. We define and implement an approach for automatic refactoring of labels following action-noun style into verb-object labels. We evaluate the proposed techniques using a collection of real-world process models—the SAP Reference Model.

1 Introduction

Business process modeling is an integral part of process management in large enterprises. Many of these companies design and maintain up to several thousand models to capture their operations [17]. Nowadays, laymen and casual modelers design a great share of these process models with detrimental consequences for model quality. In this context, there is growing demand for automatic techniques to check and improve the quality of these models.

Process model quality has been approached from different angles, including verification, error probability, and comprehension [1,13,14]. Also the small pieces of text that capture the names of activities (activity labels) have been investigated from a usability perspective. Such activity labels represent actions, which take place during the execution of a business process. Typically, an activity label captures an action and a business object on which the action is performed, like *Validate address* or *Creation of order*. In essence, three classes of activity labels have been found in practice: verb-object labels, action-noun labels, and a rest category [15]. An interesting point is that verb-object labels are superior to action-noun labels in terms of perceived ambiguity. Therefore, it is desirable that all labels follow the verb-object style.

In this paper we address the problem of automatic refactoring of action-noun labels to verb-object labels. A key challenge is the identification of actions and business objects in action-noun labels. As activity labels in process models are only sentence fragments, we use contextual information from neighboring control

C.J. Hopfe et al. (Eds.): NLDB 2010, LNCS 6177, pp. 268–276, 2010.

flow elements instead of standard natural language parsing techniques. The approach has been implemented and evaluated using the SAP Reference Model, a large collection of real-world business process models [10]. The results emphasize the potential of our approach.

The remainder of the paper is structured as follows. Section 2 illustrates the research problem and presents the main styles we identified. Section 3 defines algorithms for label substyle recognition, action and business object derivation, and refactoring methods. In Section 4 we evaluate the presented algorithms. Related work is discussed in Section 5. Section 6 concludes the paper.

2   Background

In this section we outline the problem of label refactoring, starting with an example covering different labeling styles. Then, we present the results from a study of action-noun labels in the SAP Reference Model.

2.1   Motivation

The problem of activity label quality can be motivated by the business process fragment in Fig. 1. It captures a part of a profit center planning process. One can see that it is easy to misinterpret activity label *Plan data transfer to EC-PCA from profitability analysis*. Ignoring the preceding and succeeding events, a reader might conclude that the label *Plan data transfer to EC-PCA from profitability analysis* instructs to *plan* a *data transfer*, and label *Plan integration of profit centers* advises to *plan* the *integration of profit centers*. However, event *Plan Data transferred from other Applications* reveals that the action in the activity on the left branch is given by noun *transfer*. Consequently, the activity label does not instruct to *plan* a *data transfer*, but to *transfer plan data*. This ambiguity partially stems from the style of labeling: the first word is a verb

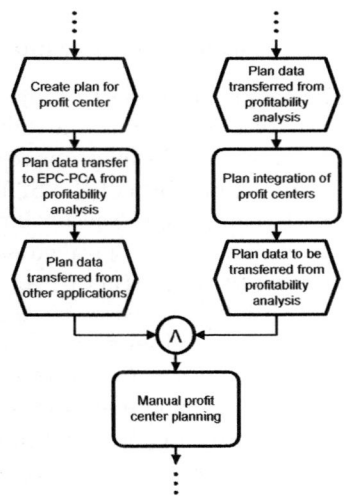

Fig. 1. Poor labeling in the SAP Reference Model

referring to an action, while in other cases the first word is a business object and the action is given as a noun.

A consistent application of the verb-object style increases understandability of labels [15]. Verb-object labels are verb phrases headed by an infinitive and succeeded by a noun phrase. The verb captures an action and the noun phrase a business object. An action-noun label states the action as a noun, which can often be confused with a business object. We propose to refactor labels of potential ambiguity by analyzing and transforming action-noun labels.

Table 1. Substyles of action-noun activity labeling style

Name	Structure	Example	Share, %
Noun phrase	NP / NN (bo) \ NN (a)	Invoice creation	78.8
Noun phrase with *of* prepositional phrase	NP / NP \ PP; NN (a), IN ('of'), NP / NN (bo)	Creation of invoice	15.0
Verb phrase (gerund)	VP / VBG (a) \ NP / NN (bo)	Creating invoice	5.1
Irregular	–	LIFO: Invoice: Creation level	1.1

2.2   Label Classification

Development of effective algorithms deriving actions and business objects from activity labels requires a thorough understanding of current labeling practices. We approached this problem in a bottom-up way by investigating the different action-noun labels of the SAP Reference Model. This model collection includes 604 Event-driven Process Chains (EPCs), each containing several activities. Table 1 shows four identified substyles of action-noun labels, each described according to their structure and illustrated by an example.

The labels of *noun phrase* style are noun phrases where a business object comes first and an action follows. A business object may be absent. The labels of *Noun phrase with* of *prepositional phrase* style are also noun phrases. However, an action is represented by a leading noun and is succeeded by a prepositional phrase. The prepositional phrase is headed by a preposition *of* and refers to a business object. The labels of noun phrase with *of* prepositional phrase style can have an optional prepositional phrase. The labels of *verb phrase (gerund)* substyle are verb phrases headed by a gerund. The action is captured as a gerund, succeeded by the business object captured as a noun. This style may have an optional prepositional phrase.

The substyles described above cover almost 99% of action-noun labels in the model collection. The remaining labels cannot be assigned to any of these substyles and are related to *irregular* style. The specific property of these labels is the use of characters, linking together parts of the label in a special way. These characters do not allow labels to qualify as any of the above named substyles. Examples are *Transfer Posting FI-LC* and *LIFO: Valuation: Pool Level*. The majority of irregular labels can be recognized by the use of the characters ':' and '–'.

Table 2. Properties of action-noun label substyles

Label class	Property
Noun phrase	none
Noun phrase with *of*	label contains a prepositional phrase
prepositional phrase	with *of* as a leading preposition
Verb phrase (gerund)	the leading gerund signifies an action
	and is followed by a business object
Phrase with coordinating	the phrase contains a coordinating
conjunction	conjunction, e.g., *and* or *or*
Irregular	label contains characters ':' or ' - '

Some labels refer to more than one business object or instruct to perform more than one action. Such labels contain a conjunction, coordinating the relations between homogeneous parts. Examples of conjunctions are *and*, *or*, comma symbol, and slash symbol. These labels can be decomposed to several labels, each capturing one action on one object.

3   Automatic Refactoring of Action-Noun Activity Labels

This section presents a stepwise approach to automatic refactoring of activity labels from action-noun style into verb-object style. The refactoring process includes (1) label style recognition, (2) derivation of an action and a business object from the label, and (3) composing a verb-noun label.

3.1   Label Style Recognition

The recognition process is driven by a set of label properties. Each label is evaluated against the set of properties and, according to evaluation results, is categorized into a particular style. Table 2 enumerates the action-noun label substyles and points to their featured properties.

Algorithm 1 formalizes label style recognition. The input of the algorithm is an action-noun label *label*, the output is *prop*—an object storing the label properties with a substyle among them. We assume that all the boolean properties in *prop* object are initialized with *false*.

First the algorithm examines, if the label contains characters that allow to classify the label as irregular (see lines 3–5). If the label contains such characters, the style of the label is irregular and the algorithm terminates. Otherwise, the algorithm continues seeking for prepositions (lines 6–8) and conjunctions (lines 9–11). If conjunctions or prepositions are found, respective flags *hasConjunctions* and *hasPrepositions* are set to *true* and the position of the first conjunction/preposition is stored in *pIndex/cIndex*.

Algorithm 1. Recognition of action-noun substyles

1: **recognizeSubstyle(Label** *label***)**
2: *prop* = new LabelProperties();
3: **if** *label* contains ':' OR ' - ' **then**
4:      *prop.style* = UNCLASSIFIED;
5:      **return** *properties*;
6: **if** *label* contains prepositions **then**
7:      *prop.hasPrepositions* = **true**;
8:      *prop.pIndex* = getFirstPrepositionIndex(*label*);
9: **if** *label* contains conjunctions **then**
10:     *prop.hasConjunctions* = **true**;
11:     *prop.cIndex* = getConjunctionIndex(*label*);
12: **if** first word in *label* has suffix 'ing' **then**
13:     *prop.hasSuffixING* = **true**;
14: *verbSize* = getVerbSize(*label*);
15: **if** *prop.hasSuffixING* **and** *label.size* > *verbSize* **and** (!*prop.hasPrepositions*
    or *prop.pIndex* > *verbSize* + 1) **then**
16:     **if** verb == action derived from label context **then**
17:         *prop.style* = GERUND;
18:         **return** *prop*;
19: **if** *prop.hasPrepositions* **and** *label*.getWordAt(*prop.pIndex*) == 'of' **then**
20:     *prop.style* = PREPOSITION_OF;
21:     **return** *prop*;
22: *prop.style* = NOUN;
23: **return** *prop*;

The algorithm proceeds checking if the label starts with a gerund (lines 12–18). It is verified if the first word of the label has an *ing* suffix. Next, Word-Net [16] is used to learn if the first word is a verb and which infinitive it has. An assessment, whether the gerund represents an action, requires a deeper investigation: if the first word of a label is a gerund, it does not imply that this word also represents the action. Consider label *Planning scenario processing*. Although *planning* is a gerund, it might also be a part of a business object. In order to resolve this ambiguity, we consider event nodes preceding and succeeding the activity with the inspected label. Returning to the example, we notice that the activity is preceded by an event labeled with *Planning scenario is processed*. A part of speech analysis of this label identifies *planning* and *scenario* as nouns and *process* as a verb. Hence, we can infer that *processing* captures an action.

If the algorithm qualifies a label to be a gerund, it terminates. In the opposite case, the algorithm proceeds checking prepositions in the label (lines 19–21). A label containing prepositions, with the first *of* preposition, is qualified as a *noun phrase with* of *prepositional phrase*. If the label is categorized to none of the enumerated substyles, the algorithm relates it to a *noun phrase* substyle.

3.2   Derivation of Action and Business Object

Recognition of a label substyle enables the derivation of an action and a business object as a next step. We provide an informal discussion of the algorithm enabling this step. The input of the algorithm is an action-noun label *label* and a corresponding *LabelProperties* object *prop* storing the properties of *label* obtained with Algorithm 1. The output of the algorithm is *prop* with *action* and *bObject* properties set.

The algorithm starts with an analysis of labels following noun phrase style. It checks for an optional prepositional phrase. If the label has a prepositional phrase, the phrase is omitted and not studied any more. If the label has only one word, e.g. *Deployment* or *Classification*, this word is recognized as an action. Otherwise, the algorithm checks if the last two words of the label constitute a phrasal verb, for instance *set up* or *carry forward*. If the first two words are recognized as a phrasal verb, this verb is perceived as an action. The rest of the label, if it exists, is recognized as a business object. If the phrasal verb is not revealed, the last word is recognized as an action, while the rest as a business object. It continues with analysis of *verb phrase (gerund)* labels. Analysis of these labels resembles the analysis of labels of noun phrase style. The key difference is that the action is expected to appear in the beginning of the label, while the business object at the end. The algorithm concludes with an analysis of activity labels following *noun phrase with prepositional phrase* style. The label part preceding preposition *of* is recognized as an action. The label part between preposition *of* and the next preposition is treated as a business object.

3.3   Label Refactoring

Refactoring aims to transform an action-noun label into a verb-object label signifying the same action performed on the same business object. Derivation of actions and business objects from activity labels enables construction of labels in verb-object style. In fact, after the analysis of the previous steps the task becomes a trivial concatenation of a verb representing an action and a noun phrase representing a business object.

4   Empirical Evaluation

We have conducted an experiment to validate how well the proposed algorithms approximate a human interpretation of activity labels. To evaluate the algorithms we have designed a test collection that includes the SAP Reference Model as a process model collection and human interpretations of activity labels. Human interpretations are captured by two mappings: one mapping from an activity label to a set of corresponding actions and another mapping from an activity to a set of business objects. Within the evaluation we compared (1) recognition of label substyles by the algorithm and by humans and (2) derivation of actions and business objects by the algorithm and by humans.

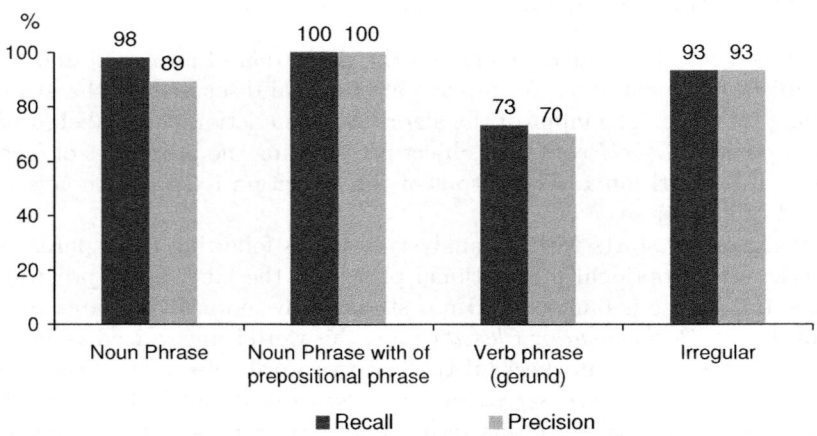

Fig. 2. Precision and recall of algorithms for label style recognition

To evaluate the substyle recognition algorithm we measured the *precision* (number of correctly recognized labels divided by retrieved labels) and *recall* (number of correctly recognized labels divided by all labels) of the algorithm [2]. Fig. 2 presents the corresponding values. The evaluation of action/business object derivation makes use of precision. The precision value for action derivation is 88%, for business object derivation 85%, and for label refactoring 85%. As the precision values are reasonably high, we conclude that the proposed algorithms are capable of automatic derivation of actions and business objects from labels.

5   Related Work

The research reported in this paper relates to guidelines for process model labeling and natural language approaches for conceptual models. The verb-object style is widely promoted in the literature for *labeling activities of process models* [12]. It has been observed though that verb-object labeling in real process models is not consistently applied. For instance, the practical guide for process modeling with ARIS [4, pp.66-70] shows models with both actions as verbs and as nouns. Our work helps to automatically refactor such action-noun labels. It also complements prior work on the automatic identification of verb-object labels in [11]. The concept of part of speech tagging is also investigated for interactive process modeling support [3]. The relationship between *process models and natural language* has been discussed and utilized in various works. In [6] the authors investigate in how far the three steps of building a conceptual model can be automated using a model for pre-design. Further text analysis approaches have been used to link activities in process models to document fragments [9] and to compare process models from a semantic perspective [5]. Most beneficiary is the verb-object style for model verbalization (see [7]), as verb-object style labels can easily be verbalized using the *You have to* prefix.

6   Conclusions and Future Work

In this paper we have proposed a method for automatic refactoring of action-noun activity labels into labels of verb-object style. We performed an evaluation of the proposed approach with process models from the SAP Reference Model. There are two directions of the future work. On the one hand, it is the improvement of the existing refactoring technique and the development of alternative methods based on natural language processing tools, e.g., [8]. As a next step, we aim to evaluate the proposed refactoring technique against other test collections. On the other hand, a number of applications calls for algorithms deriving actions and business objects from activity labels. In particular, we want to integrate the derivation technique with automatic modeling support of [18].

References

1. van der Aalst, W.M.P.: Workflow Verification: Finding Control-Flow Errors Using Petri-Net-Based Techniques. In: van der Aalst, W.M.P., Desel, J., Oberweis, A. (eds.) Business Process Management. LNCS, vol. 1806, pp. 161–183. Springer, Heidelberg (2000)
2. Baeza-Yates, R., Ribeiro-Neto, B.: Modern Information Retrieval (May 1999)
3. Becker, J., Delfmann, P., Herwig, S., Lis, L., Stein, A.: Towards Increased Comparability of Conceptual Models - Enforcing Naming Conventions through Domain Thesauri and Linguistic Grammars. In: Proceedings of ECIS (June 2009)
4. Davis, R.: Business Process Modelling with ARIS: A Practical Guide (2001)
5. Ehrig, M., Koschmider, A., Oberweis, A.: Measuring Similarity between Semantic Business Process Models. In: Proceedings of APCCM, pp. 71–80 (2007)
6. Fliedl, G., Kop, C., Mayr, H.C.: From Textual Scenarios to a Conceptual Schema. Data & Knowl. Eng. 55(1), 20–37 (2005)
7. Frederiks, P., van der Weide, T.: Information Modeling: The Process and the Required Competencies of Its Participants. Data & Knowl. Eng. 58(1), 4–20 (2006)
8. Stanford Natural Language Processing Group. Stanford Log-linear Part-of-speech Tagger, http://nlp.stanford.edu/software/tagger.shtml (accessed April 4, 2010)
9. Ingvaldsen, J.E., Gulla, J.A., Su, X., Rønneberg, H.: A Text Mining Approach to Integrating Business Process Models and Governing Documents. In: Meersman, R., Tari, Z., Herrero, P. (eds.) OTM-WS 2005. LNCS, vol. 3762, pp. 473–484. Springer, Heidelberg (2005)
10. Keller, G., Teufel, T.: SAP(R) R/3 Process Oriented Implementation: Iterative Process Prototyping. Addison-Wesley, Reading (1998)
11. Leopold, H., Smirnov, S., Mendling, J.: On Labeling Quality in Business Process Models. In: EPK. CEUR, vol. 554, pp. 42–57 (2009)
12. Malone, T.W., Crowston, K., Herman, G.A.: Organizing Business Knowledge: The MIT Process Handbook. The MIT Press, Cambridge (2003)
13. Mendling, J.: Metrics for Process Models: Empirical Foundations of Verification, Error Prediction, and Guidelines for Correctness. LNBIP, vol. 6 (1974)
14. Mendling, J., Reijers, H.A., Cardoso, J.: What Makes Process Models Understandable? Business Process Management, 48–63 (2007)

15. Mendling, J., Reijers, H.A., Recker, J.: Activity Labeling in Process Modeling: Empirical Insights and Recommendations. Information Systems (2009)
16. Miller, G.: WordNet: a lexical database for English. CACM 38(11), 39–41 (1995)
17. Rosemann, M.: Potential Pitfalls of Process Modeling: Part A. Business Process Management Journal 12(2), 249–254 (2006)
18. Smirnov, S., Weidlich, M., Mendling, J., Weske, M.: Action Patterns in Business Process Models. In: Baresi, L., Chi, C.-H., Suzuki, J. (eds.) Service-Oriented Computing. LNCS, vol. 5900, pp. 115–129. Springer, Heidelberg (2009)

Unsupervised Ontology Acquisition from Plain Texts: The *OntoGain* System

Euthymios Drymonas[1], Kalliopi Zervanou[2,*], and Euripides G.M. Petrakis[1]

[1] Intelligent Systems Laboratory, Electronic and Computer Engineering Dept.,
Technical University of Crete (TUC), Chania, Crete, Greece
{max,petrakis}@intelligence.tuc.gr
[2] Tilburg centre for Creative Computing (TiCC),
University of Tilburg, The Netherlands
k.zervanou@uvt.nl
http://www.intelligence.tuc.gr

Abstract. We propose *OntoGain*, a system for unsupervised ontology acquisition from unstructured text which relies on multi-word term extraction. For the acquisition of taxonomic relations, we exploit inherent multi-word terms' lexical information in a comparative implementation of agglomerative hierarchical clustering and formal concept analysis methods. For the detection of non-taxonomic relations, we comparatively investigate in *OntoGain* an association rules based algorithm and a probabilistic algorithm. The *OntoGain* system allows for transformation of the derived ontology into standard OWL statements. *OntoGain* results are compared to both hand-crafted ontologies, as well as to a state-of-the art system, in two different domains: the medical and computer science domains.

Keywords: ontology acquisition, formal concept analysis, term extraction, term similarity, term clustering, association rules, multi-word terms, OWL.

1   Introduction

In modern computer science, *ontologies* are formal representations of knowledge resources in terms of *concepts* and respective *relations* which describe a certain domain. Ontology development is a time and cost consuming task, requiring specialists from several fields [1]. This high development effort constitutes, in turn, an inhibiting factor in building large scale intelligent systems. Current research in the field of automatic, or semi-automatic ontology acquisition and development aims at providing methods and solutions to this problem.

A branch of current approaches to ontology acquisition relies on text analysis techniques originating from the field of information extraction, such as the extraction of named entities and relationships, based on supervised [2,3], or manually developed semantic analysis patterns [4,5]. These methods involve human

---

* This work was carried-out while the author was with TUC.

C.J. Hopfe et al. (Eds.): NLDB 2010, LNCS 6177, pp. 277–287, 2010.

intervention and effort at various degrees, either in the development of training data or analysis resources, but they achieve relatively good representations of the domain concepts and relations. Other methods approach the ontology acquisition problem by unsupervised techniques, based on statistics and basic linguistic tools [6,7,8], or unsupervised information extraction [9,10]. The principal challenge in such approaches lies in achieving a satisfactory coverage of the domain in terms of concepts and concept relations, while reducing human effort to the absolute minimum.

Within this latter framework of approaches, we present *OntoGain*, a platform for unsupervised ontology acquisition from text. In *OntoGain*, contrary to other similar approaches [6,11,12,8], initial ontology concept acquisition relies on a domain independent method for multi-word term extraction [13]. Multi-word terms constitute the majority of terminological expressions and lexicalise domain concepts such that their semantics cannot be fully inferred from their single-word constituents [14]. Furthermore, multi-word terms inherently contain classificatory information, expressed as modifiers. For these reasons, their identification is expected to provide better conceptual coverage for a given domain and contribute to the subsequent relation acquisition tasks. For the acquisition of taxonomic and non-taxonomic relationships, we comparatively investigate four approaches: agglomerative hierarchical clustering and formal concept analysis, for the taxonomy development, and association rules and conditional probabilities, for the detection of non-taxonomic relations. These methods are adapted for multi-word term concept input and comparatively assessed in two different domains, the medical and the computer science domains.

Issues relating to ontology construction and *OntoGain* resources are discussed first in Sec. 2. The method is discussed in detail in Sec. 3. Evaluation results are presented in Sec. 4 followed by conclusions and issues for further research in Sec. 5.

2   Ontology Construction Methodology

The steps in ontology development may be viewed as a layer stack, where lower layers represent the basic tasks upon which rely the more complex, higher layers [12,15,16]. Thus, one cannot define concept relations before defining concepts.

In this perspective, *term extraction* is a basic layer for unsupervised concept acquisition from text. This task aims at identifying the set of terms which are characteristic for the domain and which will form the ontology lexicon. In our approach, terms are considered multi-word compounds. Multi-word terms constitute the large majority of term expressions. Moreover, they are vested with more compact and distinctive semantics (e.g., *"right ventricular infarction"* specifies in detail a concept different from other general mentions of myocardial infarction), and they present the advantage of lexically revealing their semantic content classificatory information by means of modifiers. For example, the compound term *"right ventricular infarction"* denotes a type of *"ventricular infarction"*, which in turn is a type of *"infarction"*. The exploitation of multi-word term information in *OntoGain* allows for improved domain concept coverage by capturing full term phrases and by retaining, in concept representations, the inherent, in multi-word terms, classificatory information. Moreover, multi-word terms being more compact representations of the domain concepts, compared to long lists of single-word terms, allow for more efficient system performance in the subsequent steps of ontology acquisition and make the processing of large document collections faster, without exhausting system resources.

The subsequent layer, the *concept hierarchy*, constitutes the "backbone" of the ontology. This task aims at organising concepts into a hierarchical structure, a taxonomy, where each concept is related to its respective broader and narrower concepts. In *OntoGain*, we implement and comparatively evaluate two methods for unsupervised taxonomic relation acquisition: agglomerative hierarchical clustering and formal concept analysis. In our implementation of the hierarchical clustering approach, we exploit multi-word term lexicalised classificatory information to determine term similarity in clustering.

Concepts are also characterised by attributes and relations to other concepts in the hierarchy. This is the "relation" layer, where *non-taxonomic* relations are defined. These relationships are typically expressed by a verb relating pair of concepts [17]. For this layer in *OntoGain*, we again implement and compare two approaches: association rules and a probabilistic approach.

OntoGain exhibits the following two important advantages: (a) produces a semantically rich ontology of multi-word domain concepts, rather than an ontology of mere single-word terms and, (b) produces an ontology in standard OWL representation[1]. Moreover, it allows for experimentation and comparative assessment of four different approaches to the unsupervised acquisition of taxonomic and non-taxonomic relations.

3   The *OntoGain* Modules

The principal modules of *OntoGain* are:

1. *preprocessing* which performs the linguistic analysis tasks required by subsequent modules,
2. *concept extraction* which identifies multi-word term phrases denoting domain concepts,
3. *taxonomy construction* which hierarchically structures the discovered concepts, and
4. *non-taxonomic relation acquisition* which enriches the taxonomy with domain specific concept relationships.

In what follows, we present these modules' implementations in more detail.

3.1   Preprocessing

For this module, we use the OpenNLP[2] suite of tools for tokenisation, POS tagging and shallow parsing. For the acquisition of word lemma information, we

---

[1] Implementing the *Jena Semantic Web Framework*: http://jena.sourceforge.net
[2] http://opennlp.sourceforge.net

used the WordNet Java Library (JWNL[3]). POS tagging and lemma information are required in the concept extraction phase, whereas shallow parsing information is used in the non-taxonomic relation acquisition phase for the detection of verbal dependencies.

3.2 Concept Extraction

For its concept extraction module, *OntoGain* implements C/NC-value [13], a domain-independent method for the extraction of multi-word and nested terms.

In this approach, noun phrases are initially selected by linguistic filtering. The subsequent statistical component defines the candidate noun phrase termhood by two measures: C-value and NC-value. The first measure, the C-value, is based on the hypothesis that multi-word terms tend to consist of other terms (nested in the compound term). For example, the terms *"coronary artery"* and *"artery disease"* are nested within the term *"coronary artery disease"*. Thus, C-value is defined as the relation of the cumulative frequency of occurrence of a word sequence in the text, with the frequency of occurrence of this sequence as part of larger proposed terms in the same text. The second measure, the NC-value, is based on the hypothesis that terms tend to appear in specific context and often co-occur with other terms. Thus, NC-value refines C-value by assigning additional weights to candidate terms which tend to co-occur with specific context words.

3.3 Taxonomy Construction

Hierarchical Clustering: Hierarchical agglomerative clustering proceeds bottom-up. Initially, each term phrase is considered a cluster and, at each step, the similarity between all pairs of clusters is computed and the most similar pair is merged. The algorithm typically continues until a single cluster is formed at the top of the hierarchy. We used the group average method to compute the similarity between two clusters. In particular, the group average method computes the average similarity across all pairs of concepts within the two clusters (C_i, C_j) that will be merged:

$$sim(C_i, C_j) = \frac{\sum_{x \in C_i, y \in C_j} sim(x, y)}{|C_i| * |C_j|} \tag{1}$$

where x is a concept in cluster C_i and y in cluster C_j respectively. For the computation of term similarity among multi-word terms, we use the *lexical similarity* measure [18] which takes into consideration multi-word term constituents (head/modifier) and is computed according to a Dice-like coefficient formula. Thus, the lexical similarity sim_{lex}, between term concept x and term concept y (the heads of which are denoted by x_h and y_h respectively, and their set of constituents by C) is computed as:

$$sim_{lex}(x, y) = \frac{|C(x_h) \cap C(y_h)|}{|C(x_h)| + |C(y_h)|} + \frac{|C(x) \cap C(y)|}{|C(x)| + |C(y)|} \tag{2}$$

---

[3] http://jwordnet.sourceforge.net

where the numerators denote the number of shared constituents and the denominators the sum of all constituents.

In our implementation of agglomerative clustering for the *taxonomy construction* module, the clustering process is terminated before reaching a single, top cluster. More specifically, clustering repeats as long as the merged clusters share common term heads. Furthermore, the lexical similarity measure gives credit to the shared heads between two similar multi-word terms. For this reason the created clusters consist of terms with shared heads. This cluster characteristic is exploited by *OntoGain* in appropriately labeling the top clusters of the derived concept hierarchy.

Formal Concept Analysis (FCA): FCA [19] is a popular approach for building concept hierarchies [7,6]. It relies on the idea that objects (i.e., concepts) are associated with their attributes (i.e., characteristics). FCA takes as input a matrix specifying a set of *formal objects* and *formal attributes*. In *OntoGain*, objects are the extracted multi-word terms, whereas attributes are the associated verbs, as identified in the syntactic dependencies analysis of the shallow parser. These dependencies are used to form the *formal contexts* matrix which constitutes the input to the FCA algorithm. An example of a formal context matrix is illustrated in Table 1. Dependencies are denoted by asterisks.

Table 1. Computer Science knowledge as a formal context

	submit	test	describe	print	compute	search
html form	⋆			⋆		⋆
hierarchical clustering					⋆	⋆
text retrieval						⋆
root node		⋆	⋆		⋆	⋆
single cluster			⋆		⋆	⋆
web page				⋆		⋆

For the subsequent selection of the optimal set of concept discriminative attributes, *OntoGain* implements conditional probability measures. In a comparative study by Cimiano [6], conditional probability is reported to outperform other measures, such as pointwise mutual information (PMI) [20], and selectional strength [21].

In our experiments with conditional probability threshold values we have found that the object-attribute dependency pairs above threshold $t = 0.003$ are the optimal set of dependencies for both of our application corpora domains.

For the implementation of FCA, we used the Lindig's *colibri* Java library[4] which implements the Next-Closure algorithm [22]. Table 1 illustrates an example of the input lattice to FCA, whereas a tree illustration of the OWL output taxonomy is shown in Fig. 1 and 2.

---

[4] http://www.st.cs.uni-saarland.de/~lindig

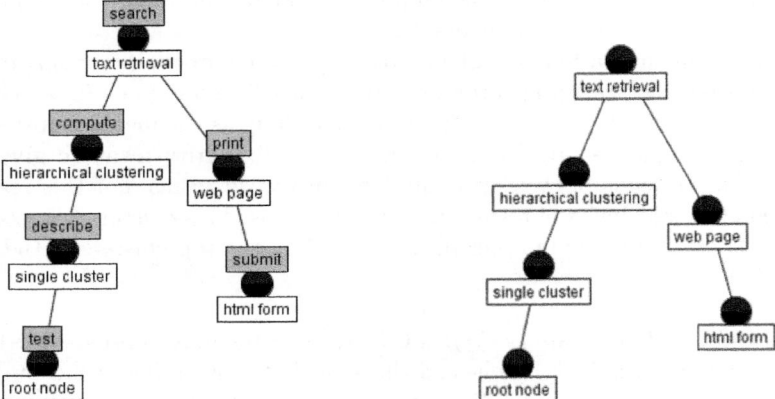

Fig. 1. Sample FCA taxonomy (Computer Science corpus)

Fig. 2. Final FCA taxonomy (without attribute labels)

3.4   Non-taxonomic Relation Acquisition

Association Rules: One of the two methods implemented in *OntoGain* for non-taxonomic relation acquisition relies on the generalised association rules algorithm, as extended for determining associations at the appropriate level of generalisation with respect to a given taxonomy [23]. In our implementation, the "subject-verb-object" dependencies set relating multi-word term concepts is enhanced with more general terms (super concepts) of the concepts it contains, based on our built taxonomy. The resulting association rules are subsequently filtered, so as to eliminate rules $X \Rightarrow Y$, where Y consists of some element in X and the rules also subsumed by $X' \Rightarrow Y'$, where X' and Y' are super concepts of X and Y respectively.

Association rules in *OntoGain* are implemented using the *predictive apriori* algorithm implementation of the Weka platform[5]. In this application, each rule is evaluated by *predictive accuracy*, a trade-off between support and confidence for each rule maximising correct predictions [24]. In our experiments with predictive accuracy threshold values, we have found that the rules above threshold $t = 0.2$ are the best for both of our application corpora domains. Table 2 illustrates sample non-taxonomic relationships output based on association rules, for the medical corpus.

Probabilistic algorithm: In this approach, reported by Cimiano et al. [25], the selection of the most appropriate non-taxonomic relationships from the set of all dependency relations found in text relies on conditional probability measures. According to this approach, first, the frequency of a dependency relation is estimated and then this frequency is propagated through the respective super

---

[5] http://www.cs.waikato.ac.nz/ml/weka

Table 2. Non-taxonomic relationships based on association rules

Domain	Range	Label
chiasmal syndrome	pituitary disproportion	cause by
medial collateral ligament	surgical treatment	need
blood transfusion	antibiotic prophylaxis	result
lipid peroxidation	cardiopulmonary bypass	lead to
prostate specific antigen	prostatectomy	follow
chronic fatigue syndrome	cardiac function	yield
right ventricular infarction	radionuclide ventriculography	analyze by
creatinine clearance	arteriovenous hemofiltration	achieve
sudden cardiac death	tachyarrhythmias	cause
cardioplegic solution	superoxide dismutase	give
bacterial translocation	antibiotic prophylaxis	decrease
accurate diagnosis	clinical suspicion	depend
ultrasound examination	clinical suspicion	give
total body oxygen consumption	epidural analgesia	attenuate by
coronary arteriography	physician assistant	perform by

concepts in a given taxonomy. The conditional probability measure is subsequently estimated as:

$$P(c|v) = \frac{f(c, v)}{f(v)} \tag{3}$$

where c is the concept, v the relation as lexicalised by a verb, and $f(c, v)$, $f(v)$ the cumulative frequencies of the concept and the relation respectively. If two or more concepts share the same conditional probability, the dependency relation of the most specific concept is selected. In our approach in *OntoGain*, dependency relations are provided by the shallow parser and concepts are multi-word terms.

4   Evaluation

In the *OntoGain* evaluation, we attempt to assess how well the extracted ontology reflects the application domains and how our *OntoGain* results compare to similar state-of-the-art approaches and hand-crafted ontologies.

We experimented with two different domain corpora, a medical and a computer science corpus. Our medical corpus was the OHSUMED[6] collection [26], containing 348,566 references from MEDLINE[7]. The computer science corpus consisted of computer science papers and articles [27].

The approach we followed for the quality assessment of the domains' representation consisted of two stages. In the first, the resulting domain ontologies are decomposed into their constituent parts (i.e., concept terms, taxonomic & non-taxonomic relations) [16] and domain experts assess each individual constituent result. In the second stage, *OntoGain* results are compared to hand-crafted

---

[6] http://ir.ohsu.edu/ohsumed/ohsumed.html
[7] http://www.nlm.nih.gov/databases/databases_medline.html

ontologies. Finally, for the comparative assessment of *OntoGain* against a similar system, we applied the Text2Onto system[8] to our domain corpora.

4.1   Evaluation of Ontology Constituent Parts

For assessing *OntoGain* in terms of *precision*, we selected the top 200 multi-word terms extracted by the concept extraction module and we proceeded to taxonomic and non-taxonomic relation extraction by all four respective methods. The domain experts then indicated the correctly acquired terms and relations. For our estimation of *recall*, domain experts examined the first 500 lines of each corpus and extracted the multi-word concepts and relations (taxonomic and non-taxonomic). These hand-crafted ontologies were compared to the respective results obtained by *OntoGain*.

The evaluation results illustrated in Table 3 show that C/NC-value performs very well in the task of identifying concepts for the ontology lexicon and in accordance with the results reported in the literature. In the taxonomy construction task, clustering clearly outperforms FCA in both corpora. Finally, association rules deliver better non-taxonomic relations when compared to probabilistic measures.

Table 3. Evaluation results for the Computer Science & OHSUMED corpora

Method	Computer Science corpus		Medical corpus	
	Precision	Recall	Precision	Recall
Concept Extraction				
C/NC-Value	86.67 %	89.6 %	89.7 %	91.4 %
Taxonomy Construction				
Formal Concept Analysis	44.2 %	48.6 %	47.1 %	41.6 %
Agglomerative Clustering	71.33 %	62.7 %	71.2 %	67.3 %
Non-Taxonomic Relation Acquisition				
Association Rules	72.85 %	61.7 %	71.8 %	67.7 %
Probabilistic algorithm	61.67 %	49.4 %	62.7 %	55.9 %

FCA results in both domains were particularly low. This was primarily due to a large number of spurious verb dependency results. Most concepts appeared with many different verbs, resulting in the formation of huge concept lattices. In FCA, concepts sharing a verb attribute are credited, so as to form taxonomic relations. However, in our application corpora, several potentially related concepts were assigned different verb attributes which resulted in either no establishment of a candidate taxonomic relation, or in the formation of erroneous and meaningless relations.

We attempted to address this problem in two ways: The first approach relies on application of conditional probability measures, so as to distinguish important verb attributes for each concept. We then experimented with various probability thresholds. The second approach clusters candidate verb attributes in synonym

---

[8] http://ontoware.org/projects/text2onto

sets, based on WordNet[9] information, so as to reduce the number of considered attributes. However, despite our efforts, hierarchical clustering outperformed FCA. Another disadvantage of FCA is the exponential time complexity $O(2^n)$, compared to the quadratic time complexity $O(N^2)$ of agglomerative clustering.

For the acquisition of non-taxonomic relations, we observe that the association rules approach in *OntoGain* is quite effective in identifying some of the most important concept relationships. The conditional probability approach attempts to establish the correct level of generalisation in the concept hierarchy. However, our results indicate that the dependency filtering process adopted in the association rules method produces highly reasonable dependencies and succeeds in pruning less significant ancestral rules and relations for modelling the domain.

4.2 Comparison to Other Methodologies

Systems similar to *OntoGain*, such as Text2Onto rely mostly on single-word term extraction for concept identification, thus resulting in the formation of huge lists of spurious terms and relations which are not descriptive of the domain. Our experiments with the C/NC value method for concept acquisition indicate that a method designed for the extraction of domain concepts and multi-word terms, rather than mere keywords and single-word terms, provides better conceptual representations, both in terms of detailed semantics, as well as in terms of domain coverage. Moreover, unlike information extraction type of approaches, the approach proposed in *OntoGain* for concept and relation acquisition is applied in an unsupervised and knowledge-poor manner. Another advantage of *OntoGain* over systems such as Text2Onto is that it outputs the results of each ontology acquisition step in OWL. The conformance of *OntoGain* output to such standards allows for easier results visualisation in any OWL compliant ontology editor, and easier ontology editing, maintenance, reuse and exchange.

In our attempt to apply the Text2Onto system to our domain corpora we were not successful. We attempted to segment our corpus but the system was running out of memory, even though we ran it on a 64-bit server reserving 3 GB of heap space. We consider that this was due to the size of input data ($\sim$250 Mbytes for OHSUMED) which resulted in storing and processing hundreds of thousands of single-word term concepts. The subsequent effort in detection of taxonomic and non-taxonomic relations from such amounts of data leads the program to crash. This observation strengthens our assumption that multi-word terms lead to more compact representations of the examined domain, yielding dense and meaningful listings of multi-word term concepts.

5 Conclusions

We introduced *OntoGain*, a platform for ontology acquisition from texts using multi-word terms. For the acquisition of taxonomic and non-taxonomic relationships, we comparatively investigate four approaches: agglomerative

---

[9] http://wordnet.princeton.edu

hierarchical clustering and formal concept analysis, for the taxonomy development, and association rules and conditional probabilities, for the detection of non-taxonomic relations. All methods are adapted for multi-word term concept input and comparatively assessed in two different domains, the medical and the computer science domains. This evaluation indicates that agglomerative clustering and association rules outperform any other method combination reaching up to 70% precision for identification of taxonomic and non-taxonomic relations respectively in both corpora.

Investigation of method combinations for each task (e.g., agglomerative clustering with formal concept analysis, or association rules with conditional probabilities), incorporating methods for extracting or learning concept attributes (e.g., "small", "large", "black", "white"), resolving term ambiguities as well as incorporating Hearst lexico-syntactic patterns for revealing additional relationships types (e.g., "part-of") are issues for further research.

References

1. Pinto, H., Martins, J.: Ontologies: How can They be Built? Knowledge and Information Systems 6(4), 441–464 (2004)
2. Suchanek, F.M., Sozio, M., Weikum, G.: SOFIE: A Self-Organizing Framework for Information Extraction. In: Proc. of the 18th Intern. World Wide Web Conf. (WWW 2009), Madrid, Spain, pp. 631–640. ACM Press, New York (2009)
3. Pantel, P., Pennacchiotti, M.: Automatically Harvesting and Ontologizing Semantic Relations. In: Proc. of the 2008 Conf. on Ontology Learning and Population: Bridging the Gap between Text and Knowledge, pp. 171–195. IOS Press, Amsterdam (2008)
4. Velardi, P., Navigli, R., Cucchiarelli, A., Neri, F.: Evaluation of OntoLearn, a Methodology for Automatic Learning of Ontologies. In: Buitelaar, P., Cimmiano, P., Magnini, B. (eds.) Ontology Learning from Text: Methods, Evaluation and Applications, pp. 569–572. IOS Press, Amsterdam (2005)
5. Buitelaar, P., Cimiano, P., Frank, A., Racioppa, S.: SOBA: SmartWeb Ontology-based Annotation. In: Proc. of the Demo Session at the Intern. Semantic Web Conference (ISWC), Athens GA, USA (November 2006)
6. Cimiano, P., Hotho, A., Staab, S.: Learning Concept Hierarchies from Text Corpora using Formal Concept Analysis. Journal of Artificial Intelligence Research (JAIR) 24, 305–339 (2005)
7. Haav, H.M.: An application of inductive concept analysis to construction of domain-specific ontologies. In: Brandenburg University of Technology at Cottbus, pp. 63–67 (2003)
8. Maedche, A., Staab, S.: Discovering Conceptual Relations from Text. In: Proc. of the 14th European Conf. on Artificial Intelligence (ECAI 2000), August 2000, pp. 321–325. IOS Press, Amsterdam (2000)
9. Ciaramita, M., Gangemi, A., Ratsch, E., Saric, J., Rojas, I.: Unsupervised Learning of Semantic Relations for Molecular Biology Ontologies. In: Buitelaar, P., Cimiano, P. (eds.) Ontology Learning and Population: Bridging the Gap between Text and Knowledge, pp. 99–104. IOS Press, Amsterdam (2008)
10. Soderland, S., Mandhani, B.: Moving from Textual Relations to Ontologized Relations. In: Proc. of the 2007 AAAI Spring Symposium on Machine Reading, pp. 85–90. AAAI Press, Menlo Park (2007)

11. Cimiano, P., Völker, J.: Text2Onto - A Framework for Ontology Learning and Data-driven Change Discovery. In: Montoyo, A., Muñoz, R., Métais, E. (eds.) NLDB 2005. LNCS, vol. 3513, pp. 227–238. Springer, Heidelberg (2005)
12. Buitelaar, P., Cimiano, P., Magnini, B.: Ontology Learning from Text: Methods, Evaluation and Applications. IOS Press, Amsterdam (2005)
13. Frantzi, K., Ananiadou, S., Mima, H.: Automatic Recognition of Multi-Word Terms: The C-Value/NC-Value Method. Intern. Journal of Digital Libraries 3(2), 117–132 (2000)
14. Witschel, H.: Terminology Extraction and Automatic Indexing – Comparison and Qualitative Evaluation of Methods. In: Proc. of Terminology and Knowledge Engineering, TKE (2005)
15. Cimiano, P.: Ontology Learning and Population from Text: Algorithms, Evaluation and Applications. Springer, Heidelberg (2006)
16. Brank, J., Grobelnik, M., Mladenic, D.: A Survey of Ontology Evaluation Techniques. In: Proc. of the Conf. on Data Mining and Data Warehouses (SiKDD 2005), Ljubljana, Slovenia (October 2005)
17. Kavalec, M., Maedche, A., Svátek, V.: Discovery of Lexical Entries for Non-taxonomic Relations in Ontology Learning. In: Van Emde Boas, P., Pokorný, J., Bieliková, M., Štuller, J. (eds.) SOFSEM 2004. LNCS, vol. 2932, pp. 249–256. Springer, Heidelberg (2004)
18. Nenadic, G., Spasic, I., Ananiadou, S.: Automatic Discovery of Term Similarities Using Pattern Mining. Intl. Journal of Terminology 10(1), 55–80 (2004)
19. Ganter, B., Wille, R.: Formal Concept Analysis: Mathematical Foundations. Springer, Heidelberg (1999)
20. Hindle, D.: Noun Classification from Predicate-Argument Structures. In: Proc. of the 28th Annual Meeting of the Association for Computational Linguistics (ACL 1990), Pittsburgh, PA, USA, June 1990, pp. 268–275 (1990)
21. Resnik, P.: Selectional Preference and Sense Disambiguation. In: Proc. of the ACL SIGLEX Workshop on Tagging Text with Lexical Semantics: Why, What, and How?, Washington, DC (1997)
22. Ganter, B., Reuter, K.: Finding all Closed Sets: A General Approach. Order 8(3), 283–290 (1991)
23. Srikant, R., Agrawal, R.: Mining Generalized Association Rules. In: Proc. of 21th Conf. on Very Large Data Bases (VLDB 1995), Zurich, Switzerland, September 1995, pp. 407–419. Morgan Kaufmann, San Francisco (1995)
24. Scheffer, T.: Finding Association Rules that Trade Support Optimally Against Confidence. Intelligent Data Analysis 9(4), 381–395 (2005)
25. Cimiano, P., Hartung, M., Ratsch, E.: Finding the Appropriate Generalization Level for Binary Relations Extracted from the Genia Corpus. In: Proc. of the Intern. Conf. on Language Resources and Evaluation (LREC 2006), ELRA, May 2006, pp. 161–169 (2006)
26. Hersh, W., Buckley, C., Leone, T., Hickam, D.: OHSUMED: An Interactive Retrieval Evaluation and New Large Test Collection for Research. In: Proc. of the 17th ACM SIGIR, Dublin, Ireland, pp. 192–201 (1994)
27. Milios, E., Zhang, Y., He, B., Dong, L.: Automatic Term Extraction and Document Similarity in Special Text Corpora. In: 6th Conf. of the Pacific Association for Computational Linguistics, Halifax, Canada, August 2003, pp. 22–25 (2003)

Identifying Writers' Background by Comparing Personal Sense Thesauri

Polina Panicheva[1,2], John Cardiff[1], and Paolo Rosso[2]

[1] Social Media Research Group, Institute of Technology Tallaght, Dublin, Ireland
{Polina.Panicheva,John.Cardiff}@ittdublin.ie
[2] Natural Language Engineering Lab, ELiRF, Universidad Politécnica de Valencia, Spain
prosso@dsic.upv.es

Abstract. Analysis of blogpost writings is an important and growing research area. Both objective and subjective characteristics of a writer are detected. Words have word meaning that is common in the language and that is represented in their usage. Another component of word meaning, "personal sense", not inherent in the language, but different for each person, reflects a meaning of words in terms of unique personal experience and carries the personal characteristics.

In our research word meaning techniques are applied to represent personal sense of words in texts by different authors. Personalized concept structures are construed and used to infer authors' perspective from text: various notions of context combined with different thesaurus similarity scales are applied to confirm that from a certain perspective similarity in the personalized thesauri with some restrictions can correspond to similarities in the occupation of the authors.

Keywords: personal sense, semantic relatedness, co-occurrence based thesauri, writer's perspective.

1 Introduction

Research on analysis of blogpost writings is an area attracting an increasing amount of attention. Not only can objective characteristics of a writer, like age or gender, be detected, but also subjective features, such as opinion, personality traits, and emotions, can be inferred. Words have word meaning that is common in the language and that is represented in their usage. We believe that another component of word meaning – "personal sense" [4] – not inherent in the language, but which is different for each person, can carry this personal information. This component reflects a meaning of the word in terms of unique experience of a person, in addition to word meaning belonging to language.

In our research the co-occurrence distributional techniques are applied to represent personal sense of words in texts by different authors, investigating four notions of 'context'. Personal concept structures, construed of the words taking into account their personal sense, are used to infer an author's perspective from the texts they write. The results confirm that similarity in the personalized thesauri can correspond to similarities in occupation of the authors.

C.J. Hopfe et al. (Eds.): NLDB 2010, LNCS 6177, pp. 288–295, 2010.
© Springer-Verlag Berlin Heidelberg 2010

The rest of the paper is organized as follows. In Section 2 the work influencing our research is discussed. Section 3 describes some previous work using the personal sense technique. In Section 4 we present the ideas underlying our work based on the notion of personal sense. In Section 5 the experiments are described in detail; the results are presented and discussed in Section 6. Additional experiments of use to support and supplement the current work are discussed in Section 7.

2 Related Work

2.1 Author's Perspective

The importance of the distinction between what belongs to the author, the reader and the text itself is underlined in [1]. In this work the authors describe the important differences in considering subjectivity in text from these points of view. The emphasis is put on the text itself and the importance of the investigation of the opinion contained in it exclusively, as opposed to observing the reader's and writer's viewpoint in the text. Our work is concerned with personal information about the author, how and why it is presented in text.

In [2] the authors present an approach to perspective determination based on concept interrelation ontology. They manually construct an ontology of the movie-production field. The semantic relatedness measure for 25 pairs of words denoting motion picture industry concepts is presented. The next step suggested is gauging the author's perspective within the motion picture industry field.

The authors of [14] describe a method for constructing a personalized thesaurus from bookmarked web pages and documents. They construct a personalized thesaurus for every user as a context distributional profile of a word and propose a scale for measuring the difference between the thesauri.

2.2 Distributional Hypothesis

The distributional hypothesis is a widely used idea in semantics. The definition of word meaning, stating that "for a *large* class of cases... the meaning of a word is its use in the language" [13], appears to be helpful for analyzing word meaning in natural language applications. Intuitively it has been used in different linguistic tasks, e.g. [11], [5]. The degree of applicability of this hypothesis to natural language is investigated in [6], testing different algorithms realizing this idea and making a crucial distinction between its applications to words and to concepts.

3 Previous Work on Personal Sense

Our initial experiments using personal sense were described in [7]. There we investigate the personal sense of the words 'movie' and 'film', and the hypothesis is that the personal sense of these words is different for the authors writing a negative and a positive review. The resulting features are reported to perform higher than the baseline, but very modestly compared to the widely used n-gram word-count features. The authors assume that the reason for this is that different writers express different personal sense for the movie-words, even if the polarity of their reviews is the same.

An experiment testing this assumption is described in [8]. The result showed that opinion mining with a bunch of different writers yielded so much noise, that a much smaller number of instances but analyzed separately for different writers gave better performance. It was concluded that polarity in texts may be expressed in an individual way by different authors.

4   Author Characterization Using Personal Sense

Personal background and other personal information about the author is revealed in text, forming a person's idiolect – "a language that can be characterized exhaustively in terms of … properties of some single person at a time" [4]. An idiolect represents a collection of personal characteristics at the same time (i.e., age, gender, social class, occupation), as well as personal traits and private states. Words have word meaning that is common in the language and that is represented in their usage. 'Personal sense', a component of word meaning not inherent in the language, but different for each person, is defined in [4]. This component reflects word meaning in terms of unique experience of a person.

4.1   Personal Sense and Co-occurrence Distribution

Our approach is based on the hypothesis that personal sense is represented in the way that words are put together to form a text and in the way they occur together with each other. We use the co-occurrence distributional method described in [6] to represent personal senses.

We investigate different notions of 'context' in order to find the one that makes the resulting co-occurrence distribution representing the meaning or personal sense of a word best: from ordered pairs of words, one word on the left and one on the right, to all the words occurring in the same sentence with the word in question.

Authors of [6] find out that for investigating semantic measures of word-to-word distance the best results are obtained by using the cosine distance measure using conditional probability weighting of coordinates. This is the measure that we use in our work to scale semantic relatedness of personal sense of words to each other.

4.2   From Personal Sense to Author's Perspective: Forming a Structure

We follow [2] in investigating the semantic inter-relatedness of words for different persons, especially for authors with different occupations. In a text by one person, the concepts that he/she uses acquire a personal sense. Depending on unique background information underlying their idiolect, the concepts represented in the text will form a unique semantic structure in terms of their meaning, as influenced by the personal sense. Whereas the authors of [2] define every profession in terms of its semantic relatedness to every other notion, suggesting a manually created ontology of the semantic field, our consideration is to derive the semantic relatedness of the notions from text, and thus infer the author's profession.

Thus, the goal of this work is as follows. Firstly, to represent personal sense of words in text by different authors using co-occurrence distributional word meaning techniques. Secondly, to form personalized thesauri based on the inter-relations between words. Thirdly, to infer relations among authors' in terms of their background, particularly their occupation, from the distance measures between their personalized thesauri. Thus, we continue exploiting the interrelation of private states expressed in text and authorship attribution, described in [8].

5 Experiments

We work with blog text corpus described in [12]. For the initial experiment we investigated 10 randomly-selected blogs annotated as 'Accounting' and 10 annotated as 'Military', with the annotation referring to the author's self-annotation in terms of their occupation or profession. We used a Natural Language Toolkit (NLTK)[1] for the Python programming language for most of the text analysis. All the texts were annotated automatically with parts-of-speech using the Treebank POS-Tagger built-in into the NLTK.

A group of nouns was selected for the experiment, consisting of 33 nouns, including the 25 most frequent nouns ([*dad, mom, person, company, movie, husband, test, parent, dinner, world, head, place, child, thing, school, way, life, family, house, work, job, car, home, girl, friend*]) and the 8 manually selected nouns ([*career, power, war, law, rule, army, man, woman*]).

Next, their personal sense was represented by the words' co-occurrence distributional vectors. Every dimension is a word occurring in the context(s) of the investigated nouns. As an example, the contexts of the target word 'woman' taken from the following sentences are presented:

-"Part of the problem for female soldiers in the Army is the existence of *women* like one of my roommates, who…"

-"SGT B, however, the *woman* I live with, has clearly not had many female friends."

We observed 4 different definitions of context:

1. Context of a word *w1* is a set of pairs of words; every pair consists of a word immediately to the left of *w1* and a word immediately to the right of *w1*:

- *(of/IN, like/IN), (the/DT, I/PR)*

2. A set of the same words as in (1), not organized in pairs, but for each word it is indicated if it belongs to the left or to the right context.

- *l_of/IN, r_like/IN, l_the/DT, r_I/PR*

3. The same set of words as in (2), but no left- or right-side context indication.

- *of/IN, like/IN, the/DT, I/PR*

4. All the words belonging to the sentences in which *w1* occurs. except *w1* itself.

- *Part/NN of/IN the/DT problem/NN for/IN female/JJ soldiers/NN in/IN the/DT Army/NN is/VB the/DT existence/NN of/IN* etc.

- *SGT/NN B/NN ,/, however/RB ,/, the/DT, I/PR live/VB with/IN ,/, has/VB clearly/RB not/RB had/VB many/JJ female/JJ friends/NN*

Every word in context acquires a weight of the conditional probability of the word given the target word *w1* (see formula 1).

$$P(w \mid w_1) = \frac{P(w \cap w_1)}{P(w_1)} = \frac{Frequency\,(w, w_1)co - occurring}{Frequency\,(w_1)} \qquad (1)$$

---

[1] Available at http://www.nltk.org/

For computing similarity between two words in a text we used the cosine similarity measure between the target words (see the Cosine for Conditional Probabilities formula, (2)).

$$CosCP(w_1, w_2) = \frac{\sum_{w \in cont(w_1) \cup cont(w_2)} (P(w \mid w_1) * P(w \mid w_2))}{\sqrt{\sum_{w \in cont(w_1)} P(w \mid w_1)^2} * \sqrt{\sum_{w \in cont(w_2)} P(w \mid w_2)^2}}$$ (2)

The experiments are dedicated to comparing the obtained personalized thesauri with the techniques similar to those described in [14]. We compared three different thesauri distance measures, using the basic formula presented in [14].

$$d = \sqrt{\frac{\sum_{i=1}^{m} \sum_{j=1}^{m} d'_{v_i w_i}^2}{q}}$$ (3)

$$d_{vw}, \begin{vmatrix} sim_{vw}^S - sim_{vw}^T \end{vmatrix}, (v, w) \in S, T$$
$$sim_{vw}^S, (v, w) \in S, \notin T$$
$$sim_{vw}^S, (v, w) \in T, \notin S$$
$$x, (v, w) \notin S, \notin T$$ (4)

The first distance measure, d_1, was exactly the same as described in [14]. That means, in equation (3) q is equal to m^2, the squared number of words in the target words vocabulary; and x in equation (4) equals to 1. For the second measure, d_2, x in equation (4) is equal to 0. In other words, when the word-pair (v, w) is not present in neither of the two thesauri S and T, we add nothing to the sum. For d_3, x is also equal to 0. But q in equation (3) represents the number of the word pairs present in at least one of the two vocabularies S and T. The difference between the thesauri distance formulas 1, 2 and 3 lies in approaching the words absent from the personal blog vocabulary in a different way. We assume that this difference will affect the results in a considerable way, because the two groups of blog authors that we are investigating manifest differences in terms of their vocabulary usage: the 'Military' authors tend not to use all of the target words in their blogs, whereas for the 'Accounting' authors this is usually not the case.

We assume that at least one notion of 'context' constitutes the personal senses of the observed words so that the personal sense and their inter-relations represent the authors' background to a sufficient degree, and the resulting average distances between personalized thesauri are bigger for authors having different occupation than for those sharing one.

6   Results and Discussion

The personalized thesauri comparison results for the 33 target words, for 4 different types of context and the 3 distance functions are presented in Table 1.

Our assumption was that the personal word sense interrelation structures would be more similar among authors belonging to the same occupational background, than between the authors belonging to different professional groups. We represented the personal sense interrelations of every author as a personalized thesauri based on their texts. The distances between the thesauri were computed for every pair of authors. We divided the pairs of thesauri into 3 types depending on their authors:

- (1) (T[military], T[military])
- (2) (T[accounting], T[accounting])
- (3) (T[military], T[accounting]), and (T[accounting], T[military]),

where T[x] represents the thesaurus of an author belonging to the occupation x.

We computed the distances between the thesauri in groups 1, 2 and 3 separately. According to our assumption, the distances in groups (1) and (2), i.e. between thesauri by authors of the same occupation, would be smaller than for the distances of group (3), representing the distances between the different occupational group thesauri. In spite of the fact that the results did not unanimously confirm our overall hypothesis, they indicate some very important tendencies and provide evidence supporting our contention that the personal sense technique is useful for discriminating between authors' occupations.

The results with high statistical significance (p-value < 0.01), obtained for the context definition 1, for distance measures 2 and 3, for the 'Military' occupation group, confirm the hypothesis and show that the inter-group distances are lower than the intra-group distances. This could also indicate the fact that the military group is easier to discriminate than the accounting group, as the personalized thesauri for these group are closely related to each other, whereas for the accounting group the structure is sparser. In most of the cases little statistical significance was achieved, indicating that more target words and more texts should be analyzed.

It is important to note that we have obtained contradictory results for different distance options, according to our expectations described in Section 5. In most of the cases distance 1 yields exactly opposite results than distances 2 and 3, in terms of the inter- and intra- group distances between the occupations. These formulas mainly differ in the way that they handle words that did not occur in the texts by specific authors. As the results for the different distance formulas suggest, the difference in approaching these words alters the results dramatically, regardless all the other options. This means that for the current data and context definition, the absence or presence of words in the authors' vocabularies outperform the personal sense differences computed for the words. This was our expectation that was confirmed by using the words that were not contained in all of the texts.

The only context definition that yielded more consistent results for different distance functions is the context number 4, containing all the words in the sentence where the target word appears. This confirms a consideration that the broader the context definition is, the more it influences the results, regardless of the different distance measures. However, the very low statistical significance in the case of such context shows that the number of target words should be increased.

Table 1. Personalized thesauri comparison results for blogs by authors of the same and different occupation

	Thesauri distance comparison		Statistical significance according to the paired t-test(p-value)	
		d_1		
Context:	'Military'vs different	'Accounting'vs different	'Military'vs different	'Accounting'vs different
1	>	<	0.74	0.2
2	>	<	0.2	0.1
3	>	<	0.1	0.1
4	>	<	0.47	0.68
		d_2		
1	<	>	*<0.001*	0.1
2	<	>	0.09	0.8
3	<	>	0.35	0.8
4	>	>	0.61	0.59
		d_3		
1	<	>	*<0.001*	*0.02*
2	<	>	0.15	0.99
3	<	<	0.56	0.99
4	>	>	0.5	0.7

7   Conclusions and Future Work

Our experiments confirm that for target words of certain number and frequency, and for a certain definition of context and a certain distance measure for thesauri comparison, the differences between the resulting thesauri can effectively represent the differences among the occupation background of the authors of texts. Personalized concept structures were constructed from texts by different authors, taking into account the personal sense of the words they used. The resulting structures were unique for every blog writer, representing the unique personal sense. The personal concept structures were used to infer the authors' perspective from their writings. The results confirmed that with certain restrictions, the method of representing the similarities in the personalized thesauri could be used to reflect similarities in occupation of the authors.

The first restriction applies to the occupation groups themselves: the results prove that different occupations are not equally easy to analyze, i.e., some of the occupations present similarities in the personalized thesauri that can be easily detected, whereas others require additional experiments with more refined options.

We see the direction of such experiments as follows. First of all, it is necessary to reduce the influence of the fact that some words are absent in some blogs, by including in the target words list only the words that appear in all of the investigated blog texts. Secondly, it is important to increase the influence of the personal sense method, i.e. of the context information, on the resulting personalized thesauri, by considering context in a broader sense and increasing the context window. However, it has been illustrated that the broader context window leads to having more noisy information, and subsequently less statistical significance. This issue will be addressed by analyzing more target words and more texts, thus acquiring higher occurrence values.

Acknowledgements

The work of the third author has been supported by the MICINN research project TEXT-ENTERPRISE 2.0 TIN2009-13391-C04-03 (Plan I+D+i).

References

1. Balahur, A., Steinberger, R.: Rethinking sentiment analysis in the news: from theory to practice and back. In: Proc. 1st Workshop on Opinion Mining and Sentiment Analysis (WOMSA), CAEPIA-TTIA Conference (2009)
2. Choudhury, S., Raymond, K., Higgs, P.L.: A rule-based metric for calculating semantic relatedness score for the motion picture industry. In: Workshop on Natural Language Processing and Ontology Engineering (2008)
3. Firth, J.R.: A synopsis of linguistic theory 1930-1955. In Studies in Linguistic Analysis. Oxford: Philological Society. Reprinted in F.R. Palmer (ed.), Selected Papers of J.R. Firth 1952-1959, London: Longman (1968)
4. Leontev, A.N.: Activity, Consciousness, and Personality. Prentice-Hall, Hillsdale (1978)
5. Mitrofanova, O., Lashevskaya, O., Panicheva, P.: Statistical Word Sense Disambiguation in Contexts for Russian Nouns Denoting Physical Objects. In: Sojka, P., Horák, A., Kopeček, I., Pala, K. (eds.) TSD 2008. LNCS (LNAI), vol. 5246, pp. 153–159. Springer, Heidelberg (2008)
6. Mohammad, S., Hirst, G.: Distributional measures of concept-distance: A task-oriented evaluation. In: Proc. EMNLP 2006 (2006)
7. Panicheva, P., Cardiff, J., Rosso, P.: A Co-occurrence Based Personal Sense Approach to Opinion Mining. In: Proc. 1st Workshop on Opinion Mining and Sentiment Analysis (WOMSA), CAEPIA-TTIA Conference (2009)
8. Panicheva, P., Cardiff, J., Rosso, P.: Personal Sense and Idiolect: Combining Authorship Attribution and Opinion Analysis. Manuscript submitted for publication (2010)
9. Pinto, D., Rosso, P., Juan, A., Jiménez, H.: A Comparative study of Clustering algorithms on Narrow-Domain abstracts. In: Sociedad Española para el Procesamiento del Lenguaje Natural (SEPLN) 37 (2006)
10. Romesburg, H.C.: Clustering Analysis for researchers. Wadsworth, Inc. (1984)
11. Rubenstein, H., Goodenough, J.B.: Contextual correlates of synonymy. Communications of the ACM 8(10) (1965)
12. Schler, J., Koppel, M., Argamon, S., Pennebaker, J.: Effects of Age and Gender on Blogging. In: Proc. 2006 AAAI Spring Symposium on Computational Approaches for Analyzing Weblogs (2006)
13. Wittgenstein, L.: Philosophical Investigations: The English Text of the 3rd Edition, translated by G.E.M. Anscombe. Macmillan Publishing, New York (1973)
14. Yoshida, S., Yukawa, T., Kuwabara, K.: Constructing and examining personalized cooccurrence-based thesauri on Web pages. In: Proc. 12th International World Wide Web Conference (2003)

Retrieval of Similar Electronic Health Records Using UMLS Concept Graphs

Laura Plaza and Alberto Díaz

Universidad Complutense de Madrid
C/Profesor José García Santesmases, s/n, Madrid 28040, Spain
{lplazam,albertodiaz}@fdi.ucm.es

Abstract. Physicians often use information from previous clinical cases in their decision-making process. However, the large amount of patient records available in hospitals makes an exhaustive search unfeasible. We propose a method for the retrieval of similar clinical cases, based on mapping the text onto UMLS concepts and representing the patient records as semantic graphs. The method also deals with the problems of negation detection and concept identification in clinical free text. To evaluate the approach, an evaluation collection has been developed. The results show that our method correlates well with the expert judgments and outperforms remarkably the traditional term-vector space model.

Keywords: Similar Case Retrieval, Graph Theory, Electronic Health Record, UMLS.

1 Introduction

In their daily work, physicians often face complex clinical cases and need to refer to similar patient cases encountered before, especially when untypical cases are presented. In fact, as stated in [1], previous knowledge about the problem is proved to be one of the most influencing factors in the decision-making process. However, as shown in [2], the excessive time required to find this knowledge may undermine its convenience if no effective access technologies are provided.

When dealing with medical records, the information is mainly stored as "free-text". This information is considerably more difficult to analyze than that presented in more formal texts, such as textbooks or scientific papers, since it exhibits unique sublanguage characteristics (e.g. verbless sentences, lack of punctuation and spelling errors). Moreover, negation detection plays an important role when trying to understand the meaning of medical information.

The task of retrieving similar medical records may be considered a particular case of Information Retrieval (IR) that may be stated as: "Given a reference record, retrieve other records from the clinical database that are similar to the reference one". To decide the criteria to assess the similarity between records is the most important and difficult issue in this task. In this paper, we propose a method for the retrieval of similar Electronic Health Records (EHRs), based on mapping the text onto UMLS concepts and representing the patient records

C.J. Hopfe et al. (Eds.): NLDB 2010, LNCS 6177, pp. 296–303, 2010.

as semantic graphs, where the vertices are UMLS concepts and the edges represent *is-a* relations between them. The method also deals with the problems of negation detection and concept identification in clinical free text.

2 Background

2.1 Related Work

Information Retrieval in medicine dates back to the middle 1960s. Early approaches use words to index the documents in the corpus. However, though term-based indexing using a vector space model is simple and powerful, concept-based indexing using controlled terminologies can improve the performance of IR. A pioneer work on IR in the medical context which makes use of these resources is SAPHIRE [3]. The aim of this project was the development of methods for indexing and retrieving medical documents from bibliographic databases.

Most NLP works in the biomedical domain have focused on journal articles rather than clinical free text. A recent initiative for the development of NLP of medical records is the CLEF forum; in particular, the ImageCLEF track[1]. The goal is to retrieve the most relevant images to a predefined set of queries, based on their clinical case descriptions. The best results in 2005 and 2006 editions came from Lacoste et al. [4], who applied a "semantic indexing" to identify key concepts in certain UMLS semantic categories. Other systems that competed in the task made use of other resources, such as MeSH, to expand the queries [5].

Apart from these few approaches, most medical IR systems do not use any external resource, but base their decisions on the terms in the text, and only a few of them use *ad hoc* lists or manual tagging to determine which of these terms are symptoms, diseases or other relevant features [6]. Ontologies and controlled vocabularies may avoid the need of producing and maintaining these lists, while capturing the semantic in the text and the relations between the terms.

2.2 The Electronic Health Record and the Domain Peculiarities

Clinical records present unique attributes that must be taken into account in any NLP system. First, the structure and content vary with the audience needs. It is a mix of highly structured information and other idiosyncratic narrative text. Sometimes, images are included. Finally, medical records differ greatly in size: from a few lines to several pages. In summary, any information that is relevant for the decision process can be part of the medical record, so that it is either impossible or surprisingly difficult to find a predictable retrieval pattern.

Second, negation detection is a fundamental issue, since it can invert the sense of a text. Negation in natural language can be extremely subtle, but medical language is much more restricted and negations are expected to be more direct [7]. A variety of approaches have addressed negation in medical texts [8,9] .

---

[1] ImageCLEF medical retrieval task, http://www.imageclef.org/2009/medical

Third, the peculiarities of the terminology and the writing practices of physicians make concept detection a very ambitious task [10]. The first challenge is the problem of *synonyms* and *homonyms*. Another handicap is the presence of *neologisms*. Finally, *elisions* and *abbreviations* complicate the automatic disambiguation of medical text.

2.3   The Use of UMLS for Concept Annotation

One of the most popular biomedical terminologies in NLP applications is the Unified Medical Language System (UMLS) [11]. UMLS consists of 3 main components: the Specialist Lexicon, the Metathesaurus and the Semantic Network. The *Metathesaurus* comprises a collection of biomedical concepts derived from more than 100 vocabulary sources and the relations among them. The *Semantic Network* consists of a set of categories (semantic types) that provides a categorization of the concepts in the Metathesaurus. Using UMLS for concept annotation presents two main advantages: first, it lists more than 15000 entries of ambiguous terms (which attenuate the problems of synonymy and homonymy); second, it contains numerous entries for elisions and abbreviations.

In order to map the text onto UMLS concepts, the *MetaMap Transfer tool (MMTx)* is used. MetaMap [12] allows mapping biomedical free-form text to Metathesaurus concepts with a high level of accuracy [13].

3   A Method for Automatic Retrieval of Similar EHRs

In this section, a concept graph-based method for retrieving similar EHRs is presented. It consists of 4 steps. Each step is discussed in the following subsections. Besides, in order to clarify how the algorithm works, the following radiology report from the CMC-NLP corpus[2] is used as working example:

> *CLINICAL HISTORY:* Eleven years old with ALL, bone narrow transplant on Jan.2, now with three day history of cough.

> *IMPRESSION:* No focal pneumonia. Likely chronic changes at the left lung base. Mild anterior wedging of the thoracic vertebral bodies.

3.1   Extraction of UMLS Concepts

In this step, the text in each EHR from the database is mapped onto UMLS concepts and semantic types using MetaMap. In order to understand what information is relevant for the retrieval of similar clinical cases, a hospital physician has been consulted. The decision apparently depends on multiple criteria (e.g. the physician's specialty and particular concerns). However, as general guidelines, two clinical records can be considered similar if: (1) the same symptom or

---

[2] Computational Medicine Center's 2007 Medical Natural Language Processsing Challenge (CMC-NLP 2007), http://www.computationalmedicine.org/challenge

Table 1. Relevant UMLS semantic types

Category	UMLS semantic types
Symptoms and signs	Sign or Symptom
	Finding
Diseases	Disease or Syndrome
	Pathologic Function
Procedures	Therapeutic or Preventive Procedure
	Diagnosis Procedure
Body parts	Body Location or Region
	Body Part, Organ, or Organ Component
Medicaments	Pharmacologic substance

sign is presented (e.g. *fever* or *5 kg weight loss*), (2) the patients have received the same diagnosis (e.g. *bacterial pneumonia*), (3) the same test or procedure is reported (e.g. *cerebral NMR* or *endoscopy biopsy*) or (4) the same medicament has been administered (e.g. *clopidogrel*).

Therefore, according to the domain expert guidelines relevant attributes to the task are: *symptoms or signs, diseases, procedures* and *medicaments*. Identifying which concepts in a medical record correspond to each of these categories is not trivial. However, the UMLS Semantic Network includes very useful information to map the concepts in the patient records to the previous categories. Then, only the concepts from a subset of UMLS semantic types are considered. Table 1 shows these semantic types along with their mappings to the categories above. A further category has been added, *body parts*, since they are often involved in the descriptions of procedures and diseases (e.g. *fractured rib*).

3.2   Negation Detection

The aim of this step is to detect negated concepts in the EHRs since, according to the domain expert, absent symptoms or diseases are not relevant for the task. Since negations within medical records usually appear in a reduced number of forms, a simple lexical scanner from regular expressions is used. According to this, we have come up with 4 negation classes. Table 2 shows the lexical patterns used to detect them, along with some examples of their occurrences within the corpora. *Concept* stands for the previously identified UMLS concepts.

3.3   Semantic Graph Representation

The next step consists of creating a graph-based representation for each EHR in the database. First, the concepts identified in the previous step are retrieved from the UMLS Metathesaurus along with their complete hierarchy of hypernyms. Second, all concept hierarchies for each category are merged, building a unique graph for each category in the EHR, where the edges represent semantic relations, and the vertices represent distinct concepts in the text. Finally, each concept is assigned a weight using equation (1), where α is the set of all the parents of the concept A, including A, and β is the set of all the parents of

Table 2. Negation classes

Lexical Pattern	Examples
no|without|rule out + adj? + concept + (or concept)*	Without focal scarring Rule out fever or cough
no|without + noun + of + concept + (or concept)*	No signs of tuberculosis No evidence of hydroureter
evaluate for + (noun|adj)? + concept + (or concept)*	Evaluate for foreign body Evaluate for abnormalities
lack|absence of + (noun|adj)? + concept + (or concept)*	Lack of kyphosis Absence of heart murmur

concept B, including B. This formula will assign greater weight to an edge as the concepts that it links become more specific. Fig. 1 shows these graphs for the record example. The edges of one of these graphs have been labeled with their weights. It can be observed how the acronym *ALL (Acute Lymphocytic Leukemia)* is expanded by MetaMap.

$$weight(A, B) = \frac{|\alpha \cap \beta|}{|\alpha \cup \beta|} \tag{1}$$

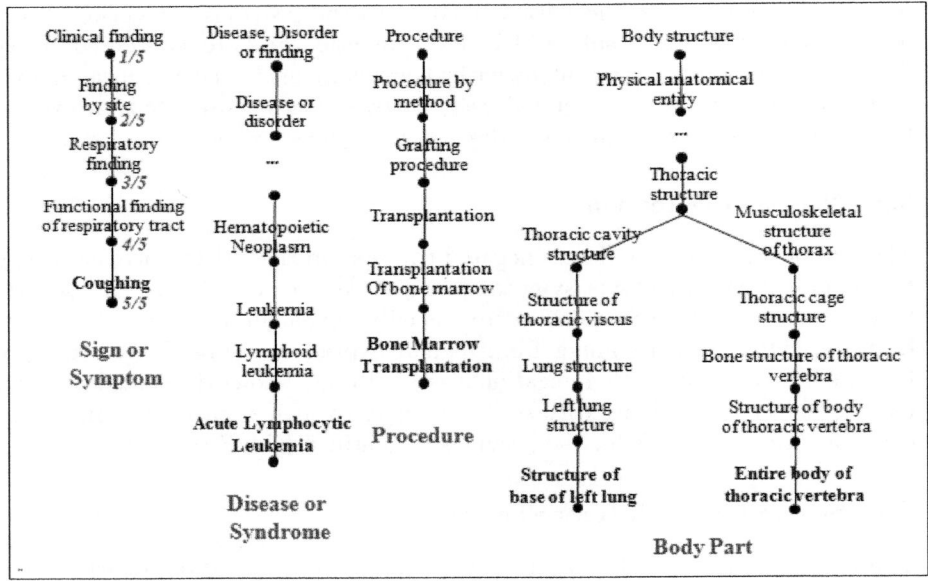

Fig. 1. An example of EHR semantic representation

3.4   Computing the Similarity between EHRs

The purpose of the last step is to compute the similarity between the graph-based representations of two patient records. To this end, a non-democratic vote mechanism is used, similar to that proposed in [14]. Given two graphs, A and B, so that the similarity of A to B has to be measured, each concept of A which is present in B assigns a score equal to the weight of that concept in the graph A, and 0 otherwise. Next, the sum of the scores for all concepts in A is computed. This result is normalized in the interval [0, *maximum similarity*]. Finally, the comparison between the representations of the patient records is accomplished by computing the similarity between the hierarchies of the graphs for each category and adding these partial results to calculate a global similarity value.

4   Evaluation Methodology

To the authors' knowledge, no corpus of relevance judgments of medical record similarity is publicly available. For this reason, a collection of radiology reports from the Cincinnati Children's Hospital Medical Center's Department of Radiology is used. This corpus was designed for being used in the CMC-NLP 2007 ICD-9-CM categorization task. To evaluate the method, 50 reports have been obtained from this corpus, using a stratified sampling, so that for each ICD category, a number of reports proportional to its size are selected. To avoid very small categories, only ten categories have been used. We will refer to this set as our 'test collection'. Next, a subset of 20 reports has been separated from the test collection, once again using a stratified sampling. We will refer to this set as our 'query collection'. Two hospital physicians were asked to select, for each record in the query collection, the most similar ones within the test collection. To measure the interjudge agreement, the Kappa statistic [15] has been calculated. An average kappa value equal to 0.7980 is obtained, which indicates that there is a substantial (0.61<= k <= 0.80) agreement.

The evaluation is done by comparing the EHRs retrieved by the system which those retrieved by the experts and calculating precision and recall metrics. As the method's output is a ranking of EHRs, two different evaluation approaches are used: first, the number of records to retrieve is set to 3 and 5 respectively; and second, the average precision at all different levels of recall is calculated [16]. Since the relevance judgments differ across judges, two experiments are presented: first, only overlapping judgments are considered; second, the union of the judgments is used.

5   Results and Discussion

In order to verify if the method proposed yields better results than the classic term vector space model, we compare the retrieval precision and recall obtained by both methods.

We first examine the results of our algorithm when 3 and 5 document cutoffs are used (Table 3: *Union-3, Intersection-3, Union-5, Intersection-5*). Note that the order of documents in the ranking is irrelevant. Our method significantly outperforms the vector space model in precision and recall, both with the union and the intersection of the relevance judgements. It can be also observed that when the 5 document cutoff is used, the precision in our method considerably decreases. The reason is that we have forced the system to retrieve 5 documents when the experts have retrieved only 3.075 documents per query on average. On the contrary, precision in the vector space model slightly increases with the number of documents retrieved. The reason is that a good number of the relevant documents retrieved are not ranked in the top 3 positions, but in positions 4 and 5. As expected, recall increases with the number of documents retrieved in both systems.

Table 3. Results of the evaluation and comparison with the vector space model

	Graph-based			Term-based		
	Precision	Recall	F-measure	Precision	Recall	F-measure
Union-3	0.717	0.676	0.696	0.467	0.476	0.471
Intersection-3	0.600	0.788	0.681	0.333	0.487	0.395
Union-5	0.530	0.765	0.626	0.470	0.708	0.564
Intersection-5	0.420	0.884	0.569	0.340	0.729	0.463
Union	0.707	–	0.692	0.594	–	0.632
Intersection	0.745	–	0.742	0.503	–	0.598

We next examine the retrieval performance when precision at all levels of recall is considered (Table 3: *Union, Intersection*), so that the position of the relevant documents in the ranking does matter now. Once again our algorithm obtains a significant improvement in precision over the term-based approach.

6   Conclusion and Future Work

In this paper, a novel approach to the automatic retrieval of similar EHRs has been presented. The method represents the medical record as a set of semantic graphs using UMLS concepts and relations. This way it gets a richer representation than the one provided by traditional models based on terms. The method achieves relatively high precision and recall, which are also well balanced, which indicates that even though some relevant records are not ranked in the top positions, most retrieved documents are relevant.

However, the intermediate results of the method have shown that the indexing is not sufficiently exhaustive, as UMLS occasionally fails to recover relevant concepts especially when these concepts are expressed in their shortened forms. Another important impairment to concept identification comes from the spelling errors that frequently occur in the clinical records.

Future work will test the method on a different evaluation collection which ideally will present longer medical records structured in different sections, so that the position of the concepts in these sections could condition whether or not these concepts are relevant for indexing. Furthermore, a user-oriented evaluation is planned which attempt to assess the system from cost-effectiveness and benefit points of view.

Acknowledgements

This research has been partially funded by the Spanish Ministerio de Ciencia e Innovación (TIN2009-14659-C03-01) and by the Spanish Ministerio de Industria, Turismo y Comercio (TIS 020100-2009-252).

References

1. Gorman, P., Helfand, M.: Information seeking in primary care: how physicians choose which clinical question to pursuit and which to leave unanswered. Medical Decision Making 15, 113–119 (1995)
2. Ely, J., Osheroff, J., Ebell, M., Chambliss, M., Vinson, D., Stevermer, J., Pifer, E.: Obstacles to answering doctors' questions about patient care with evidence: qualitative study. British Medical Journal 324, 710–713 (2002)
3. Hersh, W., Hickam, D.: Information retrieval in medicine: The Saphire experience. Journal of the American Society for Information Science 46, 743–747 (1995)
4. Lacoste, C., Lim, J., Chevallet, J., Le, D.: Medical-image retrieval based on knowledge-assisted text and image indexing. IEEE Transactions on Circuits and Systems for Video Technology 17, 889–890 (2007)
5. Navarro, S., Llopis, F., Muñoz, R.: Different Multimodal Approaches using IR-n in ImageCLEFphoto 2008. In: On-line Working Notes CLEF (2008)
6. Kwiatkowska, M., Atkins, S.: Case representation and retrieval in the diagnosis and treatment of obstructive sleep apnea: A semiofuzzy approach. In: Proceedings of the 7th ECCBR (2004)
7. Gindl, S., Kaiser, K., Miksch, S.: Syntactical negation detection in clinical practice guidelines (2008)
8. Mutalik, A., Deshpande, A., Nadkarni, P.: Use of general-purpose negation detection to augment concept indexing of medical documents. A quantitative study using the UMLS. JAIMA 8, 598–609 (2001)
9. Morante, R., Liekens, A., Daelemans, W.: Learning the scope of negation in biomedical texts. In: Proceedings of the EMNLP Conference, pp. 715–724 (2008)
10. Nadkarni, P.: Information retrieval in medicine: overview and applications. Journal of Postgraduate Medicine 46, 122–166 (2000)
11. Nelson, S., Powell, T., Humphreys, B.: The Unified Medical Language System (UMLS) Project. In: Kent, A., Hall, C.M. (eds.) Encyclopedia of Library and Information Science, Marcel Dekker, Inc., New York (2002)
12. Aronson, A.: Effective mapping of biomedical text to the umls metathesaurus: the metamap program. In: Proceedings of the AMIA Annual Symposium, pp. 17–21 (2001)
13. Pratt, W., Yetisgen-Yildiz, M.: A study of biomedical concept identification: Metamap vs. people. In: Proceedings of the AMIA Annual Symposium, pp. 529–533 (2003)
14. Yoo, I., Hu, X., Song, I.Y.: A coherent graph-based semantic clustering and summarization approach for biomedical literature and a new summarization evaluation method. BMC Bioinformatics 8(9) (2007)
15. Cohen, J.: A coefficient of agreement for nominal scales. Educational and Psychological Measurement 20, 37–46 (1960)
16. Salton, G.: Introduction to Modern Information Retrieval. McGraw-Hill, New York (1983)

Supporting the Abstraction of Clinical Practice Guidelines Using Information Extraction

Katharina Kaiser[1,2] and Silvia Miksch[2,3]

[1] Center of Medical Statistics, Informatics & Intelligent Systems
Medical University of Vienna
Spitalgasse 23, 1090 Vienna, Austria
[2] Institute of Software Technology & Interactive Systems
Vienna University of Technology
Favoritenstraße 9-11/188, 1040 Vienna, Austria
[3] Department of Information & Knowledge Engineering
Danube University Krems
Dr.-Karl-Dorrek-Straße 30, 3500 Krems, Austria

Abstract. Modelling clinical practice guidelines in a computer-interpretable format is a challenging and complex task. The modelling process involves both medical experts and computer scientists, who have to interact and communicate together. In order to support both modeller groups we propose to provide them with helpful information automatically generated using NLP methods. We identify this information using rules based on both syntactic and semantic information. The majority of the defined information extraction rules are based on semantic relationships derived from the UMLS Semantic Network. Findings in the evaluation indicate that using rules based on semantic and syntactic information provide valuable and helpful results.

1 Introduction

Clinical practice guidelines (CPGs) are important instruments to provide state-of-the-art clinical practice at point of care for the medical and clinical personnel [1]. They typically address a specific health condition and provide recommendations to the physician about issues such as who to investigate for the problem, how to investigate it, how to diagnose it, and how to treat it. It has been shown that integrating CPGs into clinical information systems can improve the quality of care [2]. For integrating CPGs into such systems, the CPGs (i.e., documents in free, narrative text) have to be translated into a computer-interpretable format. Various of such guideline representation formats have been developed (see [3] for an overview and comparison), but the translation process is still a large bottle-neck. It's a complex and challenging task requiring both medical and computer science expertise. Several methods have been developed that describe systematic approaches for guideline modelling. Some of them use intermediate formats to break the complexity of the modelling task into manageable tasks.

In various projects we use the *Many-Headed Bridge between guideline formats* (MHB) [4] as intermediate abstraction format. But still the modelling, involving both medical experts and computer scientists, is challenging and requires high training effort. Thus, we want to support the abstraction process by automatically identifying

C.J. Hopfe et al. (Eds.): NLDB 2010, LNCS 6177, pp. 304–311, 2010.

information dimensions regarding the MHB format using NLP. Our overall goal is to accomplish the abstraction process almost automatically and generate the MHB model forthwith. As a first step, we want to provide identified information dimensions to the modellers to support the further assignment to more detailed information slots.

In this paper we describe methods and an application to automatically identify information dimensions necessary for modellers of computer-interpretable guidelines using NLP.

In the next section we give a short overview on guideline modelling, knowledge-based methods for support, and describe MHB's information dimensions. In Section 3 we explain methods developed and resources used. Section 4 describes the evaluation of our system and a discussion of its results. The final section contains our conclusions.

2 Background

Modelling CPGs in a computer-interpretable and -executable format is a great challenge. Different actors are involved in this task (i.e., medical experts and computer scientists) and must find a way to communicate together.

For non-medical scientists the language in such documents is difficult to understand. On the one hand they are unfamiliar with the medical concepts, on the other hand the medical language implies a different meaning than generally assumed. For instance, the sentence *'Neonatal respiratory depression could be relieved by naloxone administration.'* means that under the condition "neonatal respiratory depression" one should perform the activity "naloxone administration" with an effect "relief". Understanding the medical language in CPGs and differentiating between explanatory information and activities to be performed are therefore the major challenge for computer scientist modellers.

As modelling is similar to programming medical scientists have difficulties with the formal concepts. Furthermore, the medical language is often vaguely formulated. To bring medical scientists to formulate concrete statements is a major challenge. In principle, MHB supports the formalization by keeping expressions used in the CPG. This makes the modelling easier for medical scientists, but still they have difficulties in detecting parameters needed and ordering of tasks.

2.1 Knowledge-Based Methods for Guideline Modelling

Several attempts have been made to support the modelling, maintenance, and shareability of computer-interpretable guidelines. For instance, Serban et al. defined linguistic patterns found in CPGs that can be used to support both the authoring and the modelling process of guidelines [5]. Furthermore, they defined an ontology out of these patterns and linked them with existing thesauri in order to use compilations of thesauri knowledge as building blocks for modeling and maintaining the content of a medical guideline [6]. In [7] we proposed a method to identify actions in ontolaryngology CPGs using a subset of the *Medical Subject Headings* (MeSH) as a thesaurus.

2.2 The *Many-Headed Bridge* between Guideline Formats

The MHB format [4] is a XML language aimed at representing an abstract representation of a CPG while translating it into the target representation language. It was designed to close the large gap between the natural language text and the formal representation of a clinical guideline, thus facilitating the formalization process. MHB provides chunks which correspond to a certain bit of information in the natural language text (e.g., a sentence, part of a sentence, or more than one sentence). These chunks are internally structured into aspects which are grouped by eight dimensions: control flow, data flow, temporal aspects, background information, evidence, resources, patient related aspects, and document structure. The most complex are the first four dimensions, which we will describe more detailed. Each dimension is then further divided into more specific information slots.

1. *Control Flow* specifies the execution order of tasks, their decomposition, and the gathering of information. It can be used to specify decisions, ordering, decomposition, and synchronization of tasks as well as actions.
2. *Data Flow* describes the data processing involved in the diagnosis and treatment.
3. *Temporal Aspects* of qualitative or quantitative nature can be specified.
4. *Background Information* can motivate the reader to follow the guideline or information to complement the statements in the recommendation part. That can be intentions, effects, relations, relations, educational information, explanations, and indicators.

Our methods are tailored to these four dimensions, which are the most challenging to be identified by modellers and also the most important ones for guideline modelling, because they contain the minimal requirements for executing guideline processes.

3   Methods

In the ReMINE project[1] we modelled a local adaptation of a guideline for natural birth for an Italian hospital. It was derived from the guideline "Induction in labour" [8]. Resulting from difficulties during the modelling process we started using the document for searching for patterns that could be used to identify several MHB dimensions. In order to generate patterns that work for our purpose, we had to use both syntactical and semantic information from the text. We defined hand-written rules to identify conditions and actions for the control dimension, data items (i.e., parameters) needed for processing conditions for the data flow dimension, temporal aspects for the time dimension, and background information.

   We use the MetaMap Transfer Program (MMTx) [9] to identify the medical concepts and their semantics in the text, the Stanford Parser [10] to generate a parse tree, and its utility Tregex [12] for matching patterns in trees.

3.1   Identifying Instances of Control Dimensions

The most challenging part of modelling is the detection of actions or activities. How can we differentiate between an action or activity and background information? The UMLS Semantic Network [11] offers amongst various semantic types also semantic

---

[1] http://www.remine-project.eu; last accessed: Jan. 12, 2010.

relationships between these types and therefore acts as an upper-level ontology. We use these semantic relationships to identify actions.

We analysed actions in our CGP and found both complete and incomplete relationships. The latter result from the fact that actions and activities in CPGs are related to tasks the health care personnel has to perform (in our specific case: gynaecologists and obstetricians, who are assigned the semantic type *Professional or Occupational Group*). As most of the actions address these users directly, they are only implicitly referred to in the text. Furthermore, passive format is frequently used, which also allows omission of the agent of the action.

In the Semantic Network there are 61 left-hand side semantic relations defined for *Professional or Occupational Group*. We use 18 of these relations that are related to activities performed by health care personnel, such as 'treats *Patient or Disabled Group*', 'performs *Therapeutic or Preventive Procedure*', and so on, and combined them to five relations of types 'diagnoses', 'interacts_with', 'performs', 'treats', and 'uses'. Furthermore, when we analysed the semantic relationships in our text, we found also new relationships that we added (e.g., 'acquires *Finding*').

The more challenging actions are those that formulate the activity implicitly. We identified 27 complete relations referring to actions, such as '*Therapeutic or Preventive Procedure* treats *Disease or Syndrome*', '*Patient or Disabled Group* performs *Therapeutic or Preventive Procedure*', or '*Therapeutic or Preventive Procedure* affects *Patient or Disabled Group*', which use relation types 'treats', 'uses', 'performs', and 'affects'. For instance, the sentence 'Women may eat a light diet in established labor.' can be described by the relation '*Population Group* performs *Therapeutic or Preventive Procedure*'.

In order to define patterns that indicate such semantic relationships we were confronted with the challenge of identifying the type of relationship. In most cases a specific verb suggests the relation type. Therefore, we tried to identify verbs for each type of relationship. We analysed our training document and assigned each relation type the verbs appearing. Afterwards we also collected synonyms of these verbs in online dictionaries. Thereby, we were able to generate syntactic rules based on semantic information. Figure 1 shows the implementation of the semantic relationships in *tregex* patterns [12] and gives examples how they work. We are now able to omit sentences not describing actions.

Actions and activities can be controlled by conditions. In many cases conditions start with 'if', 'in case of', 'when', and so on. For this kind of conditions we easily generated syntactic patterns. Other conditions could be identified based on their semantic types. For instance, the sentence 'Women with pain but no cervical changes should be reexamined ...' contains the condition 'with pain but no cervical changes'. 'Pain' is assigned the semantic type *Sign or Symptom* and 'cervical changes' is assigned *Finding*. Thus, we defined condition patterns based on semantic information, for instance, based on semantic relation '*Finding* occurs_in *Population Group*'.

3.2   Identifying Instances of Background Dimensions

The background dimension comprises different kinds of information, such as explanatory information, intentions, effects, and so on. We formulated patterns that are solely

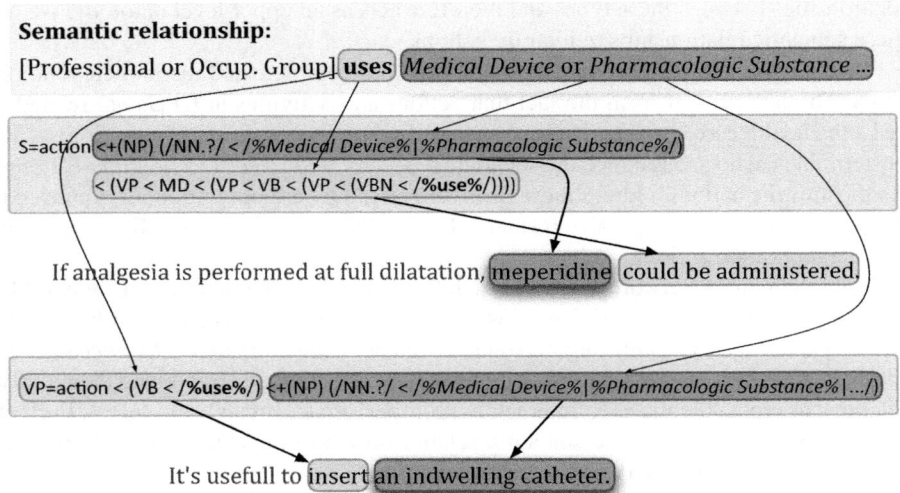

Fig. 1. *tregex* patterns implementing the relationship '[*Professional or Occupational Group*] uses *Medical Device* or *Pharmacologic Substance* or ...'

based on syntactic information. Thereby, expressions such as 'even if ...', 'in order to ...', 'because ...', or 'although ...' are used to identify different aspects of background information.

3.3   Identifying Instances of Time and Data Dimensions

Identifying temporal aspects in a CPG is a very important issue, as we can thereby suggest for connections and dependencies among actions. We generated patterns based on syntactic information and regular expressions.

Identifying data dimensions within a text has emerged to be difficult for both computer scientists and medical scientists. The data dimension refers to data items that are read from an electronic health record or asked from the personnel (i.e., 'usage'), data items that receive a certain information (e.g., newly generated information that is written into the electronic health record) (i.e., 'input'), and data items that are computed by some rules, which are defined in the document (i.e., 'abstraction'). The first ones are occurring within condition clauses, the second ones within actions, and the latter ones are appearing independently in the text. We generated patterns for input and usage of data using semantic information. Thereby, 'input' patterns are associated with certain 'action' patterns. For instance, '[*Professional or Occupation Group*] acquires *Finding* or *Organism Function* or *Body Substance* ...' also indicates for data input, which is contained in the right-hand side of the relationship.

3.4   Providing the Results

In order to provide the obtained information to the users in a simple and clearly represented way, we decided to generate a simple HTML output highlighting the different information dimensions with different colors. Due to the small amount on information dimensions the representation should be comprehensible; minimal learning effort is required. Modelling is accomplished using the DELT/A tool [13]. Thereby, the HTML document is loaded. By marking up the various dimensions the corresponding MHB model is built. Using the highlighted HTML output has the advantage that it is also displayed in the DELT/A tool and the highlighting can be used immediately for supporting the users.

4   Evaluation

We implemented a prototype system that processes the defined patterns and generates the HTML output described above. We then applied the system to another document to evaluate the performance of our system.

We used the guideline "Management of labor" [14] for evaluation. Although the CPG covers the same application area, it is structured in a different way, uses different phrasing and wording. Furthermore, it contains lots of background information and literature references and it is almost twice as large (60 pages) as the document we used to define our rules.

Due to the lack of a gold standard, we manually checked the output of our program with the help of an MHB expert. Table 1 shows the evaluation results. The overall recall and precision values are 69.3% and 79.3%, respectively. As we had expected a better result, we analysed how this outcome arose.

Table 1. Evaluation results

	CONTROL		TIME	DATA	BACKGR.	Overall
	Condition	Action				
COR	40	88	29	55	117	**329**
INC	20	27	5	11	17	**80**
PAR	5	2	5	2	2	**16**
POS	79	150	39	74	144	**486**
ACT	65	117	39	68	136	**425**
REC	53.8%	59.3%	80.8%	75.7%	81.9%	**69.3%**
PRE	65.4%	76.1%	80.8%	82.4%	86.8%	**79.3%**

COR = number of correct slot fillers generated by the system
INC = number of incorrect slot fillers generated by the system
PAR = number of partially correct slot fillers generated by the system
POS = number of slot fillers according to the key target template
ACT = number of slot fillers generated by the system
REC = (COR + PAR/2)/POS
PRE = (COR + PAR/2)/ACT

Our analysis showed that most missing assignments of actions base on incorrect parsing information and not detecting the appropriate semantic relation either due to a verb not incorporated in the knowledge base or due to not detecting the semantic type of a phrase. Incorrectly assigned actions result from wrong assignments of semantic types. The largest influence on the overall result has the erroneous output of the parser.

A minor source of errors are the completeness of semantic relationships and verbs pointing to a relation. We think that expanding our patterns by analysing more documents will improve the output.

We also provided the output of the system to students modelling the document in MHB. First feedback from medical informatics students shows that by providing information on dimensions the modelling is easier and faster accomplished. The resulting models are significantly more elaborated than those without this additional information. We are now performing experiments, for instance, to get insights which of the linguistic tasks are more critical to effectively support CPG modelling.

5   Conclusions and Future Work

In this paper we presented a method and its prototypical implementation to provide knowledge modellers of clinical practice guidelines with information about control and data flow, time dimensions, and background information. We identify this information using NLP methods based on both syntactic and semantic information. The majority of the defined information extraction rules are based on semantic relations we derived from the UMLS Semantic Network [11]. By providing this additional information the knowledge modellers see what parts they have to model and what kind of models they have to use.

Our next steps will be the further development of the extraction methods by applying the methods on documents from other clinical specialties. Thereby, we can improve and expand our rule base. Furthermore, we will try to extract more detailed information to automatically generate a MHB model. The correct interpretation of the control dimension will be the major challenge. We have to discover whether there is a single action, a decomposition of an action, or a selection of one or more alternative actions. This support can promote the application of computer-interpretable CPGs.

Acknowledgement. We'd like to thank Andreas Seyfang for his MHB support and we'd like to thank the anonymous reviewers for their valuable feedback. This work is partially supported by "Fonds zur Förderung der wissenschaftlichen Forschung FWF" (Austrian Science Fund), grants L290-N04 and TRP71-N23, and the European Community's Seventh Framework Programme (FP7/2007-2013) under grant agreement n° 2161.

References

1. Field, M.J., Lohr, K.N. (eds.): Clinical Practice Guidelines: Directions for a New Program. National Academies Press, Institute of Medicine, Washington DC (1990)
2. Quaglini, S., Ciccarese, P., Micieli, G., Cavallini, A.: Non-compliance with guidelines: Motivations and consequences in a case study. In: Kaiser, K., Miksch, S., Tu, S.W. (eds.) Computer-based Support for Clinical Guidelines and Protocols. Proceedings of the Symposium on Computerized Guidelines and Protocols (CGP 2004). Studies in Health Technology and Informatics, vol. 101, pp. 75–87. IOS Press, Prague (2004)

3. Peleg, M., Tu, S.W., Bury, J., Ciccarese, P., Fox, J., Greenes, R.A., Hall, R., Johnson, P.D., Jones, N., Kumar, A., Miksch, S., Quaglini, S., Seyfang, A., Shortliffe, E.H., Stefanelli, M.: Comparing computer-interpretable guideline models: A case-study approach. Journal of the American Medical Informatics Association (JAMIA) 10(1), 52–68 (2003)
4. Seyfang, A., Miksch, S., Marcos, M., Wittenberg, J., Polo-Conde, C., Rosenbrand, K.: Bridging the gap between informal and formal guideline representations. In: Brewka, G., Coradeschi, S., Perini, A., Traverso, P. (eds.) European Conference on Artificial Intelligence (ECAI-2006). Frontiers in Artificial Intelligence and Applications, vol. 141, pp. 447–451. IOS Press, Riva del Garda (2006)
5. Serban, R., ten Teije, A., van Harmelen, F., Marcos, M., Polo-Conde, C.: Extraction and use of linguistic patterns for modelling medical guidelines. Artificial Intelligence in Medicine 39(2), 137–149 (2007)
6. Serban, R., ten Teije, A.: Exploiting thesauri knowledge in medical guideline formalization. Methods Inf. Med. 48, 468–474 (2009)
7. Kaiser, K., Akkaya, C., Miksch, S.: How can information extraction ease formalizing treatment processes in clinical practice guidelines? A method and its evaluation. Artificial Intelligence in Medicine 39(2), 151–163 (2007)
8. National Collaborating Centre for Women's and Children's Health: Induction of labour. Clinical guideline 70, National Institute for Health and Clinical Excellence (NICE), London (UK) (July 2008)
9. Aronson, A.R.: MetaMap: Mapping text to the UMLS metathesaurus. Technical report, Lister Hill National Center for Biomedical Communications (2006)
10. Klein, D., Manning, C.D.: Fast exact inference with a factored model for natural language parsing. In: Advances in Neural Information Processing Systems 15 (NIPS 2002), pp. 3–10. MIT Press, Cambridge (2003)
11. McCray, A.T.: UMLS Semantic Network. In: Proc. of the 13th Annual Symposium on Computer Applications in Medical Care (SCAMC 1989), pp. 503–507 (1989)
12. Levy, R., Andrew, G.: Tregex and Tsurgeon: tools for querying and manipulating tree data structures. In: Fellbaum, C., Miller, G.A. (eds.) Proceedings of the fifth international conference on Language Resources and Evaluation (LREC), Genoa, Italy, ELRA, pp. 2231–2234 (2006)
13. Votruba, P., Miksch, S., Kosara, R.: Facilitating knowledge maintenance of clinical guidelines and protocols. In: [15], pp. 57–61.
14. Institute for Clinical Systems Improvement (ICSI): Management of labor. Clinical guideline, Institute for Clinical Systems Improvement (ICSI), Bloomington (MN) (May 2009)
15. Fieschi, M., Coiera, E., Li, Y.C.J. (eds.): Proceedings from the Medinfo 2004 World Congress on Medical Informatics. IOS Press, AMIA (2004)

Author Index